MATHÉMATIQUES
&
APPLICATIONS

Directeurs de la collection :
G. Allaire et M. Benaïm

66

Weijiu Liu

Elementary Feedback Stabilization of the Linear Reaction-Convection-Diffusion Equation and the Wave Equation

 Springer

Weijiu Liu
University of Central Arkansas
Department of Mathematics
201 Donaghey Avenue
Conway, AR 72035
USA
weijiul@uca.edu

ISSN 1154-483X
ISBN 978-3-642-04612-4   e-ISBN 978-3-642-04613-1
DOI 10.1007/978-3-642-04613-1
Springer Heidelberg Dordrecht London New York

Library of Congress Control Number: 2009941063

Mathematics Subject Classification (2000): 93D15, 93C20, 93B52, 93B05, 93B07, 93B55, 35B40, 35K20, 35K57, 35L05, 35L15, 35L20, 49K15, 49K20

Cover design: SPi Publisher Services

Printed on acid-free paper

Springer is part of Springer Science+Business Media (www.springer.com)

To Shi Danxia, Liu Xin, and my parents, Liu Yili and Zhang Louzhong

Preface

Mathematical control theory of applied partial differential equations is built on linear and nonlinear functional analysis and many existence theorems in control theory result from applications of theorems in functional analysis. This makes control theory inaccessible to students who do not have a background in functional analysis.

Many advanced control theory books on infinite-dimensional systems were written, using functional analysis and semigroup theory, and control theory was presented in an abstract setting. This motivates me to write this text for control theory classes in the way to present control theory by concrete examples and try to minimize the use of functional analysis. Functional analysis is not assumed and any analysis included here is elementary, using calculus such as integration by parts. The material presented in this text is just a simplification of the material from the existing advanced control books. Thus this text is accessible to senior undergraduate students and first-year graduate students in applied mathematics, who have taken linear algebra and ordinary and partial differential equations.

Elementary functional analysis is presented in Chapter 2. This material is required to present the control theory of partial differential equations. Since many control concepts and theories for partial differential equations are transplanted from finite-dimensional control systems, a brief introduction to feedback control of these systems is presented in Chapter 3. The topics covered in this chapter include controllability, observability, stabilizability, pole placement, and quadratic optimal control. Theories about the feedback stabilization of linear reaction-convection-diffusion equations are presented in Chapter 4. Both interior and boundary control problems are addressed. The methods employed to handle the problems include eigenfunction expansions, integral transforms, and optimal control. Finally, theories about feedback stabilization of linear wave equations are presented in Chapters 5 and 6. First, the one-dimensional wave equations are considered, then higher dimensional wave equations follows, since it is easier to use the one-dimensional equations to illustrate the theories and methods. The perturbed energy method is emphasized to deal with both interior and boundary control problems while the optimal control technique is used.

This text can be used as a textbook for an introductory one-semester graduate course on control theory of partial differential equations. If students do not have a background in finite-dimensional control systems, one can cover Sections 3.1-3.7 of Chapter 3, Sections 4.1-4.3 of Chapter 4, all sections of Chapter 5, and Sections 6.1- 6.3 of Chapter 6. Material from Chapter 2 can be introduced whenever needed, and all material about optimal control can be skipped. If students have had a background in both functional analysis and finite-dimensional control systems, one can cover all of Chapters 4, 5, 6 and give a brief review of Chapters 2 and 3.

Inevitably, the material is selected from the topics I have worked on or am most familiar with, and many other important topics are not covered. Control theory for other applied partial differential equations, such as elastic equations, thermoelastic equations, viscoelastic equations, Schrödinger's equations, and Navier-Stokes equations, is not included. In addition, exact and approximate controllability is not discussed since it is difficult to present it in an elementary way.

I thank my PhD advisor, Dr. Graham H. Williams, for introducing me to the area of control theory of partial differential equations. I thank Drs. George Haller, Mirsolav Krstic, and Enrique Zuazua for having directed my research and providing stimulating feedback on my work. I thank two reviewers for their hard, serious work of reading and evaluating this text and giving constructive suggestions. I thank my students, such as Jingvoon Chen, for using this text and correcting mistakes. I thank Ms. Peggy Arrigo and Dr. Danny Arrigo for their language-editing. Finally, I thank my family for their constant support.

Conway, Arkansas, USA   *Weijiu Liu*

May 2009

Contents

1 Examples of Control Systems 1
  1.1 Control of a Mass-Spring System 1
  1.2 Temperature Control of a Rod in a Chemical Reactor 3
  1.3 Control of String Vibration 6

2 Elementary Functional Analysis 9
  2.1 Banach Spaces and Hilbert Spaces 9
  2.2 Open and Closed Sets 15
  2.3 Linear Operators 17
  2.4 Sobolev Spaces 19
  2.5 Eigenvalue Problems 21
  2.6 Semigroups 29
    2.6.1 Definitions and Properties 30
    2.6.2 Formulas for Semigroup Representation 32
    2.6.3 Tests for Semigroup Generation 34
    2.6.4 Semigroup Growth Bound 38
  2.7 Abstract Cauchy Problems 44
  2.8 References and Notes 48

3 Finite Dimensional Systems 49
  3.1 General Forms of Control Systems 49
  3.2 Stability 51
  3.3 Controllability and Observability 57
  3.4 State Feedback Control 67
  3.5 PID Controllers 75
  3.6 Integral State Feedback Control 78
  3.7 Output Feedback Control 83
    3.7.1 Observer-based Output Feedback Control 83
    3.7.2 Integral Output Feedback Control 86
  3.8 Optimal Control on Finite-time Intervals 90
    3.8.1 Existence and Uniqueness 91

    3.8.2   Optimality System 95
  3.9   Optimal State Feedback Controllers 100
  3.10 Asymptotic Tracking and Disturbance Rejection.................. 111
  3.11 References and Notes 117

4   Linear Reaction-Convection-Diffusion Equation 119
  4.1   Stability 120
  4.2   Interior Feedback Stabilization 129
    4.2.1   State Feedback Controls 131
    4.2.2   Output Feedback Controls.............................. 139
  4.3   Boundary Feedback Stabilization 145
    4.3.1   State Feedback Controls 145
    4.3.2   Output Feedback Controls............................. 163
  4.4   Optimal Interior Control.................................... 173
    4.4.1   Existence and Uniqueness 174
    4.4.2   Necessary Conditions................................. 177
    4.4.3   Optimality Systems 179
    4.4.4   Optimal Interior State Feedback Controls 183
  4.5   Optimal Boundary Control................................... 193
    4.5.1   Necessary Conditions................................. 194
    4.5.2   Optimality Systems 195
    4.5.3   Existence and Uniqueness 202
    4.5.4   Optimal Boundary State Feedback Controls 205
  4.6   Generalization to Abstract Dynamical Systems 212
  4.7   References and Notes 214

5   One-dimensional Wave Equation 215
  5.1   Stability 215
  5.2   Linear Interior Feedback Stabilization 218
  5.3   Linear Boundary Feedback Stabilization 225
  5.4   References and Notes 231

6   Higher-dimensional Wave Equation 233
  6.1   Stability 233
  6.2   Linear Boundary Feedback Stabilization 235
  6.3   Nonlinear Boundary Feedback Stabilization 242
  6.4   Observability Inequalities................................... 248
  6.5   Linear Interior Feedback Stabilization 255
  6.6   Nonlinear Interior Feedback Stabilization 261
  6.7   Optimal Boundary Control................................... 271
  6.8   References and Notes 287
  References 287

Index 293

Chapter 1
Examples of Control Systems

Feedback control is a process of regulating a physical system to a desired output based on the feedback of actual outputs of the system. Feedback control systems are ubiquitous around us, including trajectory planning of a robot manipulator, guidance of a tactical missile toward a moving target, regulation of room temperature, and control of string vibrations. These control systems can be abstracted by Figure 1.1. Here a plant subject to a disturbance d is given, and a controller is to be designed so that the output y tracks the reference r. The plant can be modeled by either a system of ordinary differential equations (finite dimensional systems) or partial differential equations (infinite dimensional systems).

1.1 Control of a Mass-Spring System

Linear finite dimensional control systems have the following general form

$$\dot{\mathbf{x}} = \mathbf{Ax} + \mathbf{Bu} + \mathbf{Gd},$$
$$\mathbf{y}_m = \mathbf{Cx} + \mathbf{Du} + \mathbf{Hd},$$
$$\mathbf{y}_c = \mathbf{Ex} + \mathbf{Fu} + \mathbf{Jd},$$
$$\mathbf{x}(0) = \mathbf{x}_0,$$

where $\mathbf{x} = (x_1, x_2, \cdots, x_n)^*$ (hereafter $*$ denotes the transpose of a vector or matrix) is a state vector, $\dot{\mathbf{x}}$ denotes the derivative $\frac{d\mathbf{x}}{dt}$, $\mathbf{x}_0$ is an initial state caused by external disturbances, $\mathbf{y}_m = (y_{m1}, \cdots, y_{ml})^*$ is a measured output vector, $\mathbf{y}_c = (y_{c1}, \cdots, y_{ck})^*$ is a controlled output vector, $\mathbf{u} = (u_1, \cdots, u_m)^*$ is a control vector, and $\mathbf{A}, \mathbf{B}, \mathbf{C}, \mathbf{D}, \mathbf{E}, \mathbf{F}, \mathbf{G}, \mathbf{H}, \mathbf{J}$ are constant matrices with appropriate dimensions. Given a desired reference $\mathbf{r}$, the control objective is to design the control $\mathbf{u}$, based on the measured output $\mathbf{y}_m$, to regulate the controlled output $\mathbf{y}_c$ to $\mathbf{r}$, that is, $\mathbf{y}_c(t) \to \mathbf{r}$ as $t \to \infty$.

W. Liu, *Elementary Feedback Stabilization of the Linear Reaction-Convection-Diffusion Equation and the Wave Equation*, Mathématiques et Applications 66, DOI 10.1007/978-3-642-04613-1_1, © Springer-Verlag Berlin Heidelberg 2010

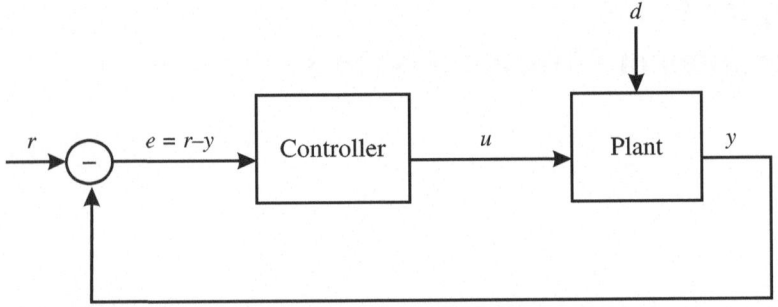

Fig. 1.1 Standard feedback control system. y is an output from a plant, d denotes a disturbance, and r is a reference input. A controller is to be designed so that the output y tracks the reference r.

As an example, we consider a mass-spring-damper system as shown in Figure 1.2. A mass with mass m is attached to a spring. If the mass is initially disturbed from its equilibrium, it will oscillate. To control the oscillation, a control actuator, a damper, is attached to the mass.

Let $y(t)$ denote the displacement of the mass from the equilibrium position. Then the system equation is given by

$$m\ddot{y} = -ky + u, \tag{1.1}$$

where m, k are positive constants, the dot $\dot{}$ denotes the time derivative, and u is the force resulting from the damper.

In analysis, sometimes it is convenient to convert a higher order differential equation to a system of first order differential equations. To do so, we define

$$x_1 = y, \quad x_2 = \dot{y}.$$

Then we have

$$\dot{x}_1 = \dot{y} = x_2,$$
$$\dot{x}_2 = \ddot{y} = -\frac{k}{m}y + \frac{1}{m}u = -\frac{k}{m}x_1 + \frac{1}{m}u.$$

This system can be written concisely in the matrix form

$$\begin{bmatrix} \dot{x}_1 \\ \dot{x}_2 \end{bmatrix} = \begin{bmatrix} 0 & 1 \\ -\frac{k}{m} & 0 \end{bmatrix} \begin{bmatrix} x_1 \\ x_2 \end{bmatrix} + \begin{bmatrix} 0 \\ \frac{1}{m} \end{bmatrix} u. \tag{1.2}$$

If our particular interest in the analysis of the system is some variables such as the displacement y, which are required to behave in a specified manner, we can then set up another equation

$$y_c = \begin{bmatrix} 1 & 0 \end{bmatrix} \begin{bmatrix} x_1 \\ x_2 \end{bmatrix} = x_1. \tag{1.3}$$

Fig. 1.2 Mass-spring-damper
system.

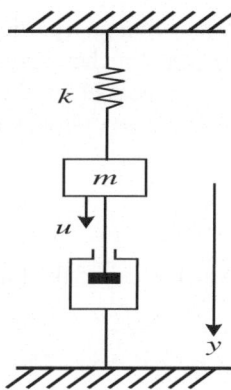

This is the controlled output. If the displacement can be measured physically, we then have another output equation

$$y_m = [1 \ 0] \begin{bmatrix} x_1 \\ x_2 \end{bmatrix} = x_1. \tag{1.4}$$

This is the measured output, which is used for feedback. The measured output can be different from the controlled output. If the displacement cannot be measured, but the velocity $x_2 = \dot{y}$ can be measured, then the measured output equation is

$$y_m = [0 \ 1] \begin{bmatrix} x_1 \\ x_2 \end{bmatrix} = x_2. \tag{1.5}$$

If the mass is expected to oscillate in a desired manner $r(t)$ or stay at a certain fixed position like the equilibrium 0, we need to design a feedback control

$$u = F(y_m)$$

such that $y_c(t) - r(t)$ converges to 0 as t tends to infinity.

1.2 Temperature Control of a Rod in a Chemical Reactor

Modern feedback control systems have become more and more complex. The stringent requirements on accuracy lead to distributed parameter systems in which partial differential equations are required to model the physical systems. Since these systems have infinite eigenfunctions, they are called infinite dimensional systems.

One class of partial differential equations consists of reaction-convection-diffusion equations

$$\frac{\partial u}{\partial t} = \mu \nabla^2 u + \nabla \cdot (u\mathbf{v}) + au.$$

Depending on a particular real problem, u can represent temperature or the concentration of a chemical species. The constant $\mu > 0$ is the diffusivity of the temperature or the species, the vector $\mathbf{v}(\mathbf{x}) = (v_1(\mathbf{x}), \cdots, v_n(\mathbf{x}))$ is the velocity field of a fluid flow, and $a(\mathbf{x})$ is a reaction rate. ∇^2 is defined by

$$\nabla^2 u = \frac{\partial^2 u}{\partial x_1^2} + \cdots + \frac{\partial^2 u}{\partial x_n^2},$$

which is called the Laplace operator. The gradient of u is denoted by

$$\nabla u = \left(\frac{\partial u}{\partial x_1}, \cdots, \frac{\partial u}{\partial x_n} \right).$$

$\nabla \cdot (u\mathbf{v}) = \mathrm{div}(u\mathbf{v}) = \sum_{i=1}^{n} \frac{\partial (uv_i)}{\partial x_i}$ denotes the divergence of the vector $u\mathbf{v}$.

For partial differential equations, we can propose two kinds of control problems: interior control and boundary control. In the problem of interior control, the control acts inside a domain:

$$\frac{\partial u}{\partial t} = \mu \nabla^2 u + \nabla \cdot (u\mathbf{v}) + au + \phi,$$

where ϕ denotes the controller. In the problem of boundary control, the control acts only on the boundary of the domain:

$$\frac{\partial u}{\partial t} = \mu \nabla^2 u + \nabla \cdot (u\mathbf{v}) + au,$$
$$u|_{\partial\Omega} = \phi,$$

where Ω denotes a bounded open set in $\mathbb{R}^n$ and $\partial\Omega$ denotes the boundary of Ω. In this control problem, the state variable is u.

The output of the system can be proposed in many ways in accordance with specific physical problems. Concentration of a chemical or a room temperature is usually desired to be homogenized. Then the controlled output is the concentration

$$o_c(u) = u(\mathbf{x}, t).$$

If the average concentration on different subsets ω_i^o ($i = 1, 2, \cdots, l$) can be measured, then the measured output is

$$\mathbf{o}_m(u) = \left(\int_{\omega_1^o} h_1(\mathbf{x}) u(\mathbf{x}, t) dV, \cdots, \int_{\omega_l^o} h_l(\mathbf{x}) u(\mathbf{x}, t) dV \right),$$

where h_i ($i = 1, 2, \cdots, l$) are given weight functions. A typical real example of such an output is the room temperature measurement where a number of thermo-sensors are placed in a room. As in the case of finite dimensional control systems, the control objective is to design the control ϕ to regulate u to a desired state, such as the room temperature of $20°C$.

Fig. 1.3 A thin rod in a chemical reactor: The reactor is fed with a pure species B and a zero-th order exothermic catalytic reaction of the form $B \rightarrow C$ takes place on the rod.

As an example of application, we consider a thin one-dimensional rod in a chemical reactor [20] as shown in Figure 1.3. The lateral surface area of the rod is not insulated. The reactor is fed with a pure species B and a zero-th order exothermic catalytic reaction of the form $B \rightarrow C$ takes place on the rod. Since the reaction is exothermic, a cooling medium that is in contact with the rod is used for cooling. We assume that the heat energy $f(x,t)$ flowing out the lateral sides per unit volume and per unit time is proportional to the temperature difference between the rod $u(x,t)$ and a known cooling medium temperature $u_m(x)$

$$f(x,t) = \alpha[u(x,t) - u_m(x)],$$

where α is a positive proportionality. We also assume that the heat energy generated by the species B per unit volume and per unit time is equal to [20]

$$\beta\left(\exp\left(-\frac{\gamma}{1 + \frac{u(x,t)}{T_0}}\right) - e^{-\gamma}\right),$$

where β is a positive proportionality, γ denotes a dimensionless activation energy, and T_0 is a reference temperature. Then the heat source is equal to

$$Q(x,t) = \beta\left(\exp\left(-\frac{\gamma}{1 + \frac{u(x,t)}{T_0}}\right) - e^{-\gamma}\right) - \alpha[u(x,t) - u_m(x)].$$

It then follows that the governing equation for the heat flow in this rod is

$$\frac{\partial u}{\partial t} = \mu\frac{\partial^2 u}{\partial x^2} + \frac{\beta}{c\rho}\left(\exp\left(-\frac{\gamma}{1 + \frac{u(x,t)}{T_0}}\right) - e^{-\gamma}\right) - \frac{\alpha}{c\rho}[u(x,t) - u_m(x)], \quad (1.6)$$

where c denotes the specific heat of the rod (the heat energy that must be supplied to a unit mass of a substance to raise its temperature one unit), and ρ denotes the density of the rod.

In this chemical engineering problem, if the cooling medium is maintained at the temperature of $u_m = 0°$, the operating equilibrium temperature $u = 0$ is desired. However, for some process parameters [20], the temperature u may grow. Thus a control mechanism is needed to regulate the temperature to the equilibrium 0. A feasible control scheme is boundary control: the temperature $u(0,t)$ at the left end $x = 0$ is fixed to be zero and the temperature $u(L,t)$ at the right end $x = L$ is controlled, that is,

$$u(L,t) = T(t).$$

If the equation (1.6) is linearized at 0, we obtain a linear reaction diffusion equation. Since the temperature distribution needs to be regulated, the controlled output is

$$o_c(x,t) = u(x,t).$$

If the temperature $u(x,t)$ can be measured, the measured output is

$$o_m(x,t) = o_c(x,t) = u(x,t).$$

We then want to design a feedback control $T(t) = F(u)$ such that $u(x,t)$ tends to zero as $t \to \infty$.

1.3 Control of String Vibration

Another important class of partial differential equations consists of linear wave equations

$$\frac{\partial^2 u}{\partial t^2} = c^2 \nabla^2 u,$$

where the positive constant c (m/s) is a wave speed. We can propose the same control problems as in the case of the reaction-convection-diffusion equations.

A typical physical example of the wave equation is the vertical vibration of a tightly stretched string. In this problem, $u = u(x,t)$ denotes the vertical displacement of the string from its equilibrium. If we are interested in the manner of vibration of the string, then the controlled output is the displacement

$$o_c(x,t) = u(x,t). \tag{1.7}$$

The measured output depends on a specific physical problem. For instance, if the velocity can be physically measured, then the measured output is

$$o_m(x,t) = \frac{\partial u}{\partial t}(x,t). \tag{1.8}$$

If we control the string at its ends, we then have a boundary control problem

$$u(0,t) = 0, \quad \frac{\partial u}{\partial x}(L,t) = f(t),$$

where f is a control force. In this case, the measured output could be the velocity at $x = L$

$$o_m(t) = \frac{\partial u}{\partial t}(L, t).$$

In a real problem, the string may be expected to stay at a certain position profile $r = r(x)$ like the equilibrium $r(x) \equiv 0$. Thus we need to find a feedback control $f = F(o_m)$ such that $u(x, t)$ converges to r as $t \to \infty$.

Chapter 2
Elementary Functional Analysis

Elementary functional analysis, such as Hilbert spaces, Sobolev spaces, and linear operators, is collected in this chapter for the reader's convenience. Since elementary stabilization results are presented without using functional analysis, readers may skip this chapter and come back when needed.

In this book, $\mathbb{N}$ denotes the set of all nonnegative natural numbers, $\mathbb{C}$ denotes the set of complex numbers, $\mathbb{R}^n$ denotes the n-dimensional Euclidean space, and $\mathbb{R} = \mathbb{R}^1$ denotes the real line. Points in $\mathbb{R}^n$ will be denoted by $\mathbf{x} = (x_1, \cdots, x_n)$, and its norm is defined by

$$\|\mathbf{x}\| = \left(\sum_{i=1}^{n} x_i^2 \right)^{\frac{1}{2}}.$$

The inner product of $\mathbf{x}$ and $\mathbf{y}$ is defined by

$$(\mathbf{x}, \mathbf{y}) = \sum_{i=1}^{n} x_i y_i.$$

2.1 Banach Spaces and Hilbert Spaces

To introduce normed linear spaces, we examine the properties of the absolute value $|x|$ of a real number x:

(1) $|x| \geq 0$ for any $x \in \mathbb{R}$, and $|x| = 0$ if and only if $x = 0$;
(2) $|ax| = |a||x|$ for any $a, x \in \mathbb{R}$;
(3) $|x + y| \leq |x| + |y|$ for any $x, y \in \mathbb{R}$.

The absolute value plays a crucial rule in Calculus. Without it, we could not describe how close two numbers are and then we could not introduce the concept of limit.

Definition 2.1. Let X be a linear space over $\mathbb{F}$ ($\mathbb{F} = \mathbb{R}$ or $\mathbb{C}$). A norm on X, denoted by $\| \cdot \|$, is a map from X to $\mathbb{R}$ that satisfies the following conditions:

W. Liu, *Elementary Feedback Stabilization of the Linear Reaction-Convection-Diffusion Equation and the Wave Equation*, Mathématiques et Applications 66, DOI 10.1007/978-3-642-04613-1_2, © Springer-Verlag Berlin Heidelberg 2010

(1) $\|x\| \geq 0$ for any $x \in X$, and $\|x\| = 0$ if and only if $x = 0$;
(2) $\|ax\| = |a| \|x\|$ for any $a \in \mathbb{F}$ and $x \in X$;
(3) $\|x + y\| \leq \|x\| + \|y\|$ for any $x, y \in X$ (the triangle inequality).

The space X endowed with the norm $\| \cdot \|$ is called a normed linear space.

According to the definition, $\mathbb{R}$ endowed with the norm $|\cdot|$, the absolute value, is a normed linear space.

The verification of condition (3) is usually difficult and needs the following Cauchy-Schwarz inequality:

$$\left| \sum_{i=1}^{n} x_i y_i \right| \leq \sqrt{\sum_{i=1}^{n} x_i^2} \sqrt{\sum_{i=1}^{n} y_i^2} \quad \text{for any } \mathbf{x}, \mathbf{y} \in \mathbb{R}^n. \tag{2.1}$$

To prove this inequality, we consider the following positive quantity:

$$\sum_{i=1}^{n} (x_i - \varepsilon y_i)^2 \geq 0$$

for any constant ε. Expanding the perfect square, we can readily derive that

$$2\varepsilon \sum_{i=1}^{n} x_i y_i \leq \sum_{i=1}^{n} x_i^2 + \varepsilon^2 \sum_{i=1}^{n} y_i^2. \tag{2.2}$$

If either $\|\mathbf{x}\| = 0$ or $\|\mathbf{y}\| = 0$, it is clear that (2.1) holds. So we can assume that $\|\mathbf{x}\| \neq 0$ and $\|\mathbf{y}\| \neq 0$. In this case, (2.1) can be derived by taking $\varepsilon = \frac{(\mathbf{x}, \mathbf{y})}{\|\mathbf{y}\|^2}$ in (2.2).

Example 2.1. The normed linear space $\mathbb{R}^n$. The linear space $\mathbb{R}^n$ endowed with the following Euclidean norm

$$\|\mathbf{x}\| = \sqrt{\sum_{i=1}^{n} x_i^2} \tag{2.3}$$

is a normed linear space. In fact, the conditions (1), (2), and (3) of Definition 2.1 are satisfied:

(1) It is clear that $\sqrt{\sum_{i=1}^{n} x_i^2} \geq 0$ and $\sqrt{\sum_{i=1}^{n} x_i^2} = 0$ if and only if $x_i = 0$ $(i = 1, 2, \cdots, n)$,
   i.e., $\mathbf{x} = \mathbf{0}$.
(2) For any number $a \in \mathbb{R}$ and a vector $\mathbf{x} \in \mathbb{R}^n$, we have

$$\|a\mathbf{x}\| = \sqrt{\sum_{i=1}^{n} (ax_i)^2} = |a| \sqrt{\sum_{i=1}^{n} x_i^2} = |a| \|\mathbf{x}\|.$$

(3) Using the Cauchy-Schwarz inequality, we derive that

$$\sum_{i=1}^{n} (x_i + y_i)^2 = \sum_{i=1}^{n} x_i^2 + 2 \sum_{i=1}^{n} x_i y_i + \sum_{i=1}^{n} y_i^2$$
$$\leq \|\mathbf{x}\|^2 + 2\|\mathbf{x}\|\|\mathbf{y}\| + \|\mathbf{y}\|^2$$
$$= (\|\mathbf{x}\| + \|\mathbf{y}\|)^2.$$

Example 2.2. The function space $L^p = L^p(\Omega)$. Let Ω be a open set in $\mathbb{R}^n$. For $p \in [1, \infty)$, let $L^p = L^p(\Omega)$ denote the vector space of all functions u defined in Ω such that

$$\| u \|_{L^p} = \left(\int_{\Omega} | u |^p \ dV \right)^{\frac{1}{p}} < \infty.$$

The integral should be understood in the sense of Lebesgue integral, but it is acceptable to treat it as the integral in Calculus for readers who do not have a background in advanced analysis. We show that L^p endowed with the above norm is a normed linear space.

It is easy to prove that $\| \cdot \|_{L^p}$ satisfies conditions (1) and (2) of Definition 2.1, but the condition (3) is difficult to prove. We need Young's inequality and Hölder's inequality.

Young's inequality. Let $1 < p, q < \infty$, $\frac{1}{p} + \frac{1}{q} = 1$. Then

$$ab \leq \frac{a^p}{p} + \frac{b^q}{q} \quad \text{for any } a, b \geq 0. \tag{2.4}$$

We do not give a detailed proof of the inequality, but provide an outline. If either $a = 0$ or $b = 0$, then (2.4) holds. Thus we can assume that $a \neq 0$. We divide the inequality by a^p to obtain

$$\frac{b}{a^{p-1}} \leq \frac{1}{p} + \frac{1}{q} \left(\frac{b}{a^{p-1}} \right)^q.$$

Here we have used that $p = (p-1)q$. This motivates us to consider the function $f(x) = \frac{1}{p} + \frac{1}{q}x^q - x$. Then the proof of Young's inequality is reduced to show that the minimum of $f(x)$ over $[0, \infty)$ is nonnegative.

Hölder's inequality. Let $1 \leq p, q < \infty$, $\frac{1}{p} + \frac{1}{q} = 1$. If $f \in L^p(\Omega)$ and $g \in L^q(\Omega)$, we have

$$\int_{\Omega} |fg| dV \leq \|f\|_{L^p} \|g\|_{L^q}. \tag{2.5}$$

Proof. If either $\|f\|_{L^p} = 0$ or $\|g\|_{L^q} = 0$, then (2.5) holds. Thus we can assume that $\|f\|_{L^p} \neq 0$ and $\|g\|_{L^q} \neq 0$. Defining $f_1 = f/\|f\|_{L^p}$ and $g_1 = g/\|g\|_{L^q}$, we have $\|f_1\|_{L^p} = \|g_1\|_{L^q} = 1$. It then follows from Young's inequality (2.4) that

$$\int_{\Omega} |f_1 g_1| dV \leq \frac{1}{p} \int_{\Omega} |f_1|^p dV + \frac{1}{q} \int_{\Omega} |g_1|^q dV = \frac{1}{p} + \frac{1}{q} = 1,$$

which implies (2.5).

Once we have Hölder's inequality, we are ready to prove the triangle inequality (3) of Definition 2.1, which, in this case, is called Minkowski's inequality.

Minkowski's inequality. Let $1 \leq p < \infty$. If $f, g \in L^p(\Omega)$, we have

$$\|f+g\|_{L^p} \leq \|f\|_{L^p} + \|g\|_{L^p}. \tag{2.6}$$

Proof. It follows from Hölder's inequality that

$$\int_\Omega |f+g|^p dV \leq \int_\Omega |f+g|^{p-1}|f| dV + \int_\Omega |f+g|^{p-1}|g| dV$$

$$\leq \left(\int_\Omega |f+g|^{(p-1)q} dV \right)^{1/q} \left(\int_\Omega |f|^p dV \right)^{1/p}$$

$$+ \left(\int_\Omega |f+g|^{(p-1)q} dV \right)^{1/q} \left(\int_\Omega |g|^p dV \right)^{1/p}$$

$$= \left(\int_\Omega |f+g|^p dV \right)^{1/q} \left(\|f\|_{L^p} + \|g\|_{L^p} \right).$$

Dividing this inequality by $\left(\int_\Omega |f+g|^p dV \right)^{1/q}$ gives (2.6).

Before we introduce the concept of inner product space, we look at the inner product in $\mathbb{C}^n$

$$(\mathbf{x}, \mathbf{y}) = \sum_{i=1}^n x_i \bar{y}_i,$$

where $\bar{}$ denotes the complex conjugate. The inner product has the following properties:

(1) $(\mathbf{x}, \mathbf{x}) \geq 0$ for any $\mathbf{x} \in \mathbb{C}^n$ and $(\mathbf{x}, \mathbf{x}) = 0$ if and only if $\mathbf{x} = 0$;
(2) $(\mathbf{x}, \mathbf{y}) = \overline{(\mathbf{y}, \mathbf{x})}$ for any $\mathbf{x}, \mathbf{y} \in \mathbb{C}^n$ ($\overline{(\mathbf{y}, \mathbf{x})}$ denoting the complex conjugate);
(3) $(a\mathbf{x} + b\mathbf{y}, \mathbf{z}) = a(\mathbf{x}, \mathbf{z}) + b(\mathbf{y}, \mathbf{z})$ for any $\mathbf{x}, \mathbf{y}, \mathbf{z} \in \mathbb{C}^n$ and any $a, b \in \mathbb{C}$.

This inner product can be extended to a general linear space.

Definition 2.2. Let X be a linear space over $\mathbb{F}$ ($\mathbb{F} = \mathbb{R}$ or $\mathbb{C}$). An inner product on X, denoted by $(\cdot, \cdot)$, is a map from $X \times X$ to $\mathbb{F}$ that satisfies the following conditions:

(1) $(x, x) \geq 0$ for any $x \in X$, and $(x, x) = 0$ if and only if $x = 0$;
(2) $(x, y) = \overline{(y, x)}$ for any $x, y \in X$;
(3) $(ax + by, z) = a(x, z) + b(y, z)$ for any $x, y, z \in X$ and any $a, b \in \mathbb{F}$.

The space X endowed with the inner product $(\cdot, \cdot)$ is called an inner product space.

Example 2.3. The inner product space $L^2(\Omega)$. For any $f, g \in L^2(\Omega)$, define

$$(f, g) = \int_\Omega f(\mathbf{x}) \bar{g}(\mathbf{x}) dV.$$

We show that $(\cdot, \cdot)$ is an inner product on $L^2(\Omega)$, and so $L^2(\Omega)$ is an inner product space.

(1) It is clear that $(f,f) = \int_\Omega f(\mathbf{x})\bar{f}(\mathbf{x})dV \geq 0$ for any $f \in L^2(\Omega)$. Moreover, $(f,f) = \int_\Omega f(\mathbf{x})\bar{f}(\mathbf{x})dV = \int_\Omega |f(\mathbf{x})|^2 dV = 0$ if and only if $f = 0$.

(2) For any $f,g \in L^2(\Omega)$, we have

$$(f,g) = \int_\Omega f(\mathbf{x})\bar{g}(\mathbf{x})dV$$
$$= \overline{\int_\Omega g(\mathbf{x})\bar{f}(\mathbf{x})dV}$$
$$= \overline{(g,f)}.$$

(3) For any $a,b \in \mathbb{C}$ and any $f,g,h \in L^2(\Omega)$, we have

$$(af+bg,h) = \int_\Omega (af(\mathbf{x})+bg(\mathbf{x}))\bar{h}(\mathbf{x})dV$$
$$= a\int_\Omega f(\mathbf{x})\bar{h}(\mathbf{x})dV + b\int_\Omega g(\mathbf{x})\bar{h}(\mathbf{x})dV$$
$$= a(f,h)+b(g,h).$$

Cauchy-Schwarz inequality (2.1) can be easily generalized to the inner product spaces

$$|(x,y)| \leq \sqrt{(x,x)}\sqrt{(y,y)}. \tag{2.7}$$

To prove it, we use the positive quantity:

$$(x-\varepsilon y, x-\varepsilon y) \geq 0$$

for any constant ε. Expanding the inner product, we can readily derive that

$$(x,x) - \bar{\varepsilon}(x,y) - \varepsilon\overline{(x,y)} + |\varepsilon|^2(y,y) \geq 0,$$

and then

$$\bar{\varepsilon}(x,y) + \varepsilon\overline{(x,y)} \leq (x,x) + |\varepsilon|^2(y,y). \tag{2.8}$$

If either $(x,x) = 0$ or $(y,y) = 0$, it is clear that (2.7) holds. So we can assume that they both are not zero. In this case, (2.7) can be derived by taking $\varepsilon = \frac{(x,y)}{(y,y)}$ in (2.8).

Using Cauchy-Schwarz inequality, we can prove the following proposition.

Proposition 2.1. *Let X be an inner product space. Then the map defined by $\|x\| = \sqrt{(x,x)}$ is a norm on X.*

The proof of this proposition is left as an exercise.

Once we define a norm in a linear space, we can introduce limit and convergence in the space.

Definition 2.3. Let X be a normed linear space. we say that a sequence $\{x_n\}$ strongly converges to $x \in X$ if

$$\lim_{n\to\infty} \|x_n - x\| = 0.$$

Definition 2.4. A sequence $\{x_n\}$ in a normed linear space X is said to a Cauchy sequence if for any $\varepsilon > 0$ there exists an integer N such that

$$\|x_n - x_m\| \leq \varepsilon \quad \text{for all} \ \ n, m \geq N.$$

Definition 2.5. A normed linear space X is said to be complete if any Cauchy sequence $\{x_n\}$ strongly converges to an $x \in X$. A complete normed linear space is called a Banach space.

Definition 2.6. Let X be an inner product space. If X under the norm $\|x\| = \sqrt{(x,x)}$ is complete, it is called a Hilbert space.

We can show that the normed linear spaces $\mathbb{R}^n$ and $L^2(\Omega)$ are Hilbert spaces and that L^p is a Banach space. The proofs of these results require advanced mathematical analysis. So we omit them and refer to the books of functional analysis [1, 34, 35, 105].

In the space $\mathbb{R}^n$, there is a basis:

$$e_1 = (1, 0, \cdots, 0), \ e_2 = (0, 1, \cdots, 0), \ e_n = (0, \cdots, 0, 1).$$

We now extend the basis concept to Hilbert spaces.

Definition 2.7. Let H be a Hilbert space.

(1) A set $\{e_i\} \subset H$ is orthogonal if

$$(e_i, e_j) = 0, \quad i \neq j.$$

(2) A set $\{e_i\} \subset H$ is orthonormal if

$$(e_i, e_j) = \delta_{ij} = \begin{cases} 1, & i = j, \\ 0, & i \neq j. \end{cases}$$

(3) An orthonormal set $\{e_i\} \subset H$ is a basis for H if

$$x = \sum_{i=1}^{\infty} (x, e_i) e_i$$

for every $x \in H$. The equation means that

$$x = \lim_{n \to \infty} \sum_{i=1}^{n} (x, e_i) e_i.$$

It is easy to verify that the set

$$\frac{1}{\sqrt{2\pi}}, \frac{1}{\sqrt{\pi}} \sin x, \frac{1}{\sqrt{\pi}} \cos x, \cdots, \frac{1}{\sqrt{\pi}} \sin(nx), \frac{1}{\sqrt{\pi}} \cos(nx), \cdots$$

is orthonormal in $L^2(0, 2\pi)$. In fact, it is also a basis, but the proof is difficult and can be found in a book of functional analysis.

Exercises 2.1

1. Let $C[a,b]$ be the linear space of all continuous functions on $[a,b]$. Define

$$\|f\|_C = \max_{a \le x \le b} |f(x)|$$

   for any $f \in C[a,b]$. Show that $\|\cdot\|_C$ is a norm on $C[a,b]$ and $C[a,b]$ endowed with this norm is a Banach space.
2. Let X be an inner product space with the inner product $(\cdot,\cdot)$. Prove that the map defined by $\|x\| = \sqrt{(x,x)}$ is a norm on X.
3. Prove Young's inequality (2.4).
4. Let

$$l^2 = \left\{ (x_1, x_2, \cdots, x_n, \cdots) \mid x_n \in \mathbb{C}, \ \sum_{n=1}^{\infty} |x_n|^2 < \infty \right\}.$$

   Show that l^2 is a Hilbert space endowed with the following inner product

$$(\mathbf{x}, \mathbf{y}) = \sum_{n=1}^{\infty} x_n \bar{y}_n.$$

2.2 Open and Closed Sets

We now extend the concept of open (closed) interval in $\mathbb{R}$ to normed linear spaces.

Definition 2.8. A subset G of a normed linear space X is said to be open if for any $x \in G$ there exists δ such that $\{y \in X \mid \|x - y\| < \delta\} \subset G$.

The open ball $B(\mathbf{c}, r) = \{\mathbf{x} \in \mathbb{R}^n \mid \|\mathbf{x} - \mathbf{c}\| < r\}$ $(r > 0)$ is an open set in $\mathbb{R}^n$ (see Figure 2.1).

Definition 2.9. Let G be a subset of a normed linear space X and $x \in X$. x is a limit point of G if there exists a sequence $\{x_n\} \subset G$ with $x_n \ne x$ such that $x_n \to x$ as $n \to \infty$.

Any points in the ball $B(\mathbf{c}, r)$ are limit points of $B(\mathbf{c}, r)$. Moreover, any points on the surface $S(\mathbf{c}, r) = \{\mathbf{x} \in \mathbb{R}^n \mid \|\mathbf{x} - \mathbf{c}\| = r\}$ of the ball are also limit points of $B(\mathbf{c}, r)$. So a limit point of G does not necessarily belong to G.

Definition 2.10. A subset G of a normed linear space X is said to be closed if it contains all its limit points.

The closed ball $\bar{B}(\mathbf{c}, r) = \{\mathbf{x} \in \mathbb{R}^n \mid \|\mathbf{x} - \mathbf{c}\| \le r\}$ $(r > 0)$ is a closed set in $\mathbb{R}^n$.

Theorem 2.1. *A subset G of a normed linear space X is closed if and only if its complement $X - G = \{x \in X \mid x \notin G\}$ is open.*

Fig. 2.1 An open ball.

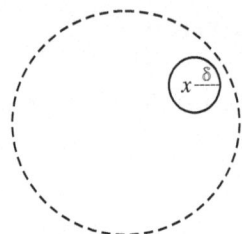

The proof of this theorem is left as an exercise. The closed ball $\bar{B}(\mathbf{c},r)$ is closed in $\mathbb{R}^n$. So its complement $\mathbb{R}^n - \bar{B}(\mathbf{c},r) = \{\mathbf{x} \in \mathbb{R}^n \mid \|\mathbf{x} - \mathbf{c}\| > r\}$ is open in $\mathbb{R}^n$.

Definition 2.11. Let G be a subset of a normed linear space X. The closure of G is the union of G with the set of all limit points of G. The closure of G is denoted by $\overline{G}$.

The closure of the open ball $B(\mathbf{c},r)$ is the closed ball $\bar{B}(\mathbf{c},r)$.

Definition 2.12. A subset G of a normed linear space X is dense in X if $\overline{G} = X$.

The set of all rational numbers is dense in $\mathbb{R}$. A set S is *countable* if there exists a one-to-one map from S to a subset of $\mathbb{N}$, the set of all nonnegative integers. The set $\{1,2,6,10\}$, the set $\mathbb{N}$, and the set of all rational numbers are countable.

Definition 2.13. A normed linear space X is separable if it contains a countable dense subset.

$\mathbb{R}$ is separable since the set of all rational numbers is countable.

Definition 2.14. A subset G of a normed linear space X is compact if for any family of open sets $\{G_\lambda, \lambda \in \Lambda\}$ with $G \subset \cup_{\lambda \in \Lambda} G_\lambda$, there exist finitely many $G_{\lambda_1}, \cdots, G_{\lambda_k}$ in this family such that $G \subset \cup_{i=1}^k G_{\lambda_i}$. G is relatively compact if the closure $\overline{G}$ is compact.

Any closed interval $[a,b]$ and the closed ball $\bar{B}(\mathbf{c},r)$ are compact. Any open interval (a,b) and open ball $B(\mathbf{c},r)$ are not compact, but relatively compact.

Exercises 2.2

1. Prove that the open ball $B(\mathbf{c},r) = \{\mathbf{x} \in \mathbb{R}^n \mid \|\mathbf{x} - \mathbf{c}\| < r\}$ $(r > 0)$ is an open set in $\mathbb{R}^n$.
2. Prove the closed ball $\bar{B}(\mathbf{c},r) = \{\mathbf{x} \in \mathbb{R}^n \mid \|\mathbf{x} - \mathbf{c}\| \leq r\}$ $(r > 0)$ is a closed set in $\mathbb{R}^n$.
3. Prove that a subset G of a normed linear space X is closed if and only if its complement $X - G = \{x \in X \mid x \notin G\}$ is open.

4. Let G be a subset of a normed linear space X. Show that

   (1) $\overline{G}$ is closed;
   (2) $\overline{G}$ is the smallest closed set containing G.

5. Prove that the closed interval $[a,b]$ and the closed ball $\bar{B}(\mathbf{c},r)$ are compact.
6. Prove that the open interval (a,b) and open ball $B(\mathbf{c},r)$ are not compact, but relatively compact.

2.3 Linear Operators

An $n \times n$ matrix $\mathbf{A}$ can be thought of as an operator in $\mathbb{R}^n$ and has the following linear property

$$\mathbf{A}(a\mathbf{x} + b\mathbf{y}) = a\mathbf{A}\mathbf{x} + b\mathbf{A}\mathbf{y} \quad \text{for any } \mathbf{x}, \mathbf{y} \in \mathbb{R}^n, a, b \in \mathbb{R}.$$

A linear operator in a normed linear space is just an extension of the matrix to the normed space.

Definition 2.15. Let X be a Banach space. A subset S of X is bounded if there exists $R > 0$ such that

$$\|x\| \leq R, \quad \forall x \in S.$$

Definition 2.16. Let X and Y be two normed linear spaces and let $D(A)$ be a subspace of X. A map $A : D(A) \rightarrow Y$ is called a linear operator if

$$A(ax + by) = aAx + bAy \quad \text{for any } x, y \in D(A),\ a, b \in \mathbb{F}.$$

The set $D(A)$ is called the domain of A. If $D(A) = X$ and A maps any bounded sets into bounded sets, that is, there exists a constant $C > 0$ such that

$$\|Ax\|_Y \leq C\|x\|_X \quad \text{for all } x \in X,$$

then A is called a bounded (or continuous) operator.

Example 2.4. For any integer $k \geq 0$, let $C^k[a,b]$ be the linear space of all k-times continuously differentiable functions on $[a,b]$. Define

$$\|f\|_{C^k} = \max_{\substack{a \leq x \leq b \\ 0 \leq i \leq k}} |f^{(i)}(x)|$$

for any $f \in C^k[a,b]$. Then $C^k[a,b]$ is a Banach space. We write $C[a,b]$ for $C^0[a,b]$. Define the differential operator

$$A = \frac{d^2}{dx^2}$$

in $C[a,b]$ with the domain $D(A) = C^2[a,b]$. It is clear that A is linear. We now show that it is unbounded in $C^2[a,b]$. Consider the set $S = \{x^n / \max\{|a|^n, |b|^n\} \mid n =$

$0,1,2,\cdots,\}$. Since $\|x^n/\max\{|a|^n,|b|^n\}\|_C = 1$, the set S is bounded. But $A(S) = \{n(n-1)x^{n-2}/\max\{|a|^n,|b|^n\} \mid n = 2,3,\cdots,\}$ is unbounded since

$$\|n(n-1)x^{n-2}/\max\{|a|^n,|b|^n\}\|_C = n(n-1)/\max\{|a|^2,|b|^2\}.$$

So A is unbounded.

Let $\mathscr{L}(X,Y)$ be the set of all linear bounded operators from X to Y. For any $a,b \in \mathbb{F}$ and $A,B \in \mathscr{L}(X,Y)$, we define $aA+bB$ as follows

$$(aA+bB)(x) = aAx+bBx \quad \text{for any } x \in X.$$

Then $\mathscr{L}(X,Y)$ is a linear space. Define

$$\|A\| = \sup_{x\neq 0} \frac{\|Ax\|}{\|x\|}.$$

We can show that $\|\cdot\|$ is a norm on $\mathscr{L}(X,Y)$ and that $\mathscr{L}(X,Y)$ is a Banach space.

If $Y = \mathbb{F}$ ($\mathbb{R}$ or $\mathbb{C}$), any $f \in \mathscr{L}(X,\mathbb{F})$ is called a linear bounded functional (or linear continuous functional) on X. We denote $X^* = \mathscr{L}(X,\mathbb{F})$ and call it the dual space of X.

Example 2.5. Let $X = L^2(0,b)$ and $b > 0$. For any function $f \in L^2(0,b)$, we define

$$F(g) = \int_0^b f(x)g(x)dx, \quad g \in L^2(0,b).$$

Then F is a linear bounded functional on $L^2(0,b)$. In fact, by Hölder's inequality, we have

$$|F(g)| \leq \int_0^b |f(x)g(x)|dx$$
$$\leq \left(\int_0^b |f(x)|^2dx\right)^{1/2} \left(\int_0^b |g(x)|^2dx\right)^{1/2}$$
$$= \|f\|_{L^2}\|g\|_{L^2}.$$

Thus F maps any bounded set in $L^2(0,b)$ into a bounded set in $\mathbb{C}$. This shows that $F \in (L^2(0,b))^*$. Since F is uniquely determined by $f(x)$, we can identify F as $f(x)$ and then say $f \in (L^2(0,b))^*$. In fact, we have $(L^2(0,b))^* = L^2(0,b)$, whose proof can be found in any functional analysis books, such as [34, 105].

Exercises 2.3

1. Show that an $n \times m$ matrix is a bounded operator from $\mathbb{R}^m$ to $\mathbb{R}^n$.
2. Let A be a linear bounded operator from X to Y. Prove that A is continuous, that is, if $x_n \to x$ in X, then $Ax_n \to Ax$ in Y.

3. Let X and Y be two normed linear spaces. Define

$$\|A\| = \sup_{x \neq 0} \frac{\|Ax\|}{\|x\|}.$$

Show that $\|\cdot\|$ is a norm on $\mathscr{L}(X,Y)$.

4. Let $k(x,y)$ be a continuous function on $[a,b] \times [a,b]$. Define the operator $K : C[a,b] \to C[a,b]$ by

$$(Kf)(x) = \int_a^b k(x,y)f(y)dy.$$

Show that K is a linear continuous operator on $C[a,b]$.

2.4 Sobolev Spaces

If $\alpha = (\alpha_1, \cdots, \alpha_n)$ is an n-tuple with $\alpha_i \in \mathbb{N}$, $i = 1, 2, \cdots, n$, we call α a multi-index, and $|\alpha| = \sum_{i=1}^n \alpha_i$ the modulus of α. We shall denote by $\mathbf{x}^\alpha$ the monomial $x_1^{\alpha_1} \cdots x_n^{\alpha_n}$, which has degree $|\alpha|$. Similarly, if $D_i = \frac{\partial}{\partial x_i}$ for $1 \leq i \leq n$, then

$$D^\alpha = D_1^{\alpha_1} \cdots D_n^{\alpha_n}$$

denotes a differential operator of order $|\alpha|$. $D^{(0,\cdots,0)}u = u$.

In what follows, Ω denotes a bounded domain (open set) in $\mathbb{R}^n$ and Γ denotes its boundary $\partial\Omega$. Given a nonnegative integer k, we let $C^k(\Omega)$ denote the vector space of all functions u, which together with all their derivatives $D^\alpha u$ of order $|\alpha| \leq k$, are continuous in Ω. We abbreviate $C^0(\Omega) = C(\Omega)$. We also set

$$C^\infty(\Omega) = \bigcap_{k=0}^\infty C^k(\Omega).$$

Moreover, the subspaces $C_0(\Omega)$ and $C_0^\infty(\Omega)$ consist of all those functions in $C(\Omega)$ and $C^\infty(\Omega)$, respectively, which have compact supports in Ω; that is, $\text{supp} u = \overline{\{\mathbf{x} \in \Omega \mid u(\mathbf{x}) \neq 0\}} \subset \Omega$. We sometimes use the notation $\mathscr{D}(\Omega)$ for $C_0^\infty(\Omega)$.

We next define $C^k(\overline{\Omega})$ ($C(\overline{\Omega})$ for $k = 0$) as the space of all those functions $u \in C^k(\Omega)$ for which $D^\alpha u$ is bounded and uniformly continuous in Ω for $0 \leq |\alpha| \leq k$. $C^k(\overline{\Omega})$ is a Banach space with the norm given by

$$\| u \|_{C^k} = \max_{0 \leq |\alpha| \leq k} \sup_{x \in \Omega} |D^\alpha u(\mathbf{x})|.$$

We now define the regularity of the boundary of a domain $\Omega \subset \mathbb{R}^n$.

Definition 2.17. The boundary $\partial\Omega$ of a domain Ω is said to be C^k if for each point $\mathbf{x}_0 \in \partial\Omega$ there exist $r > 0$ and a C^k function $f : \mathbb{R}^{n-1} \to \mathbb{R}$ such that, after relabeling and reorienting the coordinates axe if necessary, we have

$$\Omega \cap B(\mathbf{x}_0, r) = \{\mathbf{x} \in B(\mathbf{x}_0, r) \mid x_n > f(x_1, x_2, \cdots, x_{n-1})\}.$$

$\partial\Omega$ is said to be C^∞ if it is C^k for $k = 1, 2, \cdots$.

Definition 2.18. For $p \in [1, \infty)$ and a nonnegative integer m, we define

$$W^{m,p}(\Omega) = \{u \mid D^\alpha u \in L^p(\Omega) \text{ for all } |\alpha| \le m\}$$

with the following norm

$$\| u \|_{W^{m,p}} = \left(\sum_{0 \le |\alpha| \le m} \| D^\alpha u \|_{L^p}^p \right)^{\frac{1}{p}}.$$

The space $W_0^{m,p}(\Omega)$ is defined to be the closure of $C_0^\infty(\Omega)$ in the space $W^{m,p}(\Omega)$. $W^{-m,q}(\Omega) = (W_0^{m,p}(\Omega))^*$, the dual space of $W_0^{m,p}(\Omega)$, where q is the conjugate exponent to p, i.e., $\dfrac{1}{p} + \dfrac{1}{q} = 1$.

These spaces are called Sobolev spaces over Ω. In what follows, we shall write $W^{m,2}(\Omega)$ as $H^m(\Omega)$.

Theorem 2.2. *The Sobolev spaces have the following properties:*

(1) $W^{m,p}(\Omega)$ is a Banach space.
(2) For $p \in [1, \infty)$, $W^{m,p}(\Omega)$ is separable.
(3) $H^m(\Omega)$ is a Hilbert space with the inner product

$$(u, v)_{H^m} = \sum_{0 \le |\alpha| \le m} (D^\alpha u, D^\alpha v)_{L^2},$$

where $(u, v)_{L^2} = \int_\Omega u(\mathbf{x})\bar{v}(\mathbf{x}) \, dV$ is the inner product in $L^2(\Omega)$.

The proof of this theorem is referred to [1, p.45-47] or [35, Chapter 5].

Theorem 2.3. *$C_0^\infty(\Omega)$ is dense in $L^2(\Omega)$.*

The proof of this theorem is omitted and referred to [35, Chapter 5].

Exercises 2.4

1. Define the differential operator

$$A = \frac{d^2}{dx^2}$$

in $L^2(a,b)$ with the domain $D(A) = H^2(a,b)$. Show that A is unbounded in $L^2(a,b)$.

2. Show that

$$u(\mathbf{x}) = \ln\left(1 - \ln|\mathbf{x}|\right)$$

is in $H^1(B(0,1))$, where $B(0,1)$ is the unit ball in $\mathbb{R}^2$ and

$$|\mathbf{x}| = \sqrt{x_1^2 + x_2^2}.$$

2.5 Eigenvalue Problems

In matrix theory, an eigenvalue of a matrix $\mathbf{A}$ is a complex number λ such that the linear system

$$\mathbf{A}\mathbf{x} = \lambda\mathbf{x}$$

has nonzero solutions. This concept can be extended to differential equations.

Consider the eigenvalue problem

$$-\frac{d^2\varphi}{dx^2} = \lambda\varphi, \tag{2.9}$$

$$\varphi(0) = 0, \quad \varphi(L) = 0. \tag{2.10}$$

The homogeneous boundary condition $\varphi(0) = \varphi(L) = 0$ is called the *Dirichlet boundary condition*. An *eigenvalue* is a complex number λ such that (2.9)-(2.10) has nonzero solutions. The nonzero solutions are called *eigenfunctions* corresponding to the eigenvalue.

Theorem 2.4. *The eigenvalue problem (2.9)-(2.10) has eigenvalues*

$$\lambda_n = \frac{n^2\pi^2}{L^2}, \quad n = 1, 2, \cdots$$

and the corresponding eigenfunctions

$$\varphi_n = \sin\frac{n\pi x}{L}, \quad n = 1, 2, \cdots$$

Proof. The auxiliary equation for the problem is

$$m^2 = -\lambda.$$

We now discuss three cases of λ.

Case 1: $\lambda < 0$. In this case, the auxiliary equation has distinct real roots $m_1 = \sqrt{-\lambda}$ and $m_2 = -\sqrt{-\lambda}$ and then the general solution of (2.9) is

$$\varphi(x) = c_1 e^{\sqrt{-\lambda}x} + c_2 e^{-\sqrt{-\lambda}x}.$$

The boundary conditions imply that $c_1 = c_2 = 0$. So no non-zero solutions exist and therefore $\lambda < 0$ is not an eigenvalue.

Case 2: $\lambda = 0$. In this case, the auxiliary equation has repeated real roots $m_1 = m_2 = 0$ and then the general solution of (2.9) is

$$\varphi = c_1 + c_2 x.$$

The boundary conditions imply that $c_1 = c_2 = 0$. So no non-zero solutions exist and then $\lambda = 0$ is not an eigenvalue.

Case 3: $\lambda > 0$. In this case, the auxiliary equation has conjugate complex roots $m_1 = i\sqrt{\lambda}$ and $m_2 = -i\sqrt{\lambda}$ and then the general solution of (2.9) is

$$\varphi = c_1 \cos(\sqrt{\lambda}x) + c_2 \sin(\sqrt{\lambda}x).$$

$\varphi(0) = 0$ implies that $c_1 = 0$. $\varphi(L) = 0$ gives

$$\sin(\sqrt{\lambda}L) = 0.$$

So $\sqrt{\lambda}L = n\pi$ $(n = 1, 2, \cdots)$ and we obtain the eigenvalues

$$\lambda_n = \frac{n^2 \pi^2}{L^2}, \quad n = 1, 2, \cdots$$

and the corresponding eigenfunctions

$$\varphi_n = \sin\left(\frac{n\pi x}{L}\right), \quad n = 1, 2, \cdots$$

Consider the higher-dimensional eigenvalue problem

$$\mu \nabla^2 \psi + \nabla \cdot (\psi \mathbf{v}) + a\psi = \lambda \psi \quad \text{in } \Omega, \tag{2.11}$$
$$\psi = 0 \quad \text{on } \Gamma, \tag{2.12}$$

where $a(\mathbf{x})$ is a given function, $\mathbf{v}(\mathbf{x})$ is a given vector function, ∇^2 is the Laplace operator

$$\nabla^2 u = \frac{\partial^2 u}{\partial x_1^2} + \cdots + \frac{\partial^2 u}{\partial x_n^2},$$

∇ denotes the gradient operator

$$\nabla = \left(\frac{\partial}{\partial x_1}, \cdots, \frac{\partial}{\partial x_n}\right),$$

and $\nabla \cdot (\psi \mathbf{v}) = \frac{\partial(\psi v_1)}{\partial x_1} + \cdots + \frac{\partial(\psi v_n)}{\partial x_n}$, the divergence of $\psi \mathbf{v}$. The homogeneous boundary condition $\psi|_\Gamma = 0$ is called the Dirichlet boundary condition. This eigenvalue problem can be solved explicitly only for very simple geometries such as rectangles. In other situations, we have to solve it numerically.

Example 2.6. Assume that $\Omega = (0, L) \times (0, H)$, $\mathbf{v}(\mathbf{x}) = 0$, and a is a constant. Then (2.11) becomes

$$\mu \nabla^2 \psi = (\lambda - a) \psi \quad \text{in } \Omega, \tag{2.13}$$
$$\psi(0, y) = 0, \ \psi(L, y) = 0, \ \psi(x, 0) = 0, \ \psi(x, H) = 0. \tag{2.14}$$

Setting $\psi = X(x) Y(y)$ and plugging it into (2.13), we deduce that

$$\mu \frac{X''}{X} = \frac{(\lambda - a) Y - \mu Y''}{Y} = \nu,$$
$$X(0) = 0, \ X(L) = 0, \ Y(0) = 0, \ Y(H) = 0.$$

We then have two eigenvalue problems

$$\mu X'' = \nu X,$$
$$X(0) = 0, \ X(L) = 0,$$

and

$$\mu Y'' = (\lambda - a - \nu) Y,$$
$$Y(0) = 0, \ Y(H) = 0.$$

These problems have the following eigenvalues and corresponding eigenfunctions

$$\nu_n = -\mu \left(\frac{n\pi}{L} \right)^2, \quad X_n = \sin \frac{n\pi x}{L}, \quad n = 1, 2, \cdots,$$
$$\lambda_{mn} = a - \mu \left(\frac{m\pi}{H} \right)^2 - \mu \left(\frac{n\pi}{L} \right)^2, \quad Y_m = \sin \frac{m\pi y}{H}, \quad m = 1, 2, \cdots$$

Therefore (2.13)-(2.14) have the following eigenvalues and corresponding eigenfunctions

$$\lambda_{mn} = a - \mu \left(\frac{m\pi}{H} \right)^2 - \mu \left(\frac{n\pi}{L} \right)^2, \tag{2.15}$$
$$\psi_{mn} = \sin \frac{n\pi x}{L} \sin \frac{m\pi y}{H}, \quad m, n = 1, 2, \cdots \tag{2.16}$$

Example 2.7. Assume that $\Omega = (0, L) \times (0, H)$, $\mathbf{v} = (v_1, v_2)$ (v_1, v_2 are constants), and a is a constant. Then (2.11) becomes

$$\mu \nabla^2 \psi + v_1 \psi_x + v_2 \psi_y = (\lambda - a) \psi \quad \text{in } \Omega, \tag{2.17}$$
$$\psi(0, y) = 0, \ \psi(L, y) = 0, \ \psi(x, 0) = 0, \ \psi(x, H) = 0. \tag{2.18}$$

As above, setting $\psi = X(x) Y(y)$ gives two eigenvalue problems

$$\mu X_{xx} + v_1 X_x = \nu X,$$
$$X(0) = 0, \ X(L) = 0,$$

and

$$\mu Y_{yy} + v_2 Y_y = (\lambda - a - v)Y,$$
$$Y(0) = 0, \ Y(H) = 0.$$

These problems have the following eigenvalues and corresponding eigenfunctions

$$v_n = -\mu \left[\left(\frac{n\pi}{L} \right)^2 + \frac{v_1^2}{4\mu^2} \right], \quad X_n = e^{\frac{-v_1}{2\mu}x} \sin \frac{n\pi x}{L}, \quad n = 1, 2, \cdots,$$

$$\lambda_{mn} = a - \mu \left[\left(\frac{m\pi}{H} \right)^2 + \left(\frac{n\pi}{L} \right)^2 + \frac{v_1^2 + v_2^2}{4\mu^2} \right],$$

$$Y_m = e^{\frac{-v_2}{2\mu}y} \sin \frac{m\pi y}{H}, \quad m = 1, 2, \cdots$$

Therefore (2.17) has the following eigenvalues and corresponding eigenfunctions

$$\lambda_{mn} = a - \mu \left[\left(\frac{m\pi}{H} \right)^2 + \left(\frac{n\pi}{L} \right)^2 + \frac{v_1^2 + v_2^2}{4\mu^2} \right], \tag{2.19}$$

$$\psi_{mn} = e^{\frac{-1}{2\mu}(v_1 x + v_2 y)} \sin \frac{n\pi x}{L} \sin \frac{m\pi y}{H}, \quad m, n = 1, 2, \cdots \tag{2.20}$$

In general cases, although we cannot solve (2.11)-(2.12) explicitly, we can prove the following theorem.

Theorem 2.5. *Assume that* $\mathbf{v}(\mathbf{x}) = 0$. *Then the eigenvalues and corresponding eigenfunctions of (2.11)-(2.12) have the following properties:*

(1) All the eigenvalues are real.
(2) There are infinite number of eigenvalues $\lambda_1 > \lambda_2 > \cdots > \lambda_n > \cdots$ *with* $\lim\limits_{n \to \infty} \lambda_n = -\infty$.
(3) There may be many eigenfunctions corresponding to one eigenvalue.
(4) Eigenfunctions belonging to different eigenvalues λ_1 *and* λ_2 *are orthogonal, that is,*

$$\int_\Omega \psi_{\lambda_1} \psi_{\lambda_2} dV = 0.$$

Furthermore, different eigenfunctions belonging to the same eigenvalue can be made orthogonal. Thus, we may assume that all eigenfunctions are orthogonal to each other.
(5) All eigenfunctions $\{\psi_m\}_{m=1}^{\infty}$ *form a basis in* $L^2(\Omega)$, *that is, any function* $f \in L^2(\Omega)$ *can be represented by the series of the eigenfunctions*

$$f = \sum_{m=1}^{\infty} a_m \psi_m.$$

The series converges in the mean:

$$\lim_{n\to\infty} \int_{\Omega} \left| f - \sum_{m=1}^{n} a_m \psi_m \right|^2 dV = 0.$$

The proof of this theorem is referred to [31, Chapter 8] and [35, p.335, Theorem 1]. We can easily verify that the eigenvalues and eigenfunctions (2.15)-(2.16) satisfy all properties in the above theorem. If $\mathbf{v} \neq \mathbf{0}$, the eigenvalue problem is complicated and not all properties in the above theorem are true. For instance, the eigenfunctions (2.20) are not orthogonal in $L^2((0,L) \times (0,H))$. But, if we change the usual inner product of $L^2((0,L) \times (0,H))$ to the following weighted inner product

$$(u,v)_w = \int_0^L \int_0^H e^{\frac{1}{\mu}(v_1 x + v_2 y)} uv \, dx \, dy \quad u,v \in L^2((0,L) \times (0,H)), \qquad (2.21)$$

then they are orthogonal. Unfortunately, for a general $\mathbf{v}$, we do not know whether there exists a weighted inner product to make the eigenfunctions orthogonal. To state a result in this case, we introduce the concept of direct sum.

Let X_1 and X_2 be two subspaces of a vector space X. The sum of these two subspaces, denoted by $X_1 + X_2$, is the set of all the sums $x_1 + x_2$, where $x_1 \in X_1$ and $x_2 \in X_2$. If $X_1 \cap X_2 = \{0\}$, the sum is called a *direct sum* and is denoted by $X_1 \oplus X_2$. If $X = X_1 \oplus X_2$, then we say X has a *direct decomposition* and X_1 and X_2 are complementary subspaces.

As a simple example of direct sum, we consider $X = \mathbb{R}^2, X_1 = \{(x,x) \mid x \in \mathbb{R}\}$ (the line $y = x$), $X_2 = \{(x,2x) \mid x \in \mathbb{R}\}$ (the line $y = 2x$). Then $\mathbb{R}^2 = X_1 \oplus X_2$. In fact, if $(x,y) \in X_1 \cap X_2$, then $x = y = 2x$. So $x = 0$ and $y = x = 0$. In addition, for any $(x,y) \in \mathbb{R}$, let $(x,y) = (x_1,x_1) + (x_2,2x_2)$. This implies that

$$x_1 + x_2 = x,$$
$$x_1 + 2x_2 = y.$$

Solving this system gives $x_1 = 2x - y$ and $x_2 = y - x$. Thus $(x,y) = (2x - y, 2x - y) + (y - x, 2(y - x))$ with $(2x - y, 2x - y) \in X_1$ and $(y - x, 2(y - x)) \in X_2$.

Let $\lambda_1 > \lambda_2 > \cdots > \lambda_n > \cdots$ with $\lim_{n\to\infty} \lambda_n = -\infty$ be the eigenvalues of (2.11) with $\mathbf{v} = \mathbf{0}$ and let $\psi_{r1}, \psi_{r2}, \cdots, \psi_{rn_r}$ be n_r normalized (that is, $\int_{\Omega} |\psi_{r1}|^2 dV = 1$) orthogonal eigenfunctions corresponding to the eigenvalue λ_r. Let N be a positive integer. We define

$$H_N = \text{span}\{\psi_{11}, \psi_{12}, \cdots, \psi_{1n_1}; \cdots, \psi_{N1}, \psi_{N2}, \cdots, \psi_{Nn_N}\}$$

$$= \left\{ \sum_{r=1}^{N} \sum_{s=1}^{n_r} c_{rs} \psi_{rs} \mid c_{rs} \text{ are any numbers} \right\},$$

$$H_\infty = \text{span}\{\psi_{N+1,1}, \psi_{N+1,2}, \cdots, \psi_{N+1,n_{N+1}}, \cdots\}$$

$$= \left\{ \sum_{r=N+1}^{N+M} \sum_{s=1}^{n_r} c_{rs} \psi_{rs} \mid c_{rs} \text{ are any numbers and } M > 0 \text{ is an integer} \right\}.$$

As the direct result of Parts (4) and (5) of Theorem 2.5, we have the following orthogonal decomposition of $L^2(\Omega)$.

Corollary 2.1. *If* $\mathbf{v} = \mathbf{0}$, *then* $L^2(\Omega) = H_N \oplus H_\infty$. *Furthermore, the decomposition is orthogonal, that is, every function* $u_N \in H_N$ *is orthogonal to every function* $u_\infty \in H_\infty$.

For a matrix $\mathbf{A}$, if λ is not an eigenvalue of $\mathbf{A}$, then $\lambda I - \mathbf{A}$ is invertible. This concept can be extended to a linear operator in a normed space. If A is a linear, not necessarily bounded, operator in X, the *resolvent set* $\rho(A)$ of A is the set of all complex numbers λ for which $\lambda I - A$ is invertible, i.e., $(\lambda I - A)^{-1}$ is a bounded linear operator in X. The family $R(\lambda, A) = (\lambda I - A)^{-1}$, $\lambda \in \rho(A)$ of bounded linear operators is called the *resolvent* of A. The complementary set $\sigma(A)$ of $\rho(A)$ in the complex plane $\mathbb{C}$ is called the *spectrum* of A. For the matrix $\mathbf{A}$, the spectra consist of all their eigenvalues.

Definition 2.19. Let X and Y be two Banach spaces and $D(A)$ be a linear subspace of X. Let $A : D(A) \to Y$ be a linear operator.

(1) A is densely defined if $D(A)$ is dense in X.
(2) A is closed if the graph of A

$$\mathscr{G}(A) = \{(x, y) \in X \times Y \mid x \in D(A), y = Ax\}$$

is closed in $X \times Y$. Alternatively, A is closed if whenever

$$x_n \in D(A), n = 1, 2, \cdots \text{ and } \lim_{n \to \infty} x_n = x, \ \lim_{n \to \infty} Ax_n = y,$$

it follows that $x \in D(A)$ and $Ax = y$.

Consider the differential operator

$$A = \frac{d^2}{dx^2}$$

in $L^2(a, b)$ with the domain $D(A) = H^2(a, b) \cap H_0^1(a, b)$. By Theorem 2.3, we deduce that the operator is densely defined. We now show that it is closed. Let $u_n \in H^2(a, b) \cap H_0^1(a, b)$ and $u_n \to u$, $u_n'' \to v$ as $n \to \infty$. We need to prove that $v = u''$. Since $u_n \to u$ and $u_n'' \to v$, the limit and differentiation can be exchanged. So we have $v = \lim_{n \to \infty} u_n'' = \frac{d^2}{dx^2} \lim_{n \to \infty} u_n = u''$.

Definition 2.20. Let X_1 be a subspace of a Banach space X and $A : D(A) \subset X \to X$ be a linear operator. If $A(X_1 \cap D(A)) \subset X_1$, then X_1 is said to be invariant under A.

Theorem 2.6. *Assume that the spectrum* $\sigma(A)$ *of a closed operator* A *on a Banach space* X *is the union of two parts,* $\sigma_r(A)$ *and* $\sigma_l(A)$, *such that a rectifiable, simple closed curve* Γ *can be drawn so as to enclose an open set containing* $\sigma_r(A)$ *in its interior and* $\sigma_l(A)$ *in its exterior. Then the operator defined by*

$$Px = \frac{1}{2\pi i} \int_\Gamma (\lambda I - A)^{-1} x \, d\lambda \qquad (2.22)$$

is a projection (that is, $P^2 = P$), where the integral is in the counterclockwise direction along Γ. This projection induces a decomposition of the space X

$$X = X_r \oplus X_l, \quad X_r = PX, \, X_l = (I - P)X.$$

Moreover, X_r and X_l are invariant under A, the restriction A_r of A to X_r is a bounded operator on X_r with the spectrum $\sigma(A_r) = \sigma_r(A)$, and the restriction A_l of A to X_l has the spectrum $\sigma(A_l) = \sigma_l(A)$.

The proof of this theorem needs advanced complex analysis and thus is referred to [24, p.71, Lemma 2.5.7] and [44, p.178, Theorem 6.17].

Using this theorem, we can show that $L^2(\Omega)$ can be decomposed by using the eigenfunctions of the operator A defined on $L^2(\Omega)$ by

$$A\psi = \mu \nabla^2 \psi + \nabla \cdot (\psi \mathbf{v}) + a(\mathbf{x})\psi \qquad (2.23)$$

with the domain $D(A) = H^2(\Omega) \cap H^1_0(\Omega)$.

Theorem 2.7. *The eigenvalues and corresponding eigenfunctions of the operator (2.23) have the following properties:*

(1) There are infinite number of eigenvalues $\{\lambda_n\}$, $n = 1, 2, \cdots$ with $\lim_{n\to\infty} |\lambda_n| = \infty$.
   The number of eigenvalues λ_n with $\mathrm{Re}\lambda_n > r$ is finite, where r is a real number.
(2) There may be many eigenfunctions corresponding to one eigenvalue.
(3) $L^2(\Omega) = H_N \oplus H_\infty$, where H_N is a finite dimensional space containing all eigenfunctions corresponding to the eigenvalues $\lambda_1, \lambda_2, \cdots, \lambda_N$ and H_∞ is an infinite dimensional space containing all eigenfunctions corresponding to the eigenvalues $\lambda_{N+1}, \lambda_{N+2}, \cdots$. Moreover, H_N and H_∞ are invariant under the operator A defined by (2.23).

The proof of this theorem is refereed to [35, p.305, Theorem 5] and [44, p. 178, Theorem 6.17]. If $\mathbf{v} \neq \mathbf{0}$, the decomposition $L^2(\Omega) = H_N \oplus H_\infty$ is usually not orthogonal.

The following divergence theorem will be frequently used. For any continuously differentiable vector $\mathbf{F}$, we have

$$\int_\Omega \nabla \cdot \mathbf{F} \, dV = \int_\Gamma \mathbf{F} \cdot \mathbf{n} \, dS, \qquad (2.24)$$

where $\mathbf{n}$ denotes the outward normal of the boundary of Ω. For a scalar function G, we have

$$\nabla \cdot (G\mathbf{F}) = \mathbf{F} \cdot \nabla G + G \nabla \cdot \mathbf{F}.$$

It then follows from (2.24) that

$$\int_\Omega (\mathbf{F} \cdot \nabla G + G \nabla \cdot \mathbf{F}) \, dV = \int_\Omega \nabla \cdot (G\mathbf{F}) \, dV = \int_\Gamma G\mathbf{F} \cdot \mathbf{n} \, dS,$$

and then

$$\int_\Omega \mathbf{F} \cdot \nabla G \, dV = \int_\Gamma G\mathbf{F} \cdot \mathbf{n} \, dS - \int_\Omega G\nabla \cdot \mathbf{F} \, dV. \tag{2.25}$$

The equation (2.25) is called the formula of integration by parts. Applying integration by parts, we obtain the Green's identity

$$\int_\Omega v\nabla^2 u \, dV = \int_\Gamma v\frac{\partial u}{\partial \mathbf{n}} \, dS - \int_\Omega \nabla v \cdot \nabla u \, dV. \tag{2.26}$$

The following Poincaré's inequality will play a key role in estimating solutions of partial differential equations.

Lemma 2.1. *(Poincaré's inequality) Let $\lambda_1 > 0$ be the smallest eigenvalue of*

$$-\nabla^2 \psi = \lambda \psi \quad in \ \Omega, \tag{2.27}$$

$$\psi = 0 \quad on \ \Gamma. \tag{2.28}$$

Then

$$\lambda_1 \int_\Omega |f(\mathbf{x})|^2 \, dV \le \int_\Omega |\nabla f(\mathbf{x})|^2 dV \quad for \ all \ f \in H_0^1(\Omega). \tag{2.29}$$

Proof. Let $0 < \lambda_1 \le \lambda_2 \le \cdots \le \lambda_n \le \cdots$ with $\lim_{n\to\infty} \lambda_n = \infty$ be the eigenvalues of (2.27)-(2.28) and let $\psi_1, \psi_2, \cdots, \psi_n, \cdots$ be the corresponding normalized orthogonal eigenfunctions (that is, $\int_\Omega |\psi_i|^2 dV = 1$). By Theorem 2.5, we have

$$f = \sum_{m=1}^\infty a_m \psi_m,$$

where a_m $(m = 1, 2, \cdots)$ are constants. Integration by parts gives

$$\begin{aligned}
\int_\Omega |\nabla f|^2 dV &= -\int_\Omega f\nabla^2 f \, dV \\
&= -\int_\Omega \sum_{m=1}^\infty a_m \psi_m \sum_{m=1}^\infty a_m \nabla^2 \psi_m dV \\
&= \int_\Omega \sum_{m=1}^\infty a_m \psi_m \sum_{m=1}^\infty a_m \lambda_m \psi_m dV \\
&= \sum_{m=1}^\infty \lambda_m a_m^2 \\
&\ge \lambda_1 \sum_{m=1}^\infty a_m^2 \\
&= \lambda_1 \int_\Omega |f|^2 dV.
\end{aligned}$$

Exercises 2.5

1. Solve the eigenvalue problem

$$X'' = \lambda X,$$
$$X(0) = 0, \ X'(L) = 0.$$

2. Solve the eigenvalue problem

$$\nabla^2 \psi = \lambda \psi,$$
$$\frac{\partial \psi}{\partial x}(0,y) = \frac{\partial \psi}{\partial x}(L,y) = 0, \quad \frac{\partial \psi}{\partial y}(x,0) = \frac{\partial \psi}{\partial y}(x,H) = 0.$$

3. Show that the eigenfunctions of the problem

$$\mu \nabla^2 \psi + v_1 \frac{\partial \psi}{\partial x} + v_2 \frac{\partial \psi}{\partial y} = \lambda \psi \quad \text{in } \Omega,$$
$$\psi(0,y) = 0, \ \psi(L,y) = 0, \ \psi(x,0) = 0, \ \psi(x,H) = 0,$$

are orthogonal with the weighted inner product (2.21), where v_1, v_2 are constants.

4. Prove Green's identity

$$\int_\Omega v \nabla^2 u \, dV = \int_\Gamma v \frac{\partial u}{\partial \mathbf{n}} \, dS - \int_\Omega \nabla v \cdot \nabla u \, dV.$$

2.6 Semigroups

To get an idea of what a semigroup is and how it can be used to deal with differential equations, we consider the linear system

$$\frac{d\mathbf{x}}{dt} = \mathbf{A}\mathbf{x}, \quad \mathbf{x}(0) = \mathbf{x}_0, \tag{2.30}$$

where $\mathbf{x}$ is a state vector, $\mathbf{x}_0$ is an initial state, and $\mathbf{A}$ is an $n \times n$ constant matrix. The solution of this system is given by

$$\mathbf{x}(t) = e^{\mathbf{A}t}\mathbf{x}^0,$$

where

$$e^{\mathbf{A}t} = \sum_{n=0}^\infty \frac{\mathbf{A}^n t^n}{n!}. \tag{2.31}$$

The family of the matrices $e^{\mathbf{A}t}$ has the following important properties, the so-called semigroup properties,

$$e^{\mathbf{A} \cdot 0} = \mathbf{I}, \qquad e^{\mathbf{A}(t+s)} = e^{\mathbf{A}t} e^{\mathbf{A}s}. \tag{2.32}$$

Such a family of matrices is called a semigroup.

A partial differential equation can be formulated into a problem similar to (2.30). Consider the wave equation

$$\frac{\partial^2 u}{\partial t^2} = c^2 \frac{\partial^2 u}{\partial x^2} \quad \text{in } (0, L) \times (0, \infty), \tag{2.33}$$

$$u(0, t) = 0, \ u(L, t) = 0, \quad t \geq 0, \tag{2.34}$$

$$u(x, 0) = u_0(x), \quad \frac{\partial u}{\partial t}(x, 0) = u_1(x), \quad x \in (0, L). \tag{2.35}$$

We define an operator A on the Hilbert space $\mathscr{H} = H_0^1(0, 1) \times L^2(0, 1)$ by

$$A = \begin{pmatrix} 0 & I \\ c^2 \frac{d^2}{dx^2} & 0 \end{pmatrix} \tag{2.36}$$

with the domain $D(A) = (H^2(0, 1) \cap H_0^1(0, 1)) \times H_0^1(0, 1)$, where I denotes the identity operator. Set $v = \frac{\partial u}{\partial t}$ and

$$\mathbf{u} = \begin{pmatrix} u \\ v \end{pmatrix}, \quad \mathbf{u}_0 = \begin{pmatrix} u_0 \\ u_1 \end{pmatrix}.$$

Then the problem (2.33)-(2.35) can be formulated as an abstract differential equation

$$\frac{d\mathbf{u}}{dt} = A\mathbf{u}, \quad \mathbf{u}(0) = \mathbf{u}_0 \tag{2.37}$$

on $\mathscr{H}$. We call the problem (2.37) as an abstract Cauchy problem.

We could guess that the solution of (2.37) should be given by

$$\mathbf{u}(t) = e^{At} \mathbf{u}_0.$$

However, the definition (2.31) of the matrix exponential no longer makes a sense in this case and we have to explain what e^{At} means. This leads to the concept of semigroup in a Banach space.

2.6.1 Definitions and Properties

Definition 2.21. Let X be a Banach space. A one-parameter family $T(t)$, $0 \leq t < \infty$, of bounded linear operators from X into X is called a *semigroup of bounded linear operators on X* if

(1) $T(0) = I$, (I is the identity operator on X);
(2) $T(t + s) = T(t)T(s)$ for every $t, s \geq 0$ (the semigroup property).

Furthermore, if the semigroup $T(t)$ also satisfies

$$\lim_{t \to 0^+} T(t)x = x \qquad \text{for every } x \in X, \tag{2.38}$$

then $T(t)$ is called a *strongly continuous* semigroup of bounded linear operators on X. A strongly continuous semigroup of bounded linear operators on X will be called a *semigroup of class C_0* or simply a *C_0 semigroup*.

It is clear that the definition of semigroup is an extension of $e^{\mathbf{A}t}$ satisfying (2.32), where $\mathbf{A}$ is a matrix. For the matrix $\mathbf{A}$, we have

$$\mathbf{A}\mathbf{x} = \lim_{t \to 0^+} \frac{e^{\mathbf{A}t}\mathbf{x} - \mathbf{x}}{t} = \lim_{t \to 0^+} \frac{e^{\mathbf{A}t}\mathbf{x} - e^{\mathbf{A}0}\mathbf{x}}{t} = \frac{d^+ e^{\mathbf{A}t}\mathbf{x}}{dt}\bigg|_{t=0}.$$

This is a very important relationship between the matrix $\mathbf{A}$ and the semigroup $e^{\mathbf{A}t}$. We can say that the semigroup $e^{\mathbf{A}t}$ is generated by $\mathbf{A}$. This relationship can be extended to the case of C_0 semigroup.

The linear operator A defined by

$$D(A) = \left\{ x \in X \ : \ \lim_{t \to 0^+} \frac{T(t)x - x}{t} \text{ exists} \right\} \tag{2.39}$$

and

$$Ax = \lim_{t \to 0^+} \frac{T(t)x - x}{t} = \frac{d^+ T(t)x}{dt}\bigg|_{t=0} \qquad \text{for } x \in D(A) \tag{2.40}$$

is called the *infinitesimal generator* of the semigroup $T(t)$, $D(A)$ is the domain of A. We symbolically write $T(t) = e^{At}$.

Semigroups have the following important properties.

Theorem 2.8. *Let $T(t)$ be a C_0 semigroup and A its infinitesimal generator. Then*

(1) There exist constants $\omega \geq 0$ and $M \geq 1$ such that

$$\|T(t)\| \leq Me^{\omega t}, \quad \text{for } 0 \leq t < \infty. \tag{2.41}$$

(2) For every $x \in X$, $t \to T(t)x$ is a continuous function from $[0, \infty)$ into X.
(3) For $x \in D(A)$, $T(t)x \in D(A)$ and

$$\frac{d}{dt} T(t)x = AT(t)x = T(t)Ax. \tag{2.42}$$

(4) $D(A)$, the domain of A, is dense in X and A is a closed linear operator.

The proof is omitted and referred to [91, p.4, Theorems 2.2 and 2.4] because advanced results such as the uniform boundedness theorem [95, p.43, Theorem 2.5] from functional analysis are required.

The property (2.42) shows that $u(t) = T(t)x$ is the solution of the equation

$$\frac{du}{dt} = Au.$$

This is why the semigroup theory is a powerful tool to study partial differential equations.

The domain of C_0 semigroups is the real nonnegative axis. It is natural to extend the domain of the parameter to regions in the complex plane that include the real nonnegative axis. It is clear that the domain must be an additive semigroup of complex numbers to preserve the semigroup structure, such as angles around the positive real axis.

Definition 2.22. Let $Z_\Delta = \{z \in \mathbb{C} \mid \phi_1 < \arg(z) < \phi_2, \phi_1 < 0 < \phi_2\}$. Suppose $T(z)$ is a bounded linear operator on a Banach space X for $z \in Z_\Delta$. The family $T(z)$, $z \in Z_\Delta$, is an analytic semigroup in Z_Δ if

(1) $z \to T(z)$ is analytic in Z_Δ, that is, the derivative $\frac{dT}{dz}$ exists for every $z \in Z_\Delta$;
(2) $T(0) = I$ and $\lim_{\substack{z \to 0 \\ z \in Z_\Delta}} T(z)x = x$ for every $x \in X$;
(3) $T(z_1 + z_2) = T(z_1)T(z_2)$ for z_1, $z_2 \in Z_\Delta$.

Clearly, the restriction of an analytic semigroup to the real axis is a C_0 semigroup.

2.6.2 Formulas for Semigroup Representation

If A is a matrix, then the semigroup generated by A is represented by the series (2.31). This is also true for a bounded linear operator. Unfortunately, if A is unbounded, the series no longer makes a sense. Then a question arises: Is there a formula to represent the semigroup in terms of A if A is unbounded?

To answer this question, we look at Cauchy's integral formula. If $f(\lambda)$ is analytic in and on a simple and closed curve Γ, then

$$f(\lambda_0) = \frac{1}{2\pi i} \int_\Gamma \frac{f(\lambda)}{\lambda - \lambda_0} d\lambda,$$

where λ_0 is inside Γ. Letting $\lambda_0 = a$ and $f(\lambda) = e^{\lambda t}$, we obtain

$$e^{at} = \frac{1}{2\pi i} \int_\Gamma \frac{e^{\lambda t}}{\lambda - a} d\lambda = \frac{1}{2\pi i} \int_\Gamma e^{\lambda t} (\lambda - a)^{-1} d\lambda.$$

This representation of the semigroup e^{at} generated by the number a is more useful in the sense that it can be generalized to unbounded operators.

Theorem 2.9. *Let A be the infinitesimal generator of a C_0 semigroup $T(t) = e^{At}$ on a Banach space X satisfying $\|e^{At}\| \le Me^{\omega t}$. Let $\gamma > \max(0, \omega)$. If $x \in D(A^2)$, then*

$$e^{At}x = \frac{1}{2\pi i} \int_{\gamma - i\infty}^{\gamma + i\infty} e^{\lambda t} (\lambda I - A)^{-1} x d\lambda \qquad (2.43)$$

and for every $\delta > 0$, the integral converges uniformly in t for $t \in [\delta, 1/\delta]$.

The proof is omitted and referred to [91, p.29, Corollary 7.5] since advanced complex analysis is required. The integration in (2.43) is along the straight line $x = \gamma$ in the complex plane such that $\mathrm{Im}(\lambda)$ increases from $-\infty$ to ∞.

The next question is: Can we use the semigroup e^{At} to represent the resolvent $R(\lambda, A) = (\lambda I - A)^{-1}$ of A? To answer this question, we look at the equation

$$\frac{du}{dt} = Au, \quad u(0) = u_0, \tag{2.44}$$

where A is the infinitesimal generator of a C_0 semigroup e^{At} on a Banach space X satisfying $\|e^{At}\| \leq Me^{\omega t}$. By Theorem 2.8, the solution of (2.44) is given by $u(t) = e^{At}u_0$. Multiplying (2.44) by $e^{-\lambda t}$ and integrating from 0 to ∞, we obtain

$$\int_0^\infty e^{-\lambda t} \frac{du}{dt} dt = \int_0^\infty e^{-\lambda t} Au(t) dt.$$

If $\mathrm{Re}(\lambda) > \omega$, we can integrate by parts on the left hand side to obtain

$$\lambda \int_0^\infty e^{-\lambda t} e^{At} u_0 dt - u_0 = A \int_0^\infty e^{-\lambda t} e^{At} u_0 dt.$$

It therefore follows that

$$(\lambda I - A) \int_0^\infty e^{-\lambda t} e^{At} u_0 dt = u_0,$$

and then

$$R(\lambda, A)u_0 = \int_0^\infty e^{-\lambda t} e^{At} u_0 dt.$$

Hence $R(\lambda, A)u_0$ is the Laplace transform of the semigroup e^{At}. Note that this equation is really true if A is a number. If $\mathrm{Re}(\lambda) > \omega$, we deduce that

$$\|e^{-\lambda t} e^{At} u_0\| \leq Me^{(\omega - \mathrm{Re}(\lambda))t} \|u_0\|.$$

It then follows that

$$\left\| \int_0^\infty e^{-\lambda t} e^{At} u_0 dt \right\| \leq M \int_0^\infty e^{(\omega - \mathrm{Re}(\lambda))t} \|u_0\| dt = \frac{M\|u_0\|}{\mathrm{Re}(\lambda) - \omega}.$$

This is not a rigorous mathematical reasoning because the exchange between the integral and the operator A, $\int_0^\infty e^{-\lambda t} A e^{At} u_0 dt = A \int_0^\infty e^{-\lambda t} e^{At} u_0 dt$, is not guaranteed if A is unbounded. For a rigorous mathematical proof, we refer to [24, p.24, Lemma 2.1.11] or [91, p.8, Theorem 3.1].

Theorem 2.10. *Let e^{At} be a C_0 semigroup with the infinitesimal generator A on a Banach space X. Assume that there exist constants M and ω such that $\|e^{At}\| \leq Me^{\omega t}$. If $\mathrm{Re}(\lambda) > \omega$, then $\lambda \in \rho(A)$ and for all $u_0 \in X$, the following result holds:*

$$R(\lambda,A)u_0 = \int_0^\infty e^{-\lambda t} e^{At} u_0 dt \text{ and } \|R(\lambda,A)\| \le \frac{M}{\text{Re}(\lambda) - \omega}. \tag{2.45}$$

2.6.3 Tests for Semigroup Generation

In solving a partial differential equation, usually a linear and unbounded operator such as the operator (2.36) can be defined through the equation. Therefore, to make the semigroup theory useful, it is important to establish tests with which we can determine what kind of operators generate a semigroup.

Theorem 2.11. *(Hille-Yosida) A linear (unbounded) operator A is the infinitesimal generator of a C_0 semigroup e^{At} on a Banach space X satisfying $\|e^{At}\| \le Me^{\omega t}$ if and only if*

(1) A is closed and $D(A)$ is dense in X.
(2) The resolvent set $\rho(A)$ of A contains (ω, ∞) and for every $\lambda > \omega$

$$\|(R(\lambda,A))^n\| \le \frac{M}{(\lambda - \omega)^n}, \quad n = 1, 2, 3, \cdots. \tag{2.46}$$

The proof of this theorem is omitted and referred to [24, p.26, Theorem 2.1.12], or [91, p.20, Theorem 5.3].

The inequality (2.41) provides a very important growth estimate of a semigroup. If $\omega = 0$, then $\|e^{At}\| \le M$ for all $t \ge 0$ and e^{At} is called *uniformly bounded*. If $\omega = 0$ and $M = 1$, e^{At} is called a *C_0 semigroup of contractions*. If $\omega < 0$, then $\|e^{At}\|$ converges to zero exponentially as $t \to \infty$.

In application to partial differential equations, the verification of the condition (2.46) is usually difficult. So we continue to look for other tests. For an $n \times n$ matrix $\mathbf{A} = (a_{ij})$ and vectors $\mathbf{x}, \mathbf{y}$ in $\mathbb{C}^n$, we have

$$\begin{aligned}
(\mathbf{Ax}, \mathbf{y}) &= \sum_{i=1}^n \bar{y}_i \sum_{j=1}^n a_{ij} x_j \\
&= \sum_{j=1}^n x_j \overline{\sum_{i=1}^n \bar{a}_{ij} y_i} \\
&= (\mathbf{x}, \mathbf{A}^* \mathbf{y}),
\end{aligned}$$

where $\mathbf{A}^*$ denotes the conjugate transpose of $\mathbf{A}$.

Definition 2.23. Let H be a Hilbert space and A a densely defined linear operator. The adjoint operator $A^*: D(A^*) \subset H \to H$ of A is defined as follows. The domain $D(A^*)$ consists of all $y \in H$ such that there exists a $y^* \in H$ satisfying

$$(Ax, y) = (x, y^*) \quad \text{for all } x \in D(A).$$

We then define

$$A^*y = y^* \quad \text{for all } y \in D(A^*).$$

Definition 2.24. A linear operator A is called self-adjoint if $A = A^*$.

Theorem 2.12. *A self-adjoint operator is closed.*

The proof of the theorem is omitted and referred to [105].

Definition 2.25. A linear operator A on a Hilbert space H is dissipative if $\text{Re}(Ax,x)$ ≤ 0 for every $x \in D(A)$.

Theorem 2.13. *(Lumer-Phillips) Let A be a linear operator with a dense domain $D(A)$ in a Hilbert space H.*

(1) If A is dissipative and there is a $\lambda_0 > 0$ such that the range, $R(\lambda_0 I - A)$, of $\lambda_0 I - A$ is H, then A is the infinitesimal generator of a C_0 semigroup of contractions on H.
(2) If A is the infinitesimal generator of a C_0 semigroup of contractions on H, then $R(\lambda I - A) = H$ for all $\lambda > 0$ and A is dissipative.

The proof of this theorem is omitted and referred to [91, p.14, Theorem 4.3].

Corollary 2.2. *Let A be a densely defined closed linear operator on a Hilbert space H. If both A and its adjoint operator A^* are dissipative, then A is the infinitesimal generator of a C_0 semigroup of contractions on H.*

The proof of this theorem is omitted and referred to [24, p.33, Corollary 2.2.3], or [91, p. 15, Corollary 4.4]. The Lumer-Phillips theorem is very useful in solving the partial differential equations because the conditions on A can be easily verified. We now use it to prove that the wave operator defined in (2.36) generates a C_0 semigroup.

Using Poincaré's inequality (2.29), we can readily prove the following lemma.

Lemma 2.2. *The space $H_0^1(0,L)$ endowed with the following inner product*

$$(u,v)_{H_0^1} = \int_0^L c^2 \frac{du}{dx} \frac{d\bar{v}}{dx} dx \tag{2.47}$$

is a Hilbert space. Moreover, the norm induced by this inner product is equivalent to the usual norm

$$\|u\|_{H^1} = \sqrt{\int_0^L \left(u^2 + \left| \frac{du}{dx} \right|^2 \right) dx},$$

that is, there exist constants $C_1, C_2 > 0$ such that

$$C_1 \|u\|_{H^1} \leq \sqrt{\int_0^L c^2 \left| \frac{du}{dx} \right|^2 dx} \leq C_2 \|u\|_{H^1}.$$

In what follows, we will always use the inner product (2.47) for $H_0^1(0,L)$.

Theorem 2.14. *The wave operator A defined in* (2.36) *generates a C_0 semigroup of contraction on $\mathscr{H} = H_0^1(0,L) \times L^2(0,L)$.*

Proof. It suffices to verify the conditions of Lumer-Phillips Theorem 2.13. By Theorem 2.3, $D(A) = (H^2(0,L) \cap H_0^1(0,L)) \times H_0^1(0,L)$ is dense in $\mathscr{H} = H_0^1(0,L) \times L^2(0,L)$.

Next let $\lambda > 0$ and $(f_1, f_2) \in H_0^1(0,L) \times L^2(0,L)$. Consider the equation

$$(\lambda I - A) \begin{bmatrix} u \\ v \end{bmatrix} = \begin{bmatrix} f_1 \\ f_2 \end{bmatrix},$$

which is equivalent to

$$\lambda u - v = f_1, \tag{2.48}$$

$$-c^2 \frac{d^2 u}{dx^2} + \lambda v = f_2. \tag{2.49}$$

$$u(0) = u(L) = 0. \tag{2.50}$$

Substituting equation (2.48) into (2.49) gives

$$-c^2 \frac{d^2 u}{dx^2} + \lambda^2 u = f_2 + \lambda f_1, \quad u(0) = u(L) = 0.$$

Using the variation of parameters formula, we obtain the unique solution

$$\begin{aligned}
u(x) &= c_1 \exp(\lambda x/c) + c_2 \exp(-\lambda x/c) \\
&\quad + \frac{c \exp(\lambda x/c)}{2\lambda} \int_0^x \exp(-\lambda s/c)(f_2(s) + \lambda f_1(s)) ds \\
&\quad - \frac{c \exp(-\lambda x/c)}{2\lambda} \int_0^x \exp(\lambda s/c)(f_2(s) + \lambda f_1(s)) ds,
\end{aligned}$$

where c_1, c_2 are unique constants such that the boundary condition $u(0) = u(L) = 0$ is satisfied. It is clear that u and u_x belong to $L^2(0,L)$. Also $c^2 \frac{d^2 u}{dx^2} = \lambda^2 u - f_2 - \lambda f_1$ belongs to $L^2(0,L)$. Hence we deduce that $u \in H^2(0,L) \cap H_0^1(0,L)$ and then (2.48) and (2.49) have a unique solution. This proves that $R(\lambda I - A) = \mathscr{H}$.

It remains to show that A is dissipative. We note that the inner product in $H_0^1(0,L)$ is defined by

$$(f,g)_{H_0^1} = \int_0^L c^2 \frac{df}{dx} \frac{d\bar{g}}{dx} dx$$

and the inner product in $L^2(0,L)$ is defined by

$$(f,g) = \int_0^L f(x)\bar{g}(x) dx.$$

For any $(f_1, f_2) \in D(A) = (H^2(0,L) \cap H_0^1(0,L)) \times H_0^1(0,L)$, integration by parts gives (the inner product here is the one in $H_0^1(0,L) \times L^2(0,L)$)

$$\mathrm{Re}\left(A\begin{bmatrix} f_1 \\ f_2 \end{bmatrix}, \begin{bmatrix} f_1 \\ f_2 \end{bmatrix}\right)_{H_0^1 \times L^2}$$

$$= \mathrm{Re}\left(\begin{bmatrix} f_2 \\ c^2 \frac{d^2 f_1}{dx^2} \end{bmatrix}, \begin{bmatrix} f_1 \\ f_2 \end{bmatrix}\right)_{H_0^1 \times L^2}$$

$$= \mathrm{Re} \int_0^L \left(c^2 \frac{d f_2}{dx} \frac{d \bar{f}_1}{dx} + c^2 \frac{d^2 f_1}{dx^2} \bar{f}_2\right) dx$$

$$= \mathrm{Re} \int_0^L c^2 \frac{d f_2}{dx} \frac{d \bar{f}_1}{dx} dx + \mathrm{Re} \left(c^2 \frac{d f_1}{dx} \bar{f}_2 \Big|_0^L\right) - \mathrm{Re} \int_0^L c^2 \frac{d f_1}{dx} \frac{d \bar{f}_2}{dx} dx$$

$$= 0.$$

Theorem 2.15. *Let A be a densely defined closed linear operator on a Banach space. Then A is the infinitesimal generator of an analytic semigroup e^{At} satisfying $\|e^{At}\| \leq M e^{\omega t}$ if and only if there exist $0 < \delta < \frac{\pi}{2}$, $M > 0$, and a real ω such that*

$$\rho(A) \supset S_{\omega,\delta} = \left\{\lambda \mid |\arg(\lambda - \omega)| < \frac{\pi}{2} + \delta\right\} \cup \{\omega\} \tag{2.51}$$

and

$$\|(\lambda I - A)^{-1}\| \leq \frac{M}{|\lambda - \omega|} \quad \text{for all } \lambda \in S_{\omega,\delta}, \ \lambda \neq \omega. \tag{2.52}$$

Furthermore

$$\|A e^{At}\| \leq \frac{M}{t} e^{\omega t} \tag{2.53}$$

and the semigroup e^{At} is given by

$$e^{At} = \frac{1}{2\pi i} \int_\Gamma e^{\lambda t} (\lambda - A)^{-1} d\lambda, \tag{2.54}$$

where Γ is a contour in the resolvent set $\rho(A)$ with $\arg(\lambda) \to \pm\theta$ as $|\lambda| \to \infty$ for some θ in $(\pi/2, \pi)$.

The proof is omitted and referred to [42, p.20, Theorem 1.3.4], [82, p.48, Theorem 2.47], or [91, p.61, Theorem 5.2]. The operator A satisfying the conditions (2.51) and (2.52) is called a *sectorial operator*. The sectorial conditions (2.51) and (2.52) are equivalent to the condition: there exists a constant M_1 and a real ω such that for every $\lambda = \sigma + i\tau, \sigma > \omega$

$$\|(\lambda I - A)^{-1}\| \leq \frac{M_1}{|\tau|}. \tag{2.55}$$

For a proof, we refer to [91, p.61, Theorem 5.2].

Theorem 2.16. *The operator A defined in (2.23) generates an analytic semigroup on $L^2(\Omega)$.*

The proof relies on advanced estimates about the elliptic operator A, such as Gårding's inequality, and is referred to [91, p.211, Theorem 2.7].

The property of being a generator is not destroyed by the addition of a bounded operator.

Theorem 2.17. *Let X be a Banach space and let A be the infinitesimal generator of a C_0 semigroup $T(t)$ on X, satisfying $\|T(t)\| \leq Me^{\omega t}$. If B is a bounded linear operator on X , then $A + B$ is the infinitesimal generator of a C_0 semigroup $S(t)$ on X, satisfying $\|S(t)\| \leq Me^{(\omega + M\|B\|)t}$.*

The proof of the theorem is referred to [91, p.77, Theorem 1.1]. If A is the infinitesimal generator of an analytic semigroup $T(t)$ on X, then the boundedness on B can be relaxed.

Theorem 2.18. *Let A be the infinitesimal generator of an analytic semigroup $T(t)$ on a Banach space X such that*

$$\|(\lambda I - A)^{-1}\| \leq \frac{M}{1 + |\lambda|}, \quad \mathrm{Re}\lambda > \omega$$

for some $M > 0$ and $\omega \in \mathbb{R}$. Let B be a linear operator satisfying $D(A) \subset D(B)$ and

$$\|Bx\| \leq a\|Ax\| + b\|x\| \quad for \ x \in D(A),$$

where $a < 1/(1 + M)$. Then $A + B$ generates an analytic semigroup on X.

The proof of the theorem is referred to [82, p.54, Theorem 2.53] or [91, p.80, Theorem 2.1].

2.6.4 Semigroup Growth Bound

The ω in (2.41) is called the *growth rate* of the semigroup. Such ω is not unique. For example, any numbers $\gamma \geq \omega$ are also growth rates. Thus we define the least growth rate by

$$\omega_0 = \inf\{\omega \mid \text{there exist constants } \omega \text{ and } M(\omega) \text{ such that } \|T(t)\| \leq Me^{\omega t}\}. \quad (2.56)$$

The ω_0 is called the *growth bound* of the semigroup.

Theorem 2.19. *Let $T(t)$ be a C_0 semigroup with the infinitesimal generator A on a Banach space X. Then the growth bound ω_0 of the semigroup satisfies*

$$\omega_0 = \lim_{t \to \infty} \frac{\ln \|T(t)\|}{t}. \quad (2.57)$$

Proof. We first prove that the limit exists. Let $t_0 > 0$ be a fixed number and $M = \sup_{t \in [0,t_0]} \|T(t)\|$. Then for every $t \geq t_0$, there exists $n \in \mathbb{N}$ such that $nt_0 \leq t < (n+1)t_0$.
It then follows from the semigroup definition that

$$
\begin{aligned}
\frac{\ln\|T(t)\|}{t} &= \frac{\ln\|T(t-nt_0+nt_0)\|}{t} \\
&= \frac{\ln\|T(nt_0)T(t-nt_0)\|}{t} \\
&= \frac{\ln\|T^n(t_0)T(t-nt_0)\|}{t} \\
&\leq \frac{\ln(\|T(t_0)\|^n \|T(t-nt_0)\|)}{t} \\
&\leq \frac{\ln\|T(t_0)\|}{t_0} \frac{nt_0}{t} + \frac{\ln M}{t}.
\end{aligned}
$$

Taking limit and noting that $\lim\limits_{t\to\infty} \dfrac{nt_0}{t} = 1$, we obtain

$$
\limsup_{t\to\infty} \frac{\ln\|T(t)\|}{t} \leq \frac{\ln\|T(t_0)\|}{t_0} < \infty.
$$

Since t_0 is arbitrary, we can let t_0 go to ∞ to obtain

$$
\limsup_{t\to\infty} \frac{\ln\|T(t)\|}{t} \leq \liminf_{t\to\infty} \frac{\ln\|T(t)\|}{t}.
$$

Hence

$$
\lim_{t\to\infty} \frac{\ln\|T(t)\|}{t} = \limsup_{t\to\infty} \frac{\ln\|T(t)\|}{t} = \liminf_{t\to\infty} \frac{\ln\|T(t)\|}{t} < \infty.
$$

For any $\omega > \omega_0$, there exists $M(\omega)$ such that

$$
\|T(t)\| \leq M(\omega)e^{\omega t}.
$$

Consequently,

$$
\lim_{t\to\infty} \frac{\ln\|T(t)\|}{t} \leq \omega + \lim_{t\to\infty} \frac{M}{t} = \omega.
$$

Since ω is arbitrary, we have

$$
\lim_{t\to\infty} \frac{\ln\|T(t)\|}{t} \leq \omega_0.
$$

If

$$
\lim_{t\to\infty} \frac{\ln\|T(t)\|}{t} < \omega_0,
$$

then there exists ω such that

$$\lim_{t \to \infty} \frac{\ln \|T(t)\|}{t} < \omega < \omega_0.$$

This implies that there exists a t_0 such that

$$\frac{\ln \|T(t)\|}{t} \leq \omega \quad \text{for } t \geq t_0$$

and then

$$\|T(t)\| \leq e^{\omega t} \quad \text{for } t \geq t_0.$$

But

$$\|T(t)\| \leq M_0 \quad \text{for } 0 \leq t \leq t_0.$$

Taking

$$M(\omega) = \max\{M_0, M_0 e^{-\omega t_0}\},$$

we obtain

$$\|T(t)\| \leq M(\omega)e^{\omega t} \quad \text{for } t \geq 0,$$

which, combing with the definition of the growth bound, implies that $\omega_0 \leq \omega$. This is a contradiction.

The growth bound is closely related the spectrum of A.

Theorem 2.20. *Let $T(t)$ be a C_0 semigroup with the infinitesimal generator A on a Banach space X and $\sigma(A)$ be the spectrum of A. Then the growth bound ω_0 of the semigroup satisfies*

$$\omega_0 \geq \sup\{\operatorname{Re}(\lambda), \lambda \in \sigma(A)\}. \tag{2.58}$$

Proof. We argue by contradiction. If $\omega_0 < \sup\{\operatorname{Re}(\lambda), \lambda \in \sigma(A)\}$, then there exists a constant ω_1 such that $\omega_0 < \omega_1 < \sup\{\operatorname{Re}(\lambda), \lambda \in \sigma(A)\}$. By the definition of the growth bound, we deduce that there exists a constant $M(\omega_1)$ such that

$$\|T(t)\| \leq M(\omega_1)e^{\omega_1 t}.$$

Moreover, there exists a $\lambda_1 \in \sigma(A)$ such that $\operatorname{Re}(\lambda_1) > \omega_1$. But it follows from Theorem 2.10 that $\lambda_1 \in \rho(A)$. This is a contradiction.

The equality in (2.58) may not hold. For such an example, we refer to [82, p.114, Example 3.6]. Fortunately, the equality does hold for a large class of semigroups such as analytic semigroups.

Lemma 2.3. *Let $\sigma(A)$ be the spectrum of an operator A and denote $\lambda_0 = \sup\{\operatorname{Re}(\lambda), \lambda \in \sigma(A)\}$. If A satisfies the sectorial conditions (2.51) and (2.52), then for any real $\lambda_1 > \lambda_0$, there exist $C > 0$ and $0 < \varphi < \pi/2$ such that*

$$\|(\lambda I - A)^{-1}\| \leq \frac{C}{|\lambda - \lambda_1|} \quad \text{whenever } |\arg(\lambda - \lambda_1)| < \frac{\pi}{2} + \varphi. \tag{2.59}$$

Fig. 2.2 Sector in the resolvent set of A.

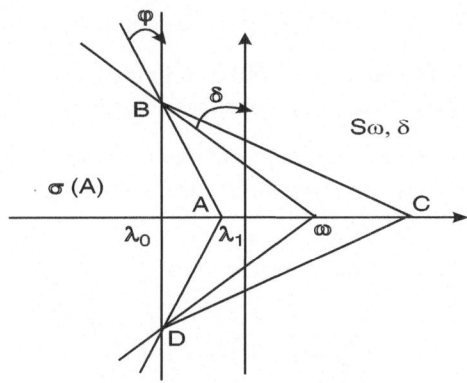

Proof. As demonstrated in Figure 2.2, there exists $0 < \varphi < \pi/2$ such that the sector $S_{\lambda_1,\varphi} \subset \rho(A)$. Let R_{ABCD} denote the region of the polygon bounded by segments AB, BC, CD, and DA. Since $R_{ABCD} \subset \rho(A)$, we have

$$M_1 = \max_{\lambda \in R_{ABCD}} |\lambda - \lambda_1| \|(\lambda I - A)^{-1}\| < \infty.$$

By the conditions (2.51) and (2.52), we derive that $|\lambda - \omega| \|(\lambda I - A)^{-1}\| \leq M$ for all $\lambda \in S_{\omega,\delta}$. It then follows that

$$
\begin{aligned}
M_2 &= \max_{\lambda \in S_{\lambda_1,\varphi} - R_{ABCD}} |\lambda - \lambda_1| \|(\lambda I - A)^{-1}\| \\
&= \max_{\lambda \in S_{\lambda_1,\varphi} - R_{ABCD}} \frac{|\lambda - \lambda_1|}{|\lambda - \omega|} |\lambda - \omega| \|(\lambda I - A)^{-1}\| \\
&\leq M \max_{\lambda \in S_{\lambda_1,\varphi} - R_{ABCD}} \frac{|\lambda - \lambda_1|}{|\lambda - \omega|} < \infty.
\end{aligned}
$$

Then (2.59) can be derived by taking $C = \max(M_1, M_2)$.

Theorem 2.21. *Let $T(t)$ be an analytic semigroup with the infinitesimal generator A on a Banach space X and $\sigma(A)$ be the spectrum of A. Then the growth bound ω_0 of the semigroup satisfies*

$$\omega_0 = \sup\{\mathrm{Re}(\lambda), \lambda \in \sigma(A)\}. \tag{2.60}$$

Proof. We argue by contradiction. If $\omega_0 > \sup\{\mathrm{Re}(\lambda), \lambda \in \sigma(A)\}$, then there exists λ_1 such that $\omega_0 > \lambda_1 > \sup\{\mathrm{Re}(\lambda), \lambda \in \sigma(A)\}$. Since A is the infinitesimal generator of the analytic semigroup, it follows from Theorem 2.15 that A satisfies the sectorial conditions (2.51) and (2.52). Then by Lemma 2.3, A satisfies the condition (2.59). It therefore follows from Theorem 2.15 that

$$\|T(t)\| \leq Ce^{\lambda_1 t}.$$

By the definition of growth bound, we deduce that $\lambda_1 \geq \omega_0$. This is a contradiction.

The equality (2.60) also holds for another class of semigroups whose eigenvectors form an orthogonal basis in a Hilbert space.

Theorem 2.22. *Suppose that the linear, closed operator A has eigenvalues $\{\lambda_n\}_{n=1}^{\infty}$ and the corresponding eigenvectors $\{\phi_n\}_{n=1}^{\infty}$ form an orthogonal basis on a Hilbert space H. Then*

(1) A has the representation

$$Ax = \sum_{n=1}^{\infty} \lambda_n (x, \phi_n) \phi_n, \quad x \in D(A) = \left\{ x \in H \mid \sum_{n=1}^{\infty} |\lambda_n|^2 |(x, \phi_n)|^2 < \infty \right\}. \quad (2.61)$$

(2) If $\sup\limits_{n \geq 1} \mathrm{Re}(\lambda_n) < \infty$, then A is the infinitesimal generator of a C_0 semigroup $T(t)$ given by

$$T(t)x = \sum_{n=1}^{\infty} (x, \phi_n) e^{\lambda_n t} \phi_n. \quad (2.62)$$

(3) The growth bound of the semigroup is given by

$$\omega_0 = \sup_{n \geq 1} \mathrm{Re}(\lambda_n). \quad (2.63)$$

Proof. For $x \in D(A)$, let

$$x = \sum_{n=1}^{\infty} (x, \phi_n) \phi_n.$$

Since A is closed, we have

$$Ax = A \left(\lim_{N \to \infty} \sum_{n=1}^{N} (x, \phi_n) \phi_n \right)$$

$$= \lim_{N \to \infty} A \left(\sum_{n=1}^{N} (x, \phi_n) \phi_n \right)$$

$$= \lim_{N \to \infty} \sum_{n=1}^{N} (x, \phi_n) A \phi_n$$

$$= \lim_{N \to \infty} \sum_{n=1}^{N} \lambda_n (x, \phi_n) \phi_n$$

$$= \sum_{n=1}^{\infty} \lambda_n (x, \phi_n) \phi_n.$$

We can easily check that $T(t)$ defined by (2.62) is a semigroup. Moreover

$$Ax = \lim_{t \to 0^+} \frac{T(t)x - x}{t} = \sum_{n=1}^{\infty} (x, \phi_n) \frac{d}{dt} \left(e^{\lambda_n t} \right) |_{t=0} \phi_n = \sum_{n=1}^{\infty} \lambda_n (x, \phi_n) \phi_n.$$

Using the orthogonality of the basis, we deduce that

$$\|T(t)x\|^2 = \sum_{n=1}^{\infty} |(x,\phi_n)|^2 e^{2\mathrm{Re}(\lambda_n)t}$$

$$\leq \exp\left(2\sup_{n\geq 1}\mathrm{Re}(\lambda_n)t\right) \sum_{n=1}^{\infty} |(x,\phi_n)|^2$$

$$= \|x\|^2 \exp\left(2\sup_{n\geq 1}\mathrm{Re}(\lambda_n)t\right).$$

This implies (2.63).

Theorem 2.22 also holds for Riesz-spectral operators whose eigenvectors form a Riesz basis in H. The Riesz basis can be obtained from an orthogonal basis by an invertible linear bounded transformation. For details, we refer to [24, Section 2.3].

There is a necessary and sufficient condition for the equality (2.60): for every $\omega > \omega_0$, the resolvent $R(\omega + \lambda, A)$ as an operator-valued function of λ is bounded on the right-half complex plane $\mathrm{Re}(\lambda) > 0$. For a proof, see [24, p.223, Theorem 5.1.6]. This result is nice, but the condition is usually difficult to verify in applications to partial differential equations.

Exercises 2.6

1. Suppose that $T(t)$ is a C_0-semigroup on the Hilbert space X.

   a. Let $\lambda \in \mathbb{C}$. Show that $e^{\lambda t}T(t)$ is also a C_0-semigroup.
   b. Prove that the infinitesimal generator of $e^{\lambda t}T(t)$ is $\lambda I + A$, where A is the infinitesimal generator of $T(t)$.

2. A linear operator A on a Hilbert space H is nonnegative if

$$(Ax,x) \geq 0 \quad \text{for all } x \in D(A).$$

Prove that if A is a self-adjoint and nonnegative operator on H, then $-A$ is the infinitesimal generator of a contraction semigroup on H.

3. Consider the heat equation with Dirichlet boundary conditions

$$\frac{\partial u}{\partial t} = k\frac{\partial^2 u}{\partial x^2},$$
$$u(0,t) = 0, \quad u(1,t) = 0,$$
$$u(x,0) = f(x),$$

where k is a positive constant. Define the differential operator A in $L^2(0,1)$ by

$$A = k\frac{d^2}{dx^2}$$

with the domain $D(A) = H^2(0,1) \cap H_0^1(0,1)$.

a. Prove that A is the infinitesimal generator of a contraction semigroup.
b. Give an explicit expression for the semigroup by solving the equation with the Fourier method (separation of variables).

4. Consider the convection-diffusion equation with Dirichlet boundary conditions

$$\frac{\partial u}{\partial t} = k\frac{\partial^2 u}{\partial x^2} + a\frac{\partial u}{\partial x},$$
$$u(0,t) = 0, \quad u(1,t) = 0,$$
$$u(x,0) = f(x),$$

where a, k are constant and $k > 0$. Define

$$A = k\frac{\partial^2 u}{\partial x^2} + a\frac{\partial u}{\partial x}$$

with the domain $D(A) = H^2(0,1) \cap H_0^1(0,1)$.

a. Show that A is self-adjoint on $L_a^2 = L^2(0,1)$ with the weighted inner product

$$(u,v)_{L_a^2} = \int_0^1 u(x)v(x)e^{ax/k}dx.$$

b. Prove that A is the infinitesimal generator of a contraction semigroup.

2.7 Abstract Cauchy Problems

Let X be a Banach space and let A be a linear operator from $D(A) \subset X$ into X. Given $x \in X$, the abstract Cauchy problem for A with an initial condition x consists of finding a solution $u(t)$ to the initial value problem

$$\frac{du(t)}{dt} = Au(t), \quad u(0) = x. \tag{2.64}$$

By a solution here we mean an X-valued function $u(t)$ such that it is continuous for $t \geq 0$, continuously differentiable, $u(t) \in D(A)$ for $t > 0$, and (2.64) is satisfied.

Theorem 2.23. *Let A be a densely defined linear operator with a nonempty resolvent set $\rho(A)$. The initial value problem (2.64) has a unique solution $u(t)$, which is continuously differentiable on $[0,\infty)$, for every initial value $x \in D(A)$ if and only if A is the infinitesimal generator of a C_0 semigroup $T(t)$.*

The proof of this theorem is omitted and refereed to [91, p.102, Theorem 1.3].

Definition 2.26. Let $T(t)$ be the C_0 semigroup generated by A. For any $x \in X$, $u(t) = T(t)x$ is called a weak solution of (2.64).

Using this theorem and Theorem 2.14, we can prove that the initial boundary value problem of the wave equation (2.33)-(2.35) has a unique solution.

Theorem 2.24. *For every initial condition* $(u_0, u_1) \in (H^2(0,L) \cap H_0^1(0,L)) \times H_0^1(0,L)$, *the initial boundary value problem of the wave equation (2.33)-(2.35) has a unique solution*

$$\begin{bmatrix} u(t) \\ \frac{\partial u}{\partial t} \end{bmatrix} = e^{At} \begin{bmatrix} u_0 \\ u_1 \end{bmatrix},$$

where the wave operator A is defined in (2.36).

This theorem just tells us that the problem (2.33)-(2.35) has a unique solution, but it does not tell us what the explicit expression of the solution is. Using the Fourier method, we can obtain the explicit solution

$$u(x,t) = \sum_{n=1}^{\infty} \left[a_n \cos\left(\frac{cn\pi t}{L}\right) + b_n \sin\left(\frac{cn\pi t}{L}\right) \right] \sin\left(\frac{n\pi x}{L}\right),$$

where

$$a_n = \frac{2}{L} \int_0^L u_0(x) \sin\left(\frac{n\pi x}{L}\right) dx,$$

$$b_n \frac{cn\pi}{L} = \frac{2}{L} \int_0^L u_1(x) \sin\left(\frac{n\pi x}{L}\right) dx.$$

Therefore, the semigroup e^{At} is given by

$$e^{At} \begin{bmatrix} u_0 \\ u_1 \end{bmatrix} = \begin{bmatrix} \sum_{n=1}^{\infty} \left[a_n \cos\left(\frac{cn\pi t}{L}\right) + b_n \sin\left(\frac{cn\pi t}{L}\right) \right] \sin\left(\frac{n\pi x}{L}\right) \\ \sum_{n=1}^{\infty} \left[-\frac{cn\pi a_n}{L} \sin\left(\frac{cn\pi t}{L}\right) + \frac{cn\pi b_n}{L} \cos\left(\frac{cn\pi t}{L}\right) \right] \sin\left(\frac{n\pi x}{L}\right) \end{bmatrix}.$$

We now consider the nonhomogeneous initial value problem

$$\frac{du(t)}{dt} = Au(t) + f(t), \quad u(0) = x, \tag{2.65}$$

where $f : [0,T) \to X$.

We denote by $L^p(a,b;X)$ $(1 \le p < +\infty)$ the space of (classes of) functions $f : (a,b) \to X$ such that

$$\| f \|_{L^p} = \left(\int_a^b \| f(t) \|_X^p \, dt \right)^{\frac{1}{p}} < +\infty.$$

$L^p(a,b;X)$ is a Banach space [32, p.469]. Let $C([0,T];X)$ denote the space of all continuous X-valued functions. Endowed with the following maximum norm

$$\|f\|_C = \max_{0 \le t \le T} \|f(t)\|,$$

$C([0,T];X)$ is a Banach space.

Definition 2.27. An X-valued function $u : [0,T) \to X$ is a (classical) solution if it is continuous on $[0,T)$, continuously differentiable $(0,T)$, $u(t) \in D(A)$ for $0 < t < T$, and (2.65) is satisfied on $[0,T)$.

If A is a matrix, the solution of (2.65) is given by the formula of variation of parameters

$$u(t) = e^{At}x + \int_0^t e^{A(t-s)}f(s)ds.$$

This is also true in general cases.

Definition 2.28. Let A be the infinitesmal generator of a C_0 semigroup $T(t)$. Let $x \in X$ and $f \in L^1(0,T;X)$. The function $u \in C([0,T];X)$ given by

$$u(t) = T(t)x + \int_0^t T(t-s)f(s)ds, \qquad 0 \le t \le T \tag{2.66}$$

is called a weak (or mild) solution of the initial value problem (2.65) on $[0,T]$.

Theorem 2.25. *Let A be the infinitesimal generator of a C_0 semigroup $T(t)$ on a Banach space. Assume that*

$$\|T(t)\| \le Me^{\omega t}, \quad \|f(t)\| \le Me^{\gamma t}. \tag{2.67}$$

Then the solution of (2.65) satisfies

$$\|u(t)\| \le C(\|u_0\| + t)e^{\omega t}, \quad \text{if } \omega = \gamma, \tag{2.68}$$

or

$$\|u(t)\| \le C(\|u_0\| + 1)e^{\tau t}, \quad \text{if } \omega \ne \gamma, \tag{2.69}$$

where $\tau = \max(\omega, \gamma)$ and C is a positive constant.

Proof. It follows from (2.66) that

$$\begin{aligned}
\|u(t)\| &= \left\| T(t)x + \int_0^t T(t-s)f(s)ds \right\| \\
&\le \|T(t)x\| + \left\| \int_0^t T(t-s)f(s)ds \right\| \\
&\le \|x\|Me^{\omega t} + M^2 \int_0^t e^{\omega(t-s)}e^{\gamma s}ds \\
&= \|x\|Me^{\omega t} + M^2 e^{\omega t} \int_0^t e^{(\gamma - \omega)s}ds\| \\
&= \|x\|Me^{\omega t} + M^2 e^{\omega t} \int_0^t e^{(\gamma - \omega)s}ds\|.
\end{aligned}$$

This implies (2.68) and (2.69).

For $f \in L^1(0,T;X)$, by definition, the problem (2.65) has a unique weak solution. The following theorem states the conditions imposed on f such that the weak solution becomes a (classical) solution for $x \in D(A)$.

Theorem 2.26. *Let A be the infinitesimal generator of a C_0 semigroup $T(t)$ on a Banach space X. If f is differentiable on $[0,T)$ and $f' \in L^1(0,T;X)$, then the initial value problem (2.65) has a solution u on $[0,T)$ for every $x \in D(A)$.*

The proof of this theorem is omitted and refereed to [91, p.109, Theorem 2.9].

Example 2.8. Consider the non-homogeneous wave equation

$$\frac{\partial^2 u}{\partial t^2} = c^2 \frac{\partial^2 u}{\partial x^2} + f(x,t) \quad \text{in } (0,L) \times (0,\infty),$$
$$u(0,t) = 0, u(L,t) = 0 \quad t \geq 0,$$
$$u(x,0) = u_0(x), \quad \frac{\partial u}{\partial t}(x,0) = u_1(x) \quad x \in (0,L).$$

If $f(x,t) \in L^2(0,L)$ for every $t \geq 0$ and is continuously differentiable in t, then it follows from Theorem 2.26 that the problem has a unique solution.

Exercises 2.7

1. Consider the wave equation with Dirichlet boundary conditions

$$\frac{\partial^2 u}{\partial t^2} = \frac{\partial^2 u}{\partial x^2} - a\frac{\partial u}{\partial x},$$
$$u(0,t) = 0, \quad u(1,t) = 0,$$
$$u(x,0) = u_0(x), \quad \frac{\partial u}{\partial t}(x,0) = u_1(x),$$

where a is a constant. Define

$$A = \frac{\partial^2 u}{\partial x^2} - a\frac{\partial u}{\partial x}$$

with the domain $D(A) = H^2(0,1) \cap H_0^1(0,1)$.

a. Formulate this problem into an abstract Cauchy problem in the state space $\mathcal{H} = H_{0,a}^1 \times L_a^2$, where $H_{0,a}^1 = H_0^1(0,1)$ with the weighted inner product

$$(h,f)_{H_{0,a}^1} = \int_0^1 e^{-ax} Ah(x)\bar{f}(x)dx,$$

and $L_a^2 = L^2(0,1)$ with the weighted inner product

$$(h,f)_{L_a^2} = \int_0^1 e^{-ax}h(x)\bar{f}(x)dx.$$

b. Prove that the Cauchy problem has a unique solution using the semigroup theory.

2.8 References and Notes

This chapter is mainly based on the references [1, 24, 30, 31, 32, 35, 39, 40, 44, 82, 91, 95, 105]. Sections 2.1, 2.2, and 2.3 are adopted from [44, 95, 105], Section 2.4 from [1, 35], Section 2.5 from [30, 31, 32, 35, 41, 44], Section 2.6 from [24, 82, 91], and Section 2.7 from [91].

Chapter 3
Finite Dimensional Systems

Because many control concepts and theories in finite dimensional systems have been transplanted to partial differential equations, we present a brief introduction to feedback control of finite dimensional systems. We start with the control systems without disturbances and address the control systems subject to a disturbance in Section 3.10.

3.1 General Forms of Control Systems

Linear finite dimensional control systems have the following general form

$$\dot{\mathbf{x}} = \mathbf{A}\mathbf{x} + \mathbf{B}\mathbf{u}, \tag{3.1}$$

$$\mathbf{y} = \mathbf{C}\mathbf{x} + \mathbf{D}\mathbf{u}, \tag{3.2}$$

$$\mathbf{x}(0) = \mathbf{x}_0, \tag{3.3}$$

where $\mathbf{x} = (x_1, x_2, \cdots, x_n)^*$ (hereafter $*$ denotes the transpose of a vector or matrix) is a state vector, $\mathbf{x}_0$ is an initial state caused by external disturbances, $\mathbf{y} = (y_1, \cdots, y_l)^*$ is an output vector, $\mathbf{u} = (u_1, \cdots, u_m)^*$ is a control vector, and $\mathbf{A}, \mathbf{B}, \mathbf{C}, \mathbf{D}$ are $n \times n$, $n \times m$, $l \times n$, $l \times m$ constant matrices, respectively. The equation (3.1) is called a *state equation*. A simple example of such control systems is the mass-spring system (1.2), which is repeated as follows

$$\begin{bmatrix} \dot{x}_1 \\ \dot{x}_2 \end{bmatrix} = \begin{bmatrix} 0 & 1 \\ -\frac{k}{m} & 0 \end{bmatrix} \begin{bmatrix} x_1 \\ x_2 \end{bmatrix} + \begin{bmatrix} 0 \\ \frac{1}{m} \end{bmatrix} u, \tag{3.4}$$

$$y = \begin{bmatrix} 1 & 0 \end{bmatrix} \begin{bmatrix} x_1 \\ x_2 \end{bmatrix}. \tag{3.5}$$

Given a constant reference output $\mathbf{r}$ of the system (3.1), the control objective is to design the control $\mathbf{u}$ to regulate the output $\mathbf{y}$ to $\mathbf{r}$, that is, $\mathbf{y}(t) \to \mathbf{r}$ as $t \to \infty$. Such

W. Liu, *Elementary Feedback Stabilization of the Linear Reaction-Convection-Diffusion Equation and the Wave Equation*, Mathématiques et Applications 66, DOI 10.1007/978-3-642-04613-1_3, © Springer-Verlag Berlin Heidelberg 2010

a control system where the reference output is constant is called a *regulator system*.

Without loss of generality, we can assume that a reference output $\mathbf{r}$ of the system (3.1)-(3.3) is zero. In fact, if $\mathbf{r} \neq \mathbf{0}$, we introduce the change of variables

$$\mathbf{w} = \mathbf{x} - \bar{\mathbf{x}}, \quad \mathbf{z} = \mathbf{y} - \mathbf{r}, \quad \mathbf{v} = \mathbf{u} - \bar{\mathbf{u}},$$

where $\bar{\mathbf{x}}$ and $\bar{\mathbf{u}}$ are the steady states satisfying

$$0 = A\bar{\mathbf{x}} + B\bar{\mathbf{u}},$$
$$\mathbf{r} = C\bar{\mathbf{x}} + D\bar{\mathbf{u}},$$

and $\bar{\mathbf{y}}$ is given by

$$\bar{\mathbf{y}} = C\bar{\mathbf{x}} + D\bar{\mathbf{u}}.$$

In most of cases, this linear system has a solution. For instance, in the mass-spring system, the coefficient matrix

$$\begin{bmatrix} A & B \\ C & D \end{bmatrix}$$

is nonsingular and then the above system has a unique solution $\bar{\mathbf{x}}$ and $\bar{\mathbf{u}}$. We then have

$$\dot{\mathbf{w}} = A\mathbf{w} + B\mathbf{v},$$
$$\mathbf{z} = C\mathbf{w} + D\mathbf{v},$$

which has a zero reference output. If we substitute $\mathbf{u} = \mathbf{v} + \bar{\mathbf{u}}$ into the original equation (3.1), we obtain

$$\dot{\mathbf{x}} = A\mathbf{x} + B(\mathbf{v} + \bar{\mathbf{u}}),$$
$$\mathbf{y} = C\mathbf{x} + D(\mathbf{v} + \bar{\mathbf{u}}),$$
$$\mathbf{x}(0) = \mathbf{x}_0,$$

which usually has a nonzero steady control $\bar{\mathbf{u}}$.

Exercises 3.1

1. Consider the system

$$\dot{x}_1 = -2x_1,$$
$$\dot{x}_2 = -11x_1 + u(t),$$
$$y(t) = x_1(t) + x_2(t).$$

Write the system in state-space form (i.e., determine the matrices $\mathbf{A}, \mathbf{B}, \mathbf{C}$, and $\mathbf{D}$).

2. Consider the system

$$\ddot{x} = 9x + 2u(t),$$
$$y(t) = x(t).$$

Write the system in state-space form (i.e., determine the matrices $\mathbf{A}, \mathbf{B}, \mathbf{C}$, and $\mathbf{D}$).

3. Consider the control system

$$\dot{\mathbf{x}} = \mathbf{A}\mathbf{x} + \mathbf{B}\mathbf{u},$$
$$\mathbf{x}(0) = \mathbf{x}_0,$$

where $\mathbf{A}$ is an $n \times n$ constant matrix and $\mathbf{B}$ is an $n \times m$ constant matrix. Derive the formula of variation of parameters

$$\mathbf{x}(t) = e^{\mathbf{A}t}\mathbf{x}_0 + \int_0^t e^{\mathbf{A}(t-s)}\mathbf{B}\mathbf{u}(s)ds. \tag{3.6}$$

3.2 Stability

Stability is one of central issues in control theory. Consider a linear system

$$\dot{\mathbf{x}} = \mathbf{A}\mathbf{x}, \tag{3.7}$$
$$\mathbf{x}(0) = \mathbf{x}_0. \tag{3.8}$$

The system has the equilibrium point $\bar{\mathbf{x}} = \mathbf{0}$, that is, $\mathbf{A}\mathbf{0} = \mathbf{0}$.

Definition 3.1. The equilibrium point $\mathbf{0}$ of (3.7) is

(1) stable if, for any $\varepsilon > 0$, there exists $\delta = \delta(\varepsilon) > 0$ such that

$$\|\mathbf{x}(t)\| < \varepsilon \quad \text{for } \|\mathbf{x}(0)\| < \delta, \, t \geq 0.$$

(2) unstable if it is not stable.

(3) exponentially stable if there exist $\sigma, K > 0$ such that for all $\mathbf{x}_0$ we have

$$\|\mathbf{x}(t)\| \leq K\|\mathbf{x}_0\|e^{-\sigma t}.$$

The stability of the equilibrium point $\mathbf{0}$ can be characterized by the locations of the eigenvalues of the matrix $\mathbf{A}$. The solution of (3.7) is given by

$$\mathbf{x}(t) = e^{\mathbf{A}t}\mathbf{x}_0.$$

From matrix theory it follows that there exists a nonsingular matrix $\mathbf{P}$ that transforms $\mathbf{A}$ into its Jordan normal form

$$\mathbf{P}^{-1}\mathbf{A}\mathbf{P} = \mathbf{J} = \begin{bmatrix} \mathbf{J}_1 & \mathbf{0} & \mathbf{0} & \cdots & \mathbf{0} & \mathbf{0} \\ \mathbf{0} & \mathbf{J}_2 & \mathbf{0} & \cdots & \mathbf{0} & \mathbf{0} \\ \vdots & \vdots & \vdots & \cdots & \vdots & \vdots \\ \mathbf{0} & \mathbf{0} & \mathbf{0} & \cdots & \mathbf{J}_{r-1} & \mathbf{0} \\ \mathbf{0} & \mathbf{0} & \mathbf{0} & \cdots & \mathbf{0} & \mathbf{J}_r \end{bmatrix},$$

where $\mathbf{J}_i$ is an $m_i \times m_i$ Jordan block associated with the eigenvalue λ_i of $\mathbf{A}$. If $m_i = 1$, then $\mathbf{J}_i = [\lambda_i]$. If $m_i > 1$, then

$$\mathbf{J}_i = \begin{bmatrix} \lambda_i & 1 & 0 & \cdots & 0 & 0 \\ 0 & \lambda_i & 1 & \cdots & 0 & 0 \\ \vdots & \vdots & \vdots & \cdots & \vdots & \vdots \\ 0 & 0 & 0 & \cdots & \lambda_i & 1 \\ 0 & 0 & 0 & \cdots & 0 & \lambda_i \end{bmatrix}.$$

To illustrate the Jordan normal form, we consider the matrix

$$\mathbf{A} = \begin{bmatrix} 1 & -2 & 2 \\ -2 & 1 & -2 \\ 2 & -2 & 1 \end{bmatrix}.$$

The characteristic equation of $\mathbf{A}$ is

$$\det(\mathbf{A} - \lambda\mathbf{I}) = -(\lambda + 1)^2(\lambda - 5).$$

The eigenvectors corresponding to the eigenvalue $\lambda_1 = \lambda_2 = -1$ are

$$\mathbf{P}_1 = \begin{bmatrix} 1 \\ 1 \\ 0 \end{bmatrix}, \quad \mathbf{P}_2 = \begin{bmatrix} 0 \\ 1 \\ 1 \end{bmatrix}.$$

The eigenvector corresponding to the eigenvalue $\lambda_3 = 5$ is

$$\mathbf{P}_3 = \begin{bmatrix} 1 \\ -1 \\ 1 \end{bmatrix}.$$

Thus we obtain the nonsingular matrix $\mathbf{P} = [\mathbf{P}_1\ \mathbf{P}_2\ \mathbf{P}_3]$ such that

$$\mathbf{P}^{-1}\mathbf{A}\mathbf{P} = \begin{bmatrix} -1 & 0 & 0 \\ 0 & -1 & 0 \\ 0 & 0 & 5 \end{bmatrix}.$$

Consider another example

$$\mathbf{A} = \begin{bmatrix} 1 & 0 & 0 \\ 2 & 2 & -1 \\ 0 & 1 & 0 \end{bmatrix}.$$

The characteristic equation is

$$\det(\mathbf{A} - \lambda\mathbf{I}) = -(\lambda - 1)^3.$$

The eigenvector corresponding to the eigenvalue $\lambda_1 = \lambda_2 = \lambda_2 = 1$ is

$$\mathbf{P}_1 = \begin{bmatrix} 0 \\ 1 \\ 1 \end{bmatrix}.$$

Solving the following system

$$(\mathbf{A} - \lambda_1\mathbf{I})\mathbf{P}_2 = \mathbf{P}_1,$$
$$(\mathbf{A} - \lambda_1\mathbf{I})\mathbf{P}_3 = \mathbf{P}_2,$$

we obtain two generalized eigenvectors

$$\mathbf{P}_2 = \begin{bmatrix} 0 \\ 1 \\ 0 \end{bmatrix}, \quad \mathbf{P}_3 = \begin{bmatrix} 1/2 \\ 0 \\ 0 \end{bmatrix}.$$

Thus we obtain the nonsingular matrix $\mathbf{P} = [\mathbf{P}_1\ \mathbf{P}_2\ \mathbf{P}_3]$ such that

$$\mathbf{P}^{-1}\mathbf{AP} = \begin{bmatrix} 1 & 1 & 0 \\ 0 & 1 & 1 \\ 0 & 0 & 1 \end{bmatrix}. \tag{3.9}$$

We rewrite the Jordan block as follows

$$
\begin{aligned}
\mathbf{J}_i &= \begin{bmatrix}
\lambda_i & 1 & 0 & \cdots & 0 & 0 \\
0 & \lambda_i & 1 & \cdots & 0 & 0 \\
\vdots & \vdots & \vdots & \cdots & \vdots & \vdots \\
0 & 0 & 0 & \cdots & \lambda_i & 1 \\
0 & 0 & 0 & \cdots & 0 & \lambda_i
\end{bmatrix} \\
&= \begin{bmatrix}
\lambda_i & 0 & 0 & \cdots & 0 & 0 \\
0 & \lambda_i & 0 & \cdots & 0 & 0 \\
\vdots & \vdots & \vdots & \cdots & \vdots & \vdots \\
0 & 0 & 0 & \cdots & \lambda_i & 0 \\
0 & 0 & 0 & \cdots & 0 & \lambda_i
\end{bmatrix} + \begin{bmatrix}
0 & 1 & 0 & \cdots & 0 & 0 \\
0 & 0 & 1 & \cdots & 0 & 0 \\
\vdots & \vdots & \vdots & \cdots & \vdots & \vdots \\
0 & 0 & 0 & \cdots & 0 & 1 \\
0 & 0 & 0 & \cdots & 0 & 0
\end{bmatrix} \\
&= \Lambda_i + \mathbf{N}.
\end{aligned}
$$

Since $\exp(\Lambda_i t) = \exp(\lambda_i t)\mathbf{I}$ and $\mathbf{N}^k = 0$ for $k \geq m_i$, we deduce that

$$
\begin{aligned}
\exp(\mathbf{J}_i t) &= \exp(\Lambda_i t)\exp(\mathbf{N}t) \\
&= \exp(\lambda_i t)\sum_{k=0}^{m_i-1}\frac{t^k\mathbf{N}^k}{k!} \\
&= \begin{bmatrix}
e^{\lambda_i t} & te^{\lambda_i t} & \frac{t^2}{2!}e^{\lambda_i t} & \cdots & \frac{t^{m_i-2}}{(m_i-2)!}e^{\lambda_i t} & \frac{t^{m_i-1}}{(m_i-1)!}e^{\lambda_i t} \\
0 & e^{\lambda_i t} & te^{\lambda_i t} & \cdots & \frac{t^{m_i-3}}{(m_i-3)!}e^{\lambda_i t} & \frac{t^{m_i-2}}{(m_i-2)!}e^{\lambda_i t} \\
\vdots & \vdots & \vdots & \cdots & \vdots & \vdots \\
0 & 0 & 0 & \cdots & e^{\lambda_i t} & te^{\lambda_i t} \\
0 & 0 & 0 & \cdots & 0 & e^{\lambda_i t}
\end{bmatrix}.
\end{aligned}
$$

It then follows that

$$\mathbf{x}(t) = e^{\mathbf{A}t}\mathbf{x}_0 = \mathbf{P}e^{\mathbf{J}t}\mathbf{P}^{-1}\mathbf{x}_0 = \sum_{i=1}^{r}\sum_{j=0}^{m_i-1}t^j e^{\lambda_i t}\mathbf{C}_{ij}\mathbf{x}_0, \tag{3.10}$$

where $\mathbf{C}_{ij}$ are constant matrices. For example, for the Jordan normal form (3.9), the solution can be expressed as

$$
\begin{aligned}
\mathbf{x}(t) &= e^{\mathbf{A}t}\mathbf{x}_0 \\
&= \mathbf{P}e^{\mathbf{J}t}\mathbf{P}^{-1}\mathbf{x}_0 \\
&= \mathbf{P}\left(\begin{bmatrix} e^t & te^t & \frac{t^2}{2}e^t \\ 0 & e^t & te^t \\ 0 & 0 & e^t \end{bmatrix}\right)\mathbf{P}^{-1}\mathbf{x}_0 \\
&= \mathbf{P}\left(\begin{bmatrix} e^t & 0 & 0 \\ 0 & e^t & 0 \\ 0 & 0 & e^t \end{bmatrix} + \begin{bmatrix} 0 & te^t & 0 \\ 0 & 0 & te^t \\ 0 & 0 & 0 \end{bmatrix} + \begin{bmatrix} 0 & 0 & \frac{t^2}{2}e^t \\ 0 & 0 & 0 \\ 0 & 0 & 0 \end{bmatrix}\right)\mathbf{P}^{-1}\mathbf{x}_0 \\
&= e^t\mathbf{x}_0 + te^t\mathbf{P}\begin{bmatrix} 0 & 1 & 0 \\ 0 & 0 & 1 \\ 0 & 0 & 0 \end{bmatrix}\mathbf{P}^{-1}\mathbf{x}_0 + \frac{t^2}{2}e^t\mathbf{P}\begin{bmatrix} 0 & 0 & 1 \\ 0 & 0 & 0 \\ 0 & 0 & 0 \end{bmatrix}\mathbf{P}^{-1}\mathbf{x}_0.
\end{aligned}
$$

Given a polynomial $f(\lambda) = (\lambda - \lambda_1)^{q_1}\cdots(\lambda - \lambda_n)^{q_n}$, the algebraic multiplicity of the root λ_i $(i = 1,\cdots,n)$ of $f(\lambda)$ is defined to be q_i. From matrix theory, the dimension m_i of $\mathbf{J}_i$ is equal to 1 if and only if the dimension of the eigenvector space $\ker(\mathbf{A} - \lambda_i\mathbf{I})$ is equal to the algebraic multiplicity q_i of the eigenvalue λ_i, and if and only if $\mathrm{rank}(\mathbf{A} - \lambda_i\mathbf{I}) = n - q_i$. Hence, from (3.10), we have the following theorem.

Theorem 3.1. *The equilibrium point $\mathbf{0}$ of (3.7) is stable if and only if all eigenvalues of $\mathbf{A}$ satisfy $\mathrm{Re}\lambda_i \leq 0$ and for every eigenvalue with $\mathrm{Re}\lambda_i = 0$ and algebraic multiplicity $q_i \geq 2$, $\mathrm{rank}(\mathbf{A} - \lambda_i\mathbf{I}) = n - q_i$, where n is the dimension of $\mathbf{x}$. The equilibrium point $\mathbf{0}$ of (3.7) is exponentially stable if and only if all eigenvalues of $\mathbf{A}$ satisfy $\mathrm{Re}\lambda_i < 0$.*

Definition 3.2. A matrix $\mathbf{A}$ is said to be Hurwitz if all eigenvalues of $\mathbf{A}$ have negative real parts.

Example 3.1. Consider the system

$$\dot{\mathbf{x}} = \mathbf{A}\mathbf{x}, \tag{3.11}$$

where

$$\mathbf{A} = \begin{bmatrix} 0 & 1 \\ -1 & 0 \end{bmatrix}.$$

The characteristic equation of $\mathbf{A}$ is

$$\det(\lambda \mathbf{I} - \mathbf{A}) = \lambda^2 + 1 = 0.$$

Solving the equation, we obtain the eigenvalues $\lambda = \pm i$. It is clear that the algebraic multiplicity of both i and $-i$ is 1 and the ranks of $i\mathbf{I} - \mathbf{A}$ and $-i\mathbf{I} - \mathbf{A}$ are equal to 1. So the conditions of Theorem 3.1 are satisfied and then the equilibrium 0 is stable. In fact, the solution of the system is given by

$$\begin{bmatrix} x_1 \\ x_2 \end{bmatrix} = c_1 \begin{bmatrix} \cos t \\ -\sin t \end{bmatrix} + c_2 \begin{bmatrix} \sin t \\ \cos t \end{bmatrix},$$

which shows that the equilibrium point 0 is stable.

Example 3.2. Consider the system

$$\dot{\mathbf{x}} = \mathbf{A}\mathbf{x}, \tag{3.12}$$

where

$$\mathbf{A} = \begin{bmatrix} 0 & 1 \\ -5 & -2 \end{bmatrix}.$$

The characteristic equation of $\mathbf{A}$ is

$$\lambda^2 + 2\lambda + 5 = 0.$$

Then the eigenvalues are

$$\lambda = -1 \pm 2i.$$

Thus, by Theorem 3.1, the equilibrium 0 is exponentially stable. In fact, the solution is

$$\begin{bmatrix} x_1 \\ x_2 \end{bmatrix} = c_1 \begin{bmatrix} \cos 2t \\ -\cos 2t - 2\sin 2t \end{bmatrix} e^{-t}$$
$$+ c_2 \begin{bmatrix} \sin 2t \\ -\sin 2t + 2\cos 2t \end{bmatrix} e^{-t},$$

which decays exponentially.

For higher order systems ($n \geq 3$), it may be impossible to solve characteristic polynomials explicitly. In this case, the Routh-Hurwitz's stability criterion is needed for determining the signs of eigenvalues without actually solving for them. The proof of the criterion is long and is referred to [21].

Theorem 3.2. *(Routh-Hurwitz's Criterion) Suppose that all the coefficients of the polynomial* $f(s) = a_0 s^n + a_1 s^{n-1} + \cdots + a_{n-1} s + a_n$ *are positive. Construct the following table:*

$$
\begin{aligned}
s^n &: a_0 \ a_2 \ a_4 \ a_6 \ \cdots \\
s^{n-1} &: a_1 \ a_3 \ a_5 \ a_7 \ \cdots \\
s^{n-2} &: b_1 \ b_2 \ b_3 \ b_4 \ \cdots \\
s^{n-3} &: c_1 \ c_2 \ c_3 \ c_4 \ \cdots \\
&\ \ \vdots \ \ \vdots \ \ \vdots \ \ \vdots \ \ \vdots \\
s^2 &: d_1 \ d_2 \\
s^1 &: e_1 \\
s^0 &: f_1
\end{aligned}
$$

where any undefined entries are set to zero and

$$
b_1 = -\frac{1}{a_1}\begin{vmatrix} a_0 & a_2 \\ a_1 & a_3 \end{vmatrix}, \ b_2 = -\frac{1}{a_1}\begin{vmatrix} a_0 & a_4 \\ a_1 & a_5 \end{vmatrix}, \ b_3 = -\frac{1}{a_1}\begin{vmatrix} a_0 & a_6 \\ a_1 & a_7 \end{vmatrix}, \ \cdots
$$

$$
c_1 = -\frac{1}{b_1}\begin{vmatrix} a_1 & a_3 \\ b_1 & b_2 \end{vmatrix}, \ c_2 = -\frac{1}{b_1}\begin{vmatrix} a_1 & a_5 \\ b_1 & b_3 \end{vmatrix}, \ c_3 = -\frac{1}{b_1}\begin{vmatrix} a_1 & a_7 \\ b_1 & b_4 \end{vmatrix}, \ \cdots
$$

$$\vdots$$

Then all zeros of f have negative real parts if and only if all entries in the first column of the table are well defined and positive.

Example 3.3. Consider the polynomial $f(s) = (s+1)(s+2)(s+3) = s^3 + 6s^2 + 11s + 6$, which has three negative zeros: -1, -2, -3. The Routh's table is as follows

$$
\begin{aligned}
s^3 &: \ 1 \ \ 11 \\
s^2 &: \ 6 \ \ 6 \\
s &: \ 10 \ \ 0 \\
s^0 &: \ 6
\end{aligned}
$$

All entries in the first column of the table are positive.

Exercises 3.2

1. Consider the system

$$\dot{x} = Ax,$$

where

$$\mathbf{A} = \begin{bmatrix} -1 & 0 & 2 \\ 0 & -1 & 1 \\ 0 & -1 & -1 \end{bmatrix}.$$

a. Calculate the eigenvalues of $\mathbf{A}$.
b. Show the first column of the Routh's array of the characteristic equation $|\lambda \mathbf{I} - \mathbf{A}| = 0$ is positive.

2. Consider the system

$$\dot{\mathbf{x}} = \mathbf{A}\mathbf{x},$$

where

$$\mathbf{A} = \begin{bmatrix} 0 & 1 & 0 \\ -b_3 & 0 & 1 \\ 0 & -b_2 & -b_1 \end{bmatrix}.$$

($\mathbf{A}$ is called Schwarz matrix). Show that the first column of the Routh's array of the characteristic equation $|\lambda \mathbf{I} - \mathbf{A}| = 0$ consists of $1, b_1, b_2$, and $b_1 b_3$.

3.3 Controllability and Observability

Controllability and observability are structural properties of a system. Consider a control system

$$\dot{\mathbf{x}} = \mathbf{A}\mathbf{x} + \mathbf{B}\mathbf{u}, \tag{3.13}$$

$$\mathbf{x}(0) = \mathbf{x}_0. \tag{3.14}$$

Definition 3.3. System (3.13) or the pair $(\mathbf{A}, \mathbf{B})$ is controllable if any initial state $\mathbf{x}_0$ and final state $\mathbf{x}_f$, there exists a control vector $\mathbf{u}$ such that $\mathbf{x}(T) = \mathbf{x}_f$ for some $T > 0$.

In deriving a test for controllability, we need Cayley-Hamilton theorem that plays a very important role in solving problems involving matrix equations.

Theorem 3.3. *(Cayley-Hamilton) Any $n \times n$ matrix $\mathbf{A}$ satisfies its characteristic equation*

$$\mathbf{A}^n + a_{n-1}\mathbf{A}^{n-1} + \cdots + a_1\mathbf{A} + a_0\mathbf{I} = \mathbf{0},$$

where

$$\lambda^n + a_{n-1}\lambda^{n-1} + \cdots + a_1\lambda + a_0 = \det(\lambda \mathbf{I} - \mathbf{A}).$$

Proof. Consider the adjugate $\mathrm{adj}(\lambda \mathbf{I} - \mathbf{A})$ of $\lambda \mathbf{I} - \mathbf{A}$:

$$\mathrm{adj}(\lambda \mathbf{I} - \mathbf{A}) = \begin{bmatrix} C_{11} & C_{21} & \cdots & C_{n1} \\ C_{12} & C_{22} & \cdots & C_{n2} \\ \vdots & \vdots & \cdots & \vdots \\ C_{1n} & C_{2n} & \cdots & C_{nn} \end{bmatrix},$$

where $C_{ji} = (-1)^{i+j} \det(\lambda \mathbf{I} - \mathbf{A})_{ji}$ and $(\lambda \mathbf{I} - \mathbf{A})_{ji}$ denotes the submatrix of $\lambda \mathbf{I} - \mathbf{A}$ formed by deleting row j and column i. It is clear that $\text{adj}(\lambda \mathbf{I} - \mathbf{A})$ is a polynomial in λ of degree $n - 1$:

$$\text{adj}(\lambda \mathbf{I} - \mathbf{A}) = \mathbf{B}_{n-1} \lambda^{n-1} + \mathbf{B}_{n-2} \lambda^{n-2} + \cdots + \mathbf{B}_1 \lambda + \mathbf{B}_0.$$

Since

$$(\lambda \mathbf{I} - \mathbf{A}) \text{adj}(\lambda \mathbf{I} - \mathbf{A}) = \det(\lambda \mathbf{I} - \mathbf{A}) \mathbf{I},$$

we obtain

$$\begin{aligned} &\left(\lambda^n + a_{n-1} \lambda^{n-1} + \cdots + a_1 \lambda + a_0 \right) \mathbf{I} \\ &= (\lambda \mathbf{I} - \mathbf{A}) \left(\mathbf{B}_{n-1} \lambda^{n-1} + \mathbf{B}_{n-2} \lambda^{n-2} + \cdots + \mathbf{B}_1 \lambda + \mathbf{B}_0 \right). \end{aligned}$$

Substituting $\mathbf{A}$ for λ gives that

$$\mathbf{A}^n + a_{n-1} \mathbf{A}^{n-1} + \cdots + a_1 \mathbf{A} + a_0 \mathbf{I} = \mathbf{0}.$$

Consider the matrix

$$\mathbf{A} = \begin{bmatrix} 1 & 2 \\ 3 & 4 \end{bmatrix}.$$

We have $\det(\lambda \mathbf{I} - \mathbf{A}) = (\lambda - 1)(\lambda - 4) - 6$ and

$$\begin{aligned} (\mathbf{A} - \mathbf{I})(\mathbf{A} - 4\mathbf{I}) - 6\mathbf{I} &= \begin{bmatrix} 0 & 2 \\ 3 & 3 \end{bmatrix} \begin{bmatrix} -3 & 2 \\ 3 & 0 \end{bmatrix} - 6 \begin{bmatrix} 1 & 0 \\ 0 & 1 \end{bmatrix} \\ &= \begin{bmatrix} 6 & 0 \\ 0 & 6 \end{bmatrix} - \begin{bmatrix} 6 & 0 \\ 0 & 6 \end{bmatrix} \\ &= \begin{bmatrix} 0 & 0 \\ 0 & 0 \end{bmatrix}. \end{aligned}$$

A square matrix $\mathbf{A}$ is *positive definite* if the inner product $(\mathbf{A}\mathbf{x}, \mathbf{x}) > 0$ for any $\mathbf{x} \neq \mathbf{0}$. Consider the matrix

$$\mathbf{A} = \begin{bmatrix} 1 & 1 \\ 1 & 2 \end{bmatrix}.$$

We have

$$\begin{aligned} (\mathbf{A}\mathbf{x}, \mathbf{x}) &= (x_1 + x_2)x_1 + (x_1 + 2x_2)x_2 \\ &= x_1^2 + 2x_1 x_2 + 2x_2^2 \\ &= (x_1 + x_2)^2 + x_2^2 \\ &> 0 \end{aligned}$$

for any $\mathbf{x} \neq \mathbf{0}$. So $\mathbf{A}$ is positive definite.

Theorem 3.4. *The following statements are equivalent:*

(1) $(\mathbf{A}, \mathbf{B})$ is controllable.

(2) The matrix

$$\mathbf{P}(t) = \int_0^t e^{\mathbf{A}(t-s)} \mathbf{B}\mathbf{B}^* e^{\mathbf{A}^*(t-s)} ds$$

is positive definite for all $t > 0$.
(3) Kalman controllability matrix defined by

$$\mathscr{C} = [\mathbf{B} \ \mathbf{A}\mathbf{B} \ \mathbf{A}^2\mathbf{B} \cdots \mathbf{A}^{n-1}\mathbf{B}]$$

has rank n.

Proof. $(1) \Rightarrow (2)$. Argue by contradiction. Assume that $\mathbf{P}(T)$ is not positive definite for some $T > 0$. Then there exists a vector $\mathbf{v} \neq 0$ such that

$$\mathbf{v}^* \mathbf{P}(T) \mathbf{v} = \int_0^T \mathbf{v}^* e^{\mathbf{A}(T-s)} \mathbf{B}\mathbf{B}^* e^{\mathbf{A}^*(T-s)} \mathbf{v} ds = 0,$$

and then

$$\mathbf{v}^* e^{\mathbf{A}t} \mathbf{B} = 0.$$

For the initial state $\mathbf{x}_0 = e^{-\mathbf{A}T} \mathbf{v}$, we have

$$\mathbf{x}(T) = e^{\mathbf{A}T} e^{-\mathbf{A}T} \mathbf{v} + \int_0^t e^{\mathbf{A}(T-s)} \mathbf{B}\mathbf{u}(s) ds,$$

and then

$$\mathbf{v}^* \mathbf{x}(T) = \mathbf{v}^* \mathbf{v} > 0.$$

Thus it is impossible to drive the system to $\mathbf{x}(T) = 0$. This contradicts with the controllability.

$(2) \Rightarrow (1)$. For any initial state $\mathbf{x}_0$ and any final state $\mathbf{x}_f$, choosing the control

$$u(t) = -\mathbf{B}^* e^{\mathbf{A}^*(T-t)} \mathbf{P}(T)^{-1} \left(e^{\mathbf{A}T} \mathbf{x}_0 - \mathbf{x}_f \right),$$

we obtain

$$\begin{aligned}
\mathbf{x}(T) &= e^{\mathbf{A}T} \mathbf{x}_0 - \int_0^T e^{\mathbf{A}(T-s)} \mathbf{B}\mathbf{B}^* e^{\mathbf{A}^*(T-s)} \mathbf{P}(T)^{-1} \left(e^{\mathbf{A}T} \mathbf{x}_0 - \mathbf{x}_f \right) ds \\
&= e^{\mathbf{A}T} \mathbf{x}_0 - \int_0^T e^{\mathbf{A}(T-s)} \mathbf{B}\mathbf{B}^* e^{\mathbf{A}^*(T-s)} ds \mathbf{P}(T)^{-1} \left(e^{\mathbf{A}T} \mathbf{x}_0 - \mathbf{x}_f \right) \\
&= e^{\mathbf{A}T} \mathbf{x}_0 - \mathbf{P}(T)\mathbf{P}(T)^{-1} \left(e^{\mathbf{A}T} \mathbf{x}_0 - \mathbf{x}_f \right) \\
&= \mathbf{x}_f.
\end{aligned}$$

$(2) \Rightarrow (3)$. Argue by contradiction. Assume that the rank of $\mathscr{C}$ is less than n. Then the row vectors of $\mathscr{C}$ are linearly dependent. So we can find a nonzero vector $\mathbf{v} = (v_1, v_2, \cdots v_n)$ such that

$$\mathbf{v}[\mathbf{B} \ \mathbf{A}\mathbf{B} \ \mathbf{A}^2\mathbf{B} \cdots \mathbf{A}^{n-1}\mathbf{B}] = [\mathbf{v}\mathbf{B} \ \mathbf{v}\mathbf{A}\mathbf{B} \ \mathbf{v}\mathbf{A}^2\mathbf{B} \cdots \mathbf{v}\mathbf{A}^{n-1}\mathbf{B}] = 0,$$

or
$$\mathbf{vB} = \mathbf{vAB} = \mathbf{vA}^2\mathbf{B} = \cdots = \mathbf{vA}^{n-1}\mathbf{B} = 0.$$

It then follows from the Cayley-Hamilton theorem that

$$-\mathbf{vA}^n\mathbf{B} = a_{n-1}\mathbf{vA}^{n-1}\mathbf{B} + \cdots + a_1\mathbf{vAB} + a_0\mathbf{vB} = 0.$$

By induction, we can show that $\mathbf{vA}^m\mathbf{B} = 0$ for $m = 0, 1, 2, \cdots$. Thus we derive that

$$\mathbf{v}e^{\mathbf{A}t}\mathbf{B} = \mathbf{v}\left(\mathbf{I} + \mathbf{A}t + \frac{1}{2!}\mathbf{A}^2 t^2 + \cdots\right)\mathbf{B} = 0,$$

and then
$$\mathbf{vP}(t) = 0.$$

This contradicts the fact that $\mathbf{P}(t)$ is positive definite.

(3) $\Rightarrow$ (2). Next we assume that rank$[\mathscr{C}] = n$, but $\mathbf{P}(t)$ is not positive definite. Then there exists a nonzero vector $\mathbf{v}$ such that

$$\mathbf{v}e^{\mathbf{A}(T-s)}\mathbf{B} = \mathbf{0}, \quad 0 \le s \le T.$$

Taking $s = T$ gives that $\mathbf{vB} = \mathbf{0}$. Differentiating it and letting $s = T$ gives $\mathbf{vAB} = \mathbf{0}$. Continuing this process, we find that

$$\mathbf{vB} = \mathbf{vAB} = \mathbf{vA}^2\mathbf{B} = \cdots = \mathbf{vA}^{n-1}\mathbf{B} = 0,$$

which contradicts the assumption that rank$[\mathscr{C}] = n$.

Example 3.4. Consider the mass-spring-damper system

$$\begin{bmatrix} \dot{x}_1 \\ \dot{x}_2 \end{bmatrix} = \begin{bmatrix} 0 & 1 \\ -\frac{k}{m} & 0 \end{bmatrix} \begin{bmatrix} x_1 \\ x_2 \end{bmatrix} + \begin{bmatrix} 0 \\ \frac{1}{m} \end{bmatrix} u(t).$$

Since the Kalman controllability matrix

$$\mathscr{C} = [\mathbf{B} \ \mathbf{AB}] = \begin{bmatrix} 0 & \frac{1}{m} \\ \frac{1}{m} & 0 \end{bmatrix}$$

has rank 2, the system is controllable.

The following example shows that not every system is controllable.

Example 3.5. The system

$$\begin{bmatrix} \dot{x}_1 \\ \dot{x}_2 \end{bmatrix} = \begin{bmatrix} 1 & 0 \\ 1 & 1 \end{bmatrix} \begin{bmatrix} x_1 \\ x_2 \end{bmatrix} + \begin{bmatrix} 0 \\ 1 \end{bmatrix} u(t)$$

is not controllable since the rank of the Kalman controllability matrix

$$\mathscr{C} = [\mathbf{B} \ \mathbf{AB}] = \begin{bmatrix} 0 & 0 \\ 1 & 1 \end{bmatrix}$$

is equal to 1.

To derive another test for controllability, we need the following technical lemma.

Lemma 3.1. *Let* $\mathbf{A}, \mathbf{B}$ *be* $n \times n$ *and* $n \times m$ *matrices, respectively. If the rank of the controllability matrix*

$$\text{rank}[\mathbf{B}\ \mathbf{AB}\ \mathbf{A}^2\mathbf{B}\cdots\mathbf{A}^{n-1}\mathbf{B}] = q < n,$$

then there exists a nonsingular matrix $\mathbf{P}$ *such that*

$$\mathbf{P}^{-1}\mathbf{AP} = \begin{bmatrix} \mathbf{A}_{11} & \mathbf{A}_{12} \\ \mathbf{0} & \mathbf{A}_{22} \end{bmatrix}, \quad \mathbf{P}^{-1}\mathbf{B} = \begin{bmatrix} \mathbf{B}_{11} \\ \mathbf{0} \end{bmatrix},$$

where $\mathbf{A}_{11}$ *is a* $q \times q$ *matrix,* $\mathbf{A}_{12}$ *is a* $q \times (n-q)$ *matrix,* $\mathbf{A}_{22}$ *is an* $(n-q) \times (n-q)$ *matrix, and* $\mathbf{B}_{11}$ *is a* $q \times m$ *matrix. Moreover, the pair* $(\mathbf{A}_{11}, \mathbf{B}_{11})$ *is controllable.*

Proof. Let $\mathbf{f}_1, \cdots, \mathbf{f}_q$ be q linearly independent column vectors of the controllability matrix and let us choose $n - q$ additional vectors $\mathbf{v}_{q+1}, \cdots, \mathbf{v}_n$ such that

$$\mathbf{P} = \begin{bmatrix} \mathbf{f}_1 & \cdots & \mathbf{f}_q\ \mathbf{v}_{q+1} & \cdots & \mathbf{v}_n \end{bmatrix}$$

is nonsingular. For each $\mathbf{f}_i$, there exists a column vector $\mathbf{b}$ of $\mathbf{B}$ and an integer $0 \leq j \leq n - 1$ such that $\mathbf{f}_i = \mathbf{A}^j\mathbf{b}$. If $j < n - 1$, then $\mathbf{A}\mathbf{f}_i = \mathbf{A}^{j+1}\mathbf{b}$ is one of the column vectors of the controllability matrix and so $\mathbf{A}\mathbf{f}_i, \mathbf{f}_1, \cdots, \mathbf{f}_q$ are linearly dependent. Thus there are $a_{i1}, \cdots, a_{iq}$ such that

$$\mathbf{A}\mathbf{f}_i = a_{i1}\mathbf{f}_1 + \cdots + a_{iq}\mathbf{f}_q. \tag{3.15}$$

If $j = n - 1$, then $\mathbf{A}\mathbf{f}_i = \mathbf{A}^n\mathbf{b}$. By the Cayley-Hamilton theorem, there exist constants $c_0, c_1, \cdots, c_{n-1}$ such that $\mathbf{A}^n = c_0\mathbf{I} + c_1\mathbf{A} + \cdots + c_{n-1}\mathbf{A}^{n-1}$. So $\mathbf{A}\mathbf{f}_i = c_0\mathbf{b} + c_1\mathbf{A}\mathbf{b} + \cdots + c_{n-1}\mathbf{A}^{n-1}\mathbf{b}$. Since each $\mathbf{A}^j\mathbf{b}$ ($0 \leq j \leq n - 1$) can be expressed in terms of $\mathbf{f}_1, \cdots, \mathbf{f}_q$, (3.15) also holds in this case. It therefore follows that

$$\begin{bmatrix} \mathbf{A}\mathbf{f}_1 & \cdots & \mathbf{A}\mathbf{f}_q\ \mathbf{A}\mathbf{v}_{q+1} & \cdots & \mathbf{A}\mathbf{v}_n \end{bmatrix} = \begin{bmatrix} \mathbf{f}_1 & \cdots & \mathbf{f}_q\ \mathbf{v}_{q+1} & \cdots & \mathbf{v}_n \end{bmatrix}\begin{bmatrix} \mathbf{A}_{11} & \mathbf{A}_{12} \\ \mathbf{0} & \mathbf{A}_{22} \end{bmatrix},$$

which implies

$$\mathbf{P}^{-1}\mathbf{AP} = \begin{bmatrix} \mathbf{A}_{11} & \mathbf{A}_{12} \\ \mathbf{0} & \mathbf{A}_{22} \end{bmatrix}.$$

In addition, all column vectors of $\mathbf{B}$ can be expressed in terms of $\mathbf{f}_1, \cdots, \mathbf{f}_q$ and so

$$\mathbf{B} = \begin{bmatrix} \mathbf{f}_1 & \cdots & \mathbf{f}_q\ \mathbf{v}_{q+1} & \cdots & \mathbf{v}_n \end{bmatrix}\begin{bmatrix} \mathbf{B}_{11} \\ \mathbf{0} \end{bmatrix},$$

which implies

$$\mathbf{P}^{-1}\mathbf{B} = \begin{bmatrix} \mathbf{B}_{11} \\ \mathbf{0} \end{bmatrix}.$$

Moreover, we can show that

$$\mathbf{P}^{-1}[\mathbf{B} \ \mathbf{AB} \ \mathbf{A}^2\mathbf{B} \cdots \mathbf{A}^{n-1}\mathbf{B}] = \begin{bmatrix} \mathbf{B}_{11} & \mathbf{A}_{11}\mathbf{B}_{11} & \mathbf{A}_{11}^2\mathbf{B}_{11} & \cdots & \mathbf{A}_{11}^{n-1}\mathbf{B}_{11} \\ \mathbf{0} & \mathbf{0} & \mathbf{0} & \cdots & \mathbf{0} \end{bmatrix}.$$

Then the rank of $[\mathbf{B}_{11} \ \mathbf{A}_{11}\mathbf{B}_{11} \ \mathbf{A}_{11}^2\mathbf{B}_{11} \cdots \mathbf{A}_{11}^{n-1}\mathbf{B}_{11}]$ is equal to q and so the pair $(\mathbf{A}_{11}, \mathbf{B}_{11})$ is controllable.

Theorem 3.5. *(Popov-Belevitch-Hautus Test) System* (3.13) *is controllable if and only if the matrix* $[(\lambda \mathbf{I} - \mathbf{A}) \ \mathbf{B}]$ *has a rank n for all complex numbers* λ.

Proof. Suppose that $[(\lambda \mathbf{I} - \mathbf{A}) \ \mathbf{B}]$ does not have rank n. Then there exists a nonzero vector $\mathbf{v}^*$ such that

$$\mathbf{v}^*[(\lambda \mathbf{I} - \mathbf{A}) \ \mathbf{B}] = 0.$$

Then

$$\lambda \mathbf{v}^* = \mathbf{v}^*\mathbf{A}, \quad \mathbf{v}^*\mathbf{B} = \mathbf{0}$$

and so

$$\begin{aligned} \mathbf{v}^*\mathscr{C} &= \mathbf{v}^*[\mathbf{B} \ \mathbf{AB} \ \mathbf{A}^2\mathbf{B} \cdots \mathbf{A}^{n-1}\mathbf{B}] \\ &= \mathbf{v}^*[\mathbf{B} \ \lambda\mathbf{B} \ \lambda^2\mathbf{B} \cdots \lambda^{n-1}\mathbf{B}] \\ &= \mathbf{0}. \end{aligned}$$

Hence, by Theorem 3.4, the system (3.13) is not controllable.

We next suppose that the system (3.13) is not controllable. Then by Lemma 3.1 there exists a nonsingular matrix $\mathbf{P}$ such that

$$\mathbf{P}^{-1}\mathbf{AP} = \begin{bmatrix} \mathbf{A}_{11} & \mathbf{A}_{12} \\ \mathbf{0} & \mathbf{A}_{22} \end{bmatrix}, \quad \mathbf{P}^{-1}\mathbf{B} = \begin{bmatrix} \mathbf{B}_{11} \\ \mathbf{0} \end{bmatrix}.$$

Let λ be an eigenvalue of $\mathbf{A}_{22}$ and $\mathbf{v}_{22}$ be the left eigenvector so that $\mathbf{v}_{22}\mathbf{A}_{22} = \lambda \mathbf{v}_{22}$. Then

$$[\mathbf{0} \ \mathbf{v}_{22}] \begin{bmatrix} \lambda \mathbf{I} - \mathbf{A}_{11} & -\mathbf{A}_{12} & \mathbf{B}_{11} \\ \mathbf{0} & \lambda \mathbf{I} - \mathbf{A}_{22} & \mathbf{0} \end{bmatrix} = \mathbf{0},$$

and so

$$[\mathbf{0} \ \mathbf{v}_{22}]\mathbf{P}^{-1}[(\lambda \mathbf{I} - \mathbf{A})\mathbf{P} \ \mathbf{B}] = \mathbf{0}.$$

Since $\mathbf{P}$ is nonsingular, we derive that

$$[\mathbf{0} \ \mathbf{v}_{22}]\mathbf{P}^{-1}(\lambda \mathbf{I} - \mathbf{A}) = \mathbf{0},$$

and then

$$[\mathbf{0} \ \mathbf{v}_{22}]\mathbf{P}^{-1}[(\lambda \mathbf{I} - \mathbf{A}) \ \mathbf{B}] = \mathbf{0}.$$

This contradicts the assumption that $[(\lambda \mathbf{I} - \mathbf{A}) \ \mathbf{B}]$ has rank n.

Consider the mass-spring system (3.4). It is clear that the rank of the matrix

$$[(\lambda \mathbf{I} - \mathbf{A}) \ \mathbf{B}] = \begin{bmatrix} \lambda & -1 & 0 \\ \frac{k}{m} & \lambda & \frac{1}{m} \end{bmatrix}$$

is equal to 2. By Popov-Belevitch-Hautus Test, the mass-spring system is controllable.

Consider the observation system

$$\dot{\mathbf{x}} = \mathbf{A}\mathbf{x}, \tag{3.16}$$

$$\mathbf{y} = \mathbf{C}\mathbf{x}, \tag{3.17}$$

$$\mathbf{x}(0) = \mathbf{x}_0, \tag{3.18}$$

where $\mathbf{y} = (y_1, y_2, \cdots, y_l)^*$ is an output vector and $\mathbf{C}$ is an $l \times n$ constant matrix.

Definition 3.4. System (3.16)-(3.17) or the pair $(\mathbf{A}, \mathbf{C})$ is observable if any initial state $\mathbf{x}_0$ can be uniquely determined by the observation $\mathbf{y}(t)$ over the interval $[0, T]$ for some $T > 0$.

We define Kalman observability matrix $\mathscr{O}$ by

$$\mathscr{O} = \begin{bmatrix} \mathbf{C} \\ \mathbf{C}\mathbf{A} \\ \mathbf{C}\mathbf{A}^2 \\ \vdots \\ \mathbf{C}\mathbf{A}^{n-1} \end{bmatrix}.$$

Theorem 3.6. *The following statements are equivalent:*

(1) $(\mathbf{A}, \mathbf{C})$ is observable.
(2) Kalman observability matrix $\mathscr{O}$ has rank n.
(3) $(\mathbf{A}^*, \mathbf{C}^*)$ is controllable.
(4) The matrix

$$\mathbf{Q}(t) = \int_0^t e^{\mathbf{A}^* s} \mathbf{C}^* \mathbf{C} e^{\mathbf{A} s} ds$$

is positive definite for all $t > 0$.
(5) The matrix

$$\begin{bmatrix} \lambda \mathbf{I} - \mathbf{A} \\ \mathbf{C} \end{bmatrix}$$

has rank n for all complex numbers λ.

Proof. $(1) \Rightarrow (2)$. We first assume that the system is observable, but the rank of $\mathscr{O}$ is less than n. Then the column vectors of $\mathscr{O}$ are linearly dependent. So we can find a nonzero vector $\mathbf{v} = (v_1, v_2, \cdots v_n)^*$ such that

$$\begin{bmatrix} \mathbf{C} \\ \mathbf{C}\mathbf{A} \\ \mathbf{C}\mathbf{A}^2 \\ \vdots \\ \mathbf{C}\mathbf{A}^{n-1} \end{bmatrix} \mathbf{v} = \begin{bmatrix} \mathbf{C}\mathbf{v} \\ \mathbf{C}\mathbf{A}\mathbf{v} \\ \mathbf{C}\mathbf{A}^2\mathbf{v} \\ \vdots \\ \mathbf{C}\mathbf{A}^{n-1}\mathbf{v} \end{bmatrix} = \mathbf{0},$$

or
$$\mathbf{Cv} = \mathbf{CAv} = \mathbf{CA}^2\mathbf{v} = \cdots = \mathbf{CA}^{n-1}\mathbf{v} = \mathbf{0}.$$

It then follows from the Cayley-Hamilton theorem that

$$-\mathbf{CA}^n\mathbf{v} = a_{n-1}\mathbf{CA}^{n-1}\mathbf{v} + \cdots + a_1\mathbf{CAv} + a_0\mathbf{Cv} = \mathbf{0}.$$

By induction, we can show that $\mathbf{CA}^m\mathbf{v} = \mathbf{0}$ for $m = 0, 1, 2, \cdots$. Thus we derive that

$$\mathbf{y}(t) = \mathbf{C}e^{\mathbf{A}t}\mathbf{v} = \mathbf{C}\left(\mathbf{I} + \mathbf{A}t + \frac{1}{2!}\mathbf{A}^2t^2 + \cdots\right)\mathbf{v} = \mathbf{0}.$$

In addition, we also have $\mathbf{y}(t) = \mathbf{C}e^{\mathbf{A}t}\mathbf{0}$. This implies that two different initial states $\mathbf{0}$ and $\mathbf{v}$ have the same output $\mathbf{0}$. So the initial state $\mathbf{x}_0$ cannot be uniquely determined by the observation $\mathbf{y}(t)$ over the interval $[0, T]$ and this contradicts the observability assumption.

$(2) \Rightarrow (1)$. Next we assume that $\text{rank}[\mathcal{O}] = n$, but the system is not observable. Then there exists a nonzero initial state $\mathbf{v}$ such that

$$\mathbf{C}e^{\mathbf{A}t}\mathbf{v} = \mathbf{0}, \quad 0 \le t \le T.$$

Taking $t = 0$ gives that $\mathbf{Cv} = \mathbf{0}$. Differentiating it and letting $t = 0$ gives $\mathbf{CAv} = \mathbf{0}$. Continuing this process, we find that

$$\mathbf{Cv} = \mathbf{CAv} = \mathbf{CA}^2\mathbf{v} = \cdots = \mathbf{CA}^{n-1}\mathbf{v} = 0,$$

which contradicts the assumption that $\text{rank}[\mathcal{O}] = n$.

The rest of the statements can be proved by using Theorem 3.4.

From Theorem 3.6, we can derive the following observability inequality.

Corollary 3.1. *System* (3.16)-(3.17) *is observable if and only if the observability inequality holds*

$$\int_0^T \|\mathbf{Cx}(t)\|^2 dt \ge K\|\mathbf{x}_0\|^2 \quad \text{for all } \mathbf{x}_0, \tag{3.19}$$

where K is a positive constant, depending on T, but independent of $\mathbf{x}_0$.

Proof. To prove the corollary, it suffices to notice that

$$
\begin{aligned}
\int_0^T \|\mathbf{Cx}(t)\|^2 dt &= \int_0^T \|\mathbf{C}e^{\mathbf{A}t}\mathbf{x}_0\|^2 dt \\
&= \int_0^T \left(\mathbf{C}e^{\mathbf{A}t}\mathbf{x}_0, \mathbf{C}e^{\mathbf{A}t}\mathbf{x}_0\right) dt \\
&= \int_0^T \left(e^{\mathbf{A}^*t}\mathbf{C}^*\mathbf{C}e^{\mathbf{A}t}\mathbf{x}_0, \mathbf{x}_0\right) dt \\
&= (\mathbf{Q}(T)\mathbf{x}_0, \mathbf{x}_0).
\end{aligned}
$$

Example 3.6. Consider the mass-spring system

$$\begin{bmatrix} \dot{x}_1 \\ \dot{x}_2 \end{bmatrix} = \begin{bmatrix} 0 & 1 \\ -\frac{k}{m} & 0 \end{bmatrix} \begin{bmatrix} x_1 \\ x_2 \end{bmatrix},$$

$$y = [0 \ 1] \begin{bmatrix} x_1 \\ x_2 \end{bmatrix}.$$

Since Kalman observability matrix

$$\mathcal{O} = \begin{bmatrix} \mathbf{C} \\ \mathbf{CA} \end{bmatrix} = \begin{bmatrix} 0 & 1 \\ -\frac{k}{m} & 0 \end{bmatrix}$$

has rank 2, the system is observable.

Example 3.7. The system

$$\begin{bmatrix} \dot{x}_1 \\ \dot{x}_2 \end{bmatrix} = \begin{bmatrix} 1 & 1 \\ 0 & 1 \end{bmatrix} \begin{bmatrix} x_1 \\ x_2 \end{bmatrix},$$

$$y = [0 \ 1] \begin{bmatrix} x_1 \\ x_2 \end{bmatrix}$$

is not observable since the rank of Kalman observability matrix

$$\mathcal{O} = \begin{bmatrix} \mathbf{C} \\ \mathbf{CA} \end{bmatrix} = \begin{bmatrix} 0 & 1 \\ 0 & 1 \end{bmatrix}$$

is equal to 1.

From Theorems 3.4 and 3.6, we can derive the following duality between controllability and observability.

Theorem 3.7. *The control system*

$$\dot{\mathbf{x}} = \mathbf{Ax} + \mathbf{B}u,$$

$$\mathbf{x}(0) = \mathbf{x}_0$$

is controllable if and only if its dual observation system

$$\dot{\mathbf{x}} = \mathbf{A}^*\mathbf{x},$$

$$y = \mathbf{B}^*\mathbf{x},$$

$$\mathbf{x}(0) = \mathbf{x}_0$$

is observable.

Exercises 3.3

1. Consider the system

$$
\begin{bmatrix} \dot{x}_1 \\ \dot{x}_2 \\ \dot{x}_3 \end{bmatrix} = \begin{bmatrix} -1 & -2 & -2 \\ 0 & -1 & 1 \\ 1 & 0 & -1 \end{bmatrix} \begin{bmatrix} x_1 \\ x_2 \\ x_3 \end{bmatrix} + \begin{bmatrix} 2 \\ 0 \\ 1 \end{bmatrix} u,
$$

$$
y = \begin{bmatrix} 1 & 1 & 0 \end{bmatrix} \begin{bmatrix} x_1 \\ x_2 \\ x_3 \end{bmatrix}.
$$

   Is the system controllable and observable?

2. Consider the system

$$
\begin{bmatrix} \dot{x}_1 \\ \dot{x}_2 \\ \dot{x}_3 \end{bmatrix} = \begin{bmatrix} 2 & 0 & 0 \\ 0 & 2 & 0 \\ 0 & 3 & 1 \end{bmatrix} \begin{bmatrix} x_1 \\ x_2 \\ x_3 \end{bmatrix},
$$

$$
y = \begin{bmatrix} 1 & 1 & 1 \end{bmatrix} \begin{bmatrix} x_1 \\ x_2 \\ x_3 \end{bmatrix}.
$$

   a. Show that the system is not observable.
   b. Show that the system is observable if the output is given by

$$
\begin{bmatrix} y_1 \\ y_2 \end{bmatrix} = \begin{bmatrix} 1 & 1 & 1 \\ 1 & 2 & 3 \end{bmatrix} \begin{bmatrix} x_1 \\ x_2 \\ x_3 \end{bmatrix}.
$$

3. ([86]) Define for any nonzero p

$$
\mathbf{A} = \begin{bmatrix} p & 1 & 0 \\ 0 & p & 1 \\ 0 & 0 & p \end{bmatrix}.
$$

   Show that $(\mathbf{A}, \mathbf{B})$ is not controllable for any $\mathbf{B} = [b_1 \ b_2 \ 0]^*$. Find a condition on the observation matrix $\mathbf{C}$ such that $(\mathbf{A}, \mathbf{C})$ is observable.

4. ([86]) The dynamics of a hot air balloon is modeled by

$$
\frac{d}{dt} \begin{bmatrix} T \\ v \\ h \end{bmatrix} = \begin{bmatrix} -k_1 & 0 & 0 \\ \sigma & -k_3 & 0 \\ 0 & 1 & 0 \end{bmatrix} \begin{bmatrix} T \\ v \\ h \end{bmatrix} + \begin{bmatrix} k_2 & 0 \\ 0 & k_3 \\ 0 & 0 \end{bmatrix} \begin{bmatrix} u \\ w \end{bmatrix},
$$

   where T is the temperature of the balloon (relative to equilibrium temperature), u is heat added, v is velocity, h is height, and w is wind speed. The parameters k_1, k_2, k_3, and σ are all positive constants.

a. If the output $y(t) = h(t)$, is the system observable?
b. If the output $y(t) = T(t)$, is the system observable?
c. Is this system controllable?
d. Is this system controllable by u (heat) only? Why?
e. Is this system controllable by w (wind) only? Why?

5. Show that if an $n \times n$ matrix $\mathbf{A}$ is positive definite, then there exists a positive constant C such that
$$(\mathbf{Ax}, \mathbf{x}) \geq C\|\mathbf{x}\|^2$$
for all $\mathbf{x} \in \mathbb{R}^n$.

6. Show that if a matrix $\mathbf{A}$ is positive definite, then it is nonsingular.
7. Prove that $\text{rank}(\mathbf{AB}) \leq \min\{\text{rank}(\mathbf{A}), \text{rank}(\mathbf{B})\}$.
8. Prove that $\text{rank}(\mathbf{AB}) = \text{rank}(\mathbf{A})$ if $\mathbf{B}$ is nonsingular.

3.4 State Feedback Control

We assume that all state variables are available for feedback and design a control of the form
$$\mathbf{u} = -\mathbf{Kx}. \tag{3.20}$$

Such a scheme is called a state feedback. The $m \times n$ matrix $\mathbf{K}$ is called a state feedback gain matrix. Substituting equation (3.20) into (3.1) gives

$$\dot{\mathbf{x}} = (\mathbf{A} - \mathbf{BK})\mathbf{x}, \quad \mathbf{x}(0) = \mathbf{x}_0. \tag{3.21}$$

Definition 3.5. The pair $(\mathbf{A}, \mathbf{B})$ or the system (3.1) is stabilizable if there exists $\mathbf{K}$ such that the solution $\mathbf{x}(t)$ of (3.21) converges to zero exponentially as $t \to \infty$ for any initial state $\mathbf{x}_0$. The matrix $\mathbf{K}$ is called the feedback matrix.

If the pair $(\mathbf{A}, \mathbf{B})$ is stabilizable, then the solution $\mathbf{x}(t)$ tends to zero. Thus we have that $\mathbf{u}(t) = -\mathbf{Kx}(t) \to 0$ and $\mathbf{y}(t) = \mathbf{Cx}(t) + \mathbf{Du}(t) \to 0$. Therefore the problem of regulating the output to zero is transformed into the stabilization of the pair $(\mathbf{A}, \mathbf{B})$.

Example 3.8. Consider the mass-spring system (3.4). Let $\mathbf{K} = [k_1 \ k_2]$. The gain matrix $\mathbf{K}$ stabilizes the equilibrium point 0 if and only if all eigenvalues of the matrix $\mathbf{A} - \mathbf{BK}$ have negative real parts. Let μ_1, μ_2 be the desired eigenvalues. Then we set

$$\det(\lambda \mathbf{I} - \mathbf{A} + \mathbf{BK}) = \lambda^2 + \frac{k_2}{m}\lambda + \frac{k + k_1}{m} = (\lambda - \mu_1)(\lambda - \mu_2).$$

Equating the coefficients gives

$$k_1 = m\mu_1\mu_2 - k, \quad k_2 = -m(\mu_1 + \mu_2).$$

We then have the following control law

$$u = -k_1 x_1 - k_2 x_2 = -k_1 y - k_2 \dot{y}.$$

For the observation system, we introduce the output injection system

$$\dot{\mathbf{x}} = \mathbf{A}\mathbf{x} - \mathbf{K}\mathbf{y} = (\mathbf{A} - \mathbf{K}\mathbf{C})\mathbf{x}, \quad \mathbf{x}(0) = \mathbf{x}_0. \tag{3.22}$$

Definition 3.6. The pair $(\mathbf{A}, \mathbf{C})$ or the system (3.16)-(3.17) is detectable if there exists $\mathbf{K}$ such that the solution $\mathbf{x}(t)$ of (3.22) converges to zero exponentially as $t \to \infty$ for any initial state $\mathbf{x}_0$. The matrix $\mathbf{K}$ is called the output injection matrix.

Evidently, $(\mathbf{A}, \mathbf{C})$ is detectable if and only if $(\mathbf{A}^*, \mathbf{C}^*)$ is stabilizable.

Consider the control system

$$\dot{x}_1 = x_1 + x_2 + u,$$
$$\dot{x}_2 = x_2.$$

It is clear that this system is not stabilizable since x_2 is not controlled by u. We can readily check that the system is not controllable. Thus it can be expected that controllability plays a key role in determining stabilizability of a system. In fact, we now show that if a system is controllable, the eigenvalues of the system can be placed at any location by a feedback.

Lemma 3.2. *Let* $\mathbf{B}$ *be an* $n \times 1$ *vector and* $(\mathbf{A}, \mathbf{B})$ *be controllable. Then there exists a nonsingular matrix* $\mathbf{T}$ *that transforms* $(\mathbf{A}, \mathbf{B})$ *into the following controllable canonical form:*

$$\mathbf{T}^{-1}\mathbf{A}\mathbf{T} = \begin{bmatrix} 0 & 1 & 0 & \cdots & 0 & 0 \\ 0 & 0 & 1 & \cdots & 0 & 0 \\ 0 & 0 & 0 & \cdots & 0 & 0 \\ \vdots & \vdots & \vdots & \cdots & \vdots & \vdots \\ 0 & 0 & 0 & \cdots & 0 & 1 \\ -a_0 & -a_1 & -a_2 & \cdots & -a_{n-2} & -a_{n-1} \end{bmatrix}, \quad \mathbf{T}^{-1}\mathbf{B} = \begin{bmatrix} 0 \\ 0 \\ 0 \\ \vdots \\ 0 \\ 1 \end{bmatrix}.$$

Proof. Because $(\mathbf{A}, \mathbf{B})$ is controllable, the vectors

$$\mathbf{B}, \mathbf{A}\mathbf{B}, \cdots, \mathbf{A}^{n-1}\mathbf{B}$$

are linearly independent. Let the characteristic polynomial of $\mathbf{A}$ be

$$\det(s\mathbf{I} - \mathbf{A}) = s^n + a_{n-1}s^{n-1} + \cdots + a_1 s + a_0.$$

Define

$$f_0(s) = s^n + a_{n-1}s^{n-1} + \cdots + a_1 s + a_0,$$
$$f_1(s) = s^{n-1} + a_{n-1}s^{n-2} + \cdots + a_2 s + a_1,$$
$$\vdots$$
$$f_{n-1}(s) = s + a_{n-1},$$
$$f_n(s) = 1.$$

Then f_i satisfies the iteration relation

$$sf_i(s) = f_{i-1}(s) - a_{i-1}f_n(s).\tag{3.23}$$

Let

$$\mathbf{v}_i = f_i(\mathbf{A})\mathbf{B}, \quad \mathbf{T} = [\mathbf{v}_1, \cdots, \mathbf{v}_n].$$

Then $\mathbf{v}_1, \cdots, \mathbf{v}_n$ are linearly independent and $\mathbf{T}$ is a nonsingular matrix. Replacing s by $\mathbf{A}$ in (3.23) and operating on $\mathbf{B}$, we obtain

$$\mathbf{A}\mathbf{v}_i = \mathbf{v}_{i-1} - a_{i-1}\mathbf{v}_n, \quad \mathbf{B} = \mathbf{v}_n,$$

and then

$$\mathbf{AT} = \mathbf{T}\begin{bmatrix} 0 & 1 & 0 & \cdots & 0 & 0 \\ 0 & 0 & 1 & \cdots & 0 & 0 \\ 0 & 0 & 0 & \cdots & 0 & 0 \\ \vdots & \vdots & \vdots & \cdots & \vdots & \vdots \\ 0 & 0 & 0 & \cdots & 0 & 1 \\ -a_0 & -a_1 & -a_2 & \cdots & -a_{n-2} & -a_{n-1} \end{bmatrix}, \quad \mathbf{B} = \mathbf{T}\begin{bmatrix} 0 \\ 0 \\ 0 \\ \vdots \\ 0 \\ 1 \end{bmatrix}.$$

Example 3.9. Let

$$\mathbf{A} = \begin{bmatrix} 1 & 2 \\ 3 & 4 \end{bmatrix}, \quad \mathbf{B} = \begin{bmatrix} 1 \\ 2 \end{bmatrix}.$$

Then

$$\begin{vmatrix} s-1 & -2 \\ -3 & s-4 \end{vmatrix} = s^2 - 5s - 2$$

and

$$f_1(s) = s - 5, \quad f_2(s) = 1.$$

Thus

$$\mathbf{v}_1 = f_1(\mathbf{A})\mathbf{B} = (\mathbf{A} - 5\mathbf{I})\mathbf{B} = \begin{bmatrix} 0 \\ 1 \end{bmatrix}, \quad \mathbf{v}_2 = \mathbf{B}, \quad \mathbf{T} = \begin{bmatrix} 0 & 1 \\ 1 & 2 \end{bmatrix}$$

and

$$\mathbf{AT} = \begin{bmatrix} 2 & 5 \\ 4 & 11 \end{bmatrix} = \mathbf{T}\begin{bmatrix} 0 & 1 \\ 2 & 5 \end{bmatrix}, \quad \mathbf{T}\begin{bmatrix} 0 \\ 1 \end{bmatrix} = \mathbf{B}.$$

The following lemma [104] indicates that feedback control of a multiple input system can be reduced to a single input system.

Lemma 3.3. *Let $(\mathbf{A}, \mathbf{B})$ be controllable and $\mathbf{b} = \mathbf{B}\mathbf{u} \neq 0$, where $\mathbf{u}$ is a vector. Then there exists a matrix $\mathbf{F}$ such that $(\mathbf{A} + \mathbf{BF}, \mathbf{b})$ is controllable.*

Proof. Let $\mathbf{b}_1 = \mathbf{b}$, and let n_1 be the largest integer such that

$$\mathbf{b}_1, \mathbf{A}\mathbf{b}_1, \cdots, \mathbf{A}^{n_1}\mathbf{b}_1$$

are linearly independent. We can easily show that

$$\mathbf{b}_1, \mathbf{A}\mathbf{b}_1 + \mathbf{b}_1, \cdots, \mathbf{A}^{n_1}\mathbf{b}_1 + \mathbf{A}^{n_1-1}\mathbf{b}_1 + \cdots + \mathbf{b}_1$$

are also linearly independent. If $n_1 < n$, then there exists

$$\mathbf{b}_2 = \mathbf{B}\mathbf{u}_2 \notin \text{span}(\mathbf{b}_1, \mathbf{A}\mathbf{b}_1, \cdots, \mathbf{A}^{n_1}\mathbf{b}_1).$$

If not, then for any $\mathbf{u}$, we have

$$\mathbf{B}\mathbf{u} = c_0\mathbf{b}_1 + c_1\mathbf{A}\mathbf{b}_1 + \cdots + c_{n_1}\mathbf{A}^{n_1}\mathbf{b}_1 \in \text{span}(\mathbf{b}_1, \mathbf{A}\mathbf{b}_1, \cdots, \mathbf{A}^{n_1}\mathbf{b}_1).$$

It then follows that $\mathbf{A}^i\mathbf{B}\mathbf{u} \in \text{span}(\mathbf{b}_1, \mathbf{A}\mathbf{b}_1, \cdots, \mathbf{A}^{n_1}\mathbf{b}_1)$ for $i = 0, 1, 2, \cdots n$. This implies that the rank of

$$[\mathbf{B} \ \mathbf{A}\mathbf{B} \ \mathbf{A}^{n-1}\mathbf{B}]$$

is equal to $n_1 < n$ and then contradicts with the controllability assumption. Let n_2 be the largest integer such that

$$\mathbf{b}_2, \mathbf{A}\mathbf{b}_2, \cdots, \mathbf{A}^{n_2}\mathbf{b}_2$$

are linearly independent. If $n_1 + n_2 < n$, we can repeat the above procedure and obtain n linearly independent vectors:

$$\mathbf{b}_1, \mathbf{A}\mathbf{b}_1, \cdots, \mathbf{A}^{n_1}\mathbf{b}_1;$$
$$\mathbf{b}_2, \mathbf{A}\mathbf{b}_2, \cdots, \mathbf{A}^{n_2}\mathbf{b}_2;$$
$$\vdots$$
$$\mathbf{b}_k, \mathbf{A}\mathbf{b}_k, \cdots, \mathbf{A}^{n_k}\mathbf{b}_k.$$

Define

$$\mathbf{v}_1 = \mathbf{b}_1,$$
$$\mathbf{v}_i = \mathbf{A}\mathbf{v}_{i-1} + \mathbf{b}_1, \quad 1 < i \leq n_1,$$
$$\mathbf{v}_{n_j+i} = \mathbf{A}\mathbf{v}_{n_j+i-1} + \mathbf{b}_{j+1}, \quad 1 \leq i \leq n_{j+1}, 1 \leq j \leq k-1.$$

We can show that $\mathbf{v}_1, \cdots, \mathbf{v}_n$ are linearly independent. Define

$$\bar{\mathbf{b}}_i = \mathbf{b}_1 = \mathbf{B}\mathbf{u}_1, \quad \bar{\mathbf{u}}_i = \mathbf{u}_1, \quad 1 \leq i \leq n_1,$$
$$\bar{\mathbf{b}}_{n_j+i} = \mathbf{b}_{j+1} = \mathbf{B}\mathbf{u}_{j+1}, \quad \bar{\mathbf{u}}_{n_j+i} = \mathbf{u}_{j+1}, \quad 1 \leq i \leq n_{j+1}, 1 \leq j \leq k-1.$$

Then we have

$$\mathbf{v}_i = \mathbf{A}\mathbf{v}_{i-1} + \bar{\mathbf{b}}_{i-1}, \quad 2 \leq i \leq n.$$

Since $\mathbf{v}_1, \cdots, \mathbf{v}_n$ are linearly independent, there exists a matrix $\mathbf{F}$ such that

$$\mathbf{F}\mathbf{v}_i = \bar{\mathbf{u}}_i.$$

In fact, $\mathbf{F}$ is given by

$$\mathbf{F} = [\bar{\mathbf{u}}_1, \cdots, \bar{\mathbf{u}}_n][\mathbf{v}_1, \cdots, \mathbf{v}_n]^{-1}.$$

It therefore follows that

$$
\begin{aligned}
\mathbf{v}_i &= \mathbf{A}\mathbf{v}_{i-1} + \mathbf{B}\bar{\mathbf{u}}_{i-1} \\
&= \mathbf{A}\mathbf{v}_{i-1} + \mathbf{B}\mathbf{F}\mathbf{v}_{i-1} \\
&= (\mathbf{A} + \mathbf{B}\mathbf{F})^{i-1}\mathbf{b}, \quad 2 \le i \le n.
\end{aligned}
$$

This implies that the rank of

$$[\mathbf{b} \ (\mathbf{A} + \mathbf{B}\mathbf{F})\mathbf{b} \ (\mathbf{A} + \mathbf{B}\mathbf{F})^{n-1}\mathbf{b}]$$

is equal to n and so $(\mathbf{A} + \mathbf{B}\mathbf{F}, \mathbf{b})$ is controllable.

Theorem 3.8. *A system* $(\mathbf{A}, \mathbf{B})$ *of order n is controllable if and only if for any set* $S = \{p_1, \cdots, p_n\}$ *of complex numbers with the property that if* $p \in S$, *then* $\bar{p} \in S$, *there exists a matrix* $\mathbf{K}$ *such that* $\mathbf{A} - \mathbf{B}\mathbf{K}$ *has eigenvalues* $p_1, \cdots, p_n$.

Proof. Let $\lambda_1, \lambda_2, \cdots \lambda_n$ be all eigenvalues of $\mathbf{A}$ and let the set $S = \{p_1, \cdots p_n\}$ have the stated property in the theorem such that $S \cap \{\lambda_1, \lambda_2, \cdots \lambda_n\} = \emptyset$. Suppose that there exists a matrix $\mathbf{K}$ such that $\mathbf{A} - \mathbf{B}\mathbf{K}$ has eigenvalues $p_1, \cdots p_n$. We want to prove that $\operatorname{rank}[(s\mathbf{I} - \mathbf{A}) \ \mathbf{B}] = n$ for all complex numbers s.

For all λ_i ($i = 1, 2, \cdots, n$), the matrix $\lambda_i \mathbf{I} - \mathbf{A} + \mathbf{B}\mathbf{K}$ is nonsingular. For any $\mathbf{x} \in \mathbb{R}^n$, we define $\bar{\mathbf{x}} = (\lambda_i \mathbf{I} - \mathbf{A} + \mathbf{B}\mathbf{K})^{-1}\mathbf{x}$ and $\bar{\mathbf{u}} = \mathbf{K}(\lambda_i \mathbf{I} - \mathbf{A} + \mathbf{B}\mathbf{K})^{-1}\mathbf{x}$. Then

$$
\begin{aligned}
[(\lambda_i \mathbf{I} - \mathbf{A}) \ \mathbf{B}] \begin{bmatrix} \bar{\mathbf{x}} \\ \bar{\mathbf{u}} \end{bmatrix} &= (\lambda_i \mathbf{I} - \mathbf{A})\bar{\mathbf{x}} + \mathbf{B}\bar{\mathbf{u}} \\
&= (\lambda_i \mathbf{I} - \mathbf{A} + \mathbf{B}\mathbf{K})(\lambda_i \mathbf{I} - \mathbf{A} + \mathbf{B}\mathbf{K})^{-1}\mathbf{x} \\
&= \mathbf{x}.
\end{aligned}
$$

This shows that $[(\lambda_i \mathbf{I} - \mathbf{A}) \ \mathbf{B}]$ has rank n. For all other complex numbers s, the rank of $[(s\mathbf{I} - \mathbf{A}) \ \mathbf{B}]$ is n since the rank of $s\mathbf{I} - \mathbf{A}$ is n. Thus, by Theorem 3.5, $(\mathbf{A}, \mathbf{B})$ is controllable.

Next assume that $(\mathbf{A}, \mathbf{B})$ is controllable. Let $\mathbf{b} = \mathbf{B}\mathbf{v} \ne 0$, where $\mathbf{v}$ is a vector. Then there exists a matrix $\mathbf{F}$ such that $(\mathbf{A} + \mathbf{B}\mathbf{F}, \mathbf{b})$ is controllable. Denote $\mathbf{A}_F = \mathbf{A} + \mathbf{B}\mathbf{F}$. Then there exists a nonsingular matrix $\mathbf{T}$ that transforms $\mathbf{A}_F$ and $\mathbf{b}$ into controllable canonical form

$$\mathbf{T}^{-1}\mathbf{A}_F\mathbf{T} = \mathbf{A}_c, \quad \mathbf{T}^{-1}\mathbf{b} = \mathbf{b}_c,$$

where

$$
\mathbf{A}_c = \begin{bmatrix}
0 & 1 & 0 & \cdots & 0 & 0 \\
0 & 0 & 1 & \cdots & 0 & 0 \\
0 & 0 & 0 & \cdots & 0 & 0 \\
\vdots & \vdots & \vdots & \cdots & \vdots & \vdots \\
0 & 0 & 0 & \cdots & 0 & 1 \\
-a_0 & -a_1 & -a_2 & \cdots & -a_{n-2} & -a_{n-1}
\end{bmatrix}, \quad
\mathbf{b}_c = \begin{bmatrix} 0 \\ 0 \\ 0 \\ \vdots \\ 0 \\ 1 \end{bmatrix}.
$$

Let the state feedback

$$\mathbf{K}_c = [k_0\ k_1\ \cdots\ k_{n-1}].$$

Then

$$\det(s\mathbf{I} - \mathbf{A}_c + \mathbf{b}_c\mathbf{K}_c)$$

$$= \det \begin{bmatrix} s & -1 & 0 & \cdots & 0 & 0 \\ 0 & s & -1 & \cdots & 0 & 0 \\ 0 & 0 & s & \cdots & 0 & 0 \\ \vdots & \vdots & \vdots & \cdots & \vdots & \vdots \\ 0 & 0 & 0 & \cdots & s & -1 \\ a_0 + k_0 & a_1 + k_1 & a_2 + k_2 & \cdots & a_{n-2} + k_{n-2} & s + a_{n-1} + k_{n-1} \end{bmatrix}$$

$$= s^n + (a_{n-1} + k_{n-1})s^{n-1} + \cdots + (a_1 + k_1)s + a_0 + k_0.$$

Let

$$(s - p_1) \cdots (s - p_n) = s^n + b_{n-1}s^{n-1} + \cdots + b_1 s + b_0.$$

Choosing $k_0, k_1, \cdots, k_{n-1}$ such that

$$a_{n-1} + k_{n-1} = b_{n-1}, \cdots, a_0 + k_0 = b_0,$$

we obtain

$$\det(s\mathbf{I} - \mathbf{A}_c + \mathbf{b}_c\mathbf{K}_c) = (s - p_1) \cdots (s - p_n).$$

Since

$$\det(s\mathbf{I} - \mathbf{A}_c + \mathbf{b}_c\mathbf{K}_c) = \det(s\mathbf{I} - \mathbf{T}^{-1}\mathbf{A}_F\mathbf{T} + \mathbf{T}^{-1}\mathbf{b}\mathbf{K}_c)$$
$$= \det[\mathbf{T}^{-1}(s\mathbf{I} - \mathbf{A}_F + \mathbf{b}\mathbf{K}_c\mathbf{T}^{-1})\mathbf{T}]$$
$$= \det(s\mathbf{I} - \mathbf{A}_F + \mathbf{b}\mathbf{K}_c\mathbf{T}^{-1}),$$

we have

$$\det(s\mathbf{I} - \mathbf{A}_F + \mathbf{b}\mathbf{K}_c\mathbf{T}^{-1}) = (s - p_1) \cdots (s - p_n),$$

and then

$$\det(s\mathbf{I} - \mathbf{A} + \mathbf{B}(-\mathbf{F} + \mathbf{v}\mathbf{K}_c\mathbf{T}^{-1})) = (s - p_1) \cdots (s - p_n).$$

For a system $(\mathbf{A}, \mathbf{B})$, an eigenvalue of $\mathbf{A}$ is called a pole of the system. Thus Theorem 3.8 is referred as the pole placement theorem.

The above constructive proofs provide an algorithm for computing the feedback gain matrix of the system $(\mathbf{A}, \mathbf{B})$. Since the computation of the feedback $\mathbf{F}$ for multiple input systems is complex, we summarize the algorithm for a single input system as follows:

1. Computation of transformation matrix $\mathbf{T}$.

$$\det(s\mathbf{I} - \mathbf{A}) = s^n + a_{n-1}s^{n-1} + \cdots + a_1 s + a_0,$$
$$f_1(s) = s^{n-1} + a_{n-1}s^{n-2} + \cdots + a_2 s + a_1,$$

$$\vdots$$

$$f_{n-1}(s) = s + a_{n-1},$$
$$f_n(s) = 1,$$
$$\mathbf{v}_i = f_i(\mathbf{A})\mathbf{B}, \; i = 1, 2, \cdots, n,$$
$$\mathbf{T} = [\mathbf{v}_1, \cdots, \mathbf{v}_n].$$

2. Computation of the designated polynomial:

$$(s - p_1) \cdots (s - p_n) = s^n + b_{n-1} s^{n-1} + \cdots + b_1 s + b_0.$$

3. Computation of the gain matrix:

$$\mathbf{K} = [b_0 - a_0 \; \cdots b_{n-1} - a_{n-1}] \mathbf{T}^{-1}.$$

Example 3.10. Let

$$\mathbf{A} = \begin{bmatrix} 1 & 2 \\ 3 & 4 \end{bmatrix}, \quad \mathbf{B} = \begin{bmatrix} 1 \\ 2 \end{bmatrix}.$$

Find the feedback gain matrix $\mathbf{K}$ such that $\mathbf{A} - \mathbf{BK}$ has eigenvalues 1 and 2.

Solution. The characteristic polynomial is

$$\begin{vmatrix} s-1 & -2 \\ -3 & s-4 \end{vmatrix} = s^2 - 5s - 2.$$

So $a_0 = -2$ and $a_1 = -5$. Since

$$(s-1)(s-2) = s^2 - 3s + 2,$$

we obtain $b_0 = 2$ and $b_1 = -3$. In Example 3.9, we have already obtained

$$\mathbf{T} = \begin{bmatrix} 0 & 1 \\ 1 & 2 \end{bmatrix}, \quad \mathbf{T}^{-1} = \begin{bmatrix} -2 & 1 \\ 1 & 0 \end{bmatrix}.$$

So the feedback matrix is

$$\mathbf{K} = [4 \; 2] \mathbf{T}^{-1} = [-6 \; 4].$$

Note that a stabilizable system is not necessarily controllable. For instance, the following simple system

$$\begin{bmatrix} \dot{x}_1 \\ \dot{x}_2 \end{bmatrix} = \begin{bmatrix} -1 & 0 \\ 0 & -1 \end{bmatrix} \begin{bmatrix} x_1 \\ x_2 \end{bmatrix} + \begin{bmatrix} 0 \\ 1 \end{bmatrix} u(t)$$

is stabilizable, but not controllable.

Using Theorem 3.8, we prove the Popov-Belevitch-Hautus Test for stabilizability.

Theorem 3.9. *(Popov-Belevitch-Hautus Test) The pair* $(\mathbf{A}, \mathbf{B})$ *or the system* (3.1) *is stabilizable if and only if the matrix* $[(\lambda\mathbf{I} - \mathbf{A}) \;\; \mathbf{B}]$ *has a rank n for all complex numbers* λ *with nonnegative real parts.*

Proof. Assume that the pair $(\mathbf{A}, \mathbf{B})$ is stabilizable. If $[(\lambda\mathbf{I} - \mathbf{A}) \;\; \mathbf{B}]$ did not have rank n for some complex number λ with a nonnegative real part, then there would exist a nonzero vector $\mathbf{v}^*$ such that

$$\mathbf{v}^*[(\lambda\mathbf{I} - \mathbf{A}) \;\; \mathbf{B}] = 0.$$

Then

$$\lambda\mathbf{v}^* = \mathbf{v}^*\mathbf{A}, \quad \mathbf{v}^*\mathbf{B} = \mathbf{0}$$

and so for any $\mathbf{K}$

$$\mathbf{v}^*(\mathbf{A} - \mathbf{B}\mathbf{K}) = \lambda\mathbf{v}^*.$$

This implies that $\mathbf{A} - \mathbf{B}\mathbf{K}$ has the eigenvalue λ with a nonnegative real part. This contradicts with the fact that the pair $(\mathbf{A}, \mathbf{B})$ is stabilizable.

We next assume that the matrix $[(\lambda\mathbf{I} - \mathbf{A}) \;\; \mathbf{B}]$ has a rank n for all complex numbers λ with nonnegative real parts. If the pair $(\mathbf{A}, \mathbf{B})$ was not stabilizable, it follows from Theorem 3.8 that it would not be controllable. Then by Lemma 3.1 there exists a nonsingular matrix $\mathbf{P}$ such that

$$\mathbf{P}^{-1}\mathbf{A}\mathbf{P} = \begin{bmatrix} \mathbf{A}_{11} & \mathbf{A}_{12} \\ \mathbf{0} & \mathbf{A}_{22} \end{bmatrix}, \quad \mathbf{P}^{-1}\mathbf{B} = \begin{bmatrix} \mathbf{B}_{11} \\ \mathbf{0} \end{bmatrix}.$$

Because the pair $(\mathbf{A}, \mathbf{B})$ was not stabilizable and by Lemma 3.1 the $(\mathbf{A}_{11}, \mathbf{B}_{11})$ is controllable, $\mathbf{A}_{22}$ must have an eigenvalue λ with a nonnegative real part. Let $\mathbf{v}_{22}$ be the left eigenvector so that $\mathbf{v}_{22}\mathbf{A}_{22} = \lambda\mathbf{v}_{22}$. Then

$$[\mathbf{0} \;\; \mathbf{v}_{22}] \begin{bmatrix} \lambda\mathbf{I} - \mathbf{A}_{11} & -\mathbf{A}_{12} & \mathbf{B}_{11} \\ \mathbf{0} & \lambda\mathbf{I} - \mathbf{A}_{22} & \mathbf{0} \end{bmatrix} = 0,$$

and so

$$[\mathbf{0} \;\; \mathbf{v}_{22}]\mathbf{P}^{-1}[(\lambda\mathbf{I} - \mathbf{A})\mathbf{P} \;\; \mathbf{B}] = 0.$$

Since $\mathbf{P}$ is nonsingular, we derive that

$$[\mathbf{0} \;\; \mathbf{v}_{22}]\mathbf{P}^{-1}(\lambda\mathbf{I} - \mathbf{A}) = 0,$$

and then

$$[\mathbf{0} \;\; \mathbf{v}_{22}]\mathbf{P}^{-1}[(\lambda\mathbf{I} - \mathbf{A}) \;\; \mathbf{B}] = 0.$$

This contradict the assumption that $[(\lambda\mathbf{I} - \mathbf{A}) \;\; \mathbf{B}]$ has rank n.

Exercises 3.4

1. Consider the system
$$\dot{\mathbf{x}} = \mathbf{A}\mathbf{x} + \mathbf{B}u,$$

   where
   $$\mathbf{A} = \begin{bmatrix} 0 & 1 & 0 \\ 0 & 0 & 1 \\ -1 & -5 & -6 \end{bmatrix}, \quad \mathbf{B} = \begin{bmatrix} 0 \\ 1 \\ 1 \end{bmatrix}.$$

   a. Show that the pair $(\mathbf{A}, \mathbf{B})$ is controllable.
   b. Determine the state feedback gain matrix $\mathbf{K}$ such that the closed-loop system with the feedback control $u = -\mathbf{K}\mathbf{x}$ has the closed-loop poles (eigenvalues) at $\lambda = -2 \pm 4i$ and $\lambda = -10$.

2. Consider the system
   $$\begin{bmatrix} \dot{x}_1 \\ \dot{x}_2 \end{bmatrix} = \begin{bmatrix} 0 & 1 \\ 0 & 2 \end{bmatrix} \begin{bmatrix} x_1 \\ x_2 \end{bmatrix} + \begin{bmatrix} 1 \\ 0 \end{bmatrix} u.$$

   a. Show that the system is not controllable.
   b. Show that this system cannot be stabilized by the state feedback control $u = -\mathbf{K}\mathbf{x}$, whatever matrix $\mathbf{K}$ is chosen.

3. Consider
   $$\mathbf{A} = \begin{bmatrix} 3 & 4 \\ 5 & 6 \end{bmatrix}, \quad \mathbf{B} = \begin{bmatrix} 1 \\ 0 \end{bmatrix}.$$

   Find the nonsingular matrix $\mathbf{T}$ that transforms $(\mathbf{A}, \mathbf{B})$ into the controllable canonical form. Then find the feedback gain matrix $\mathbf{K}$ such that $\mathbf{A} - \mathbf{B}\mathbf{K}$ has eigenvalues $1 - i$ and $1 + i$.

4. Consider the system
   $$\dot{\mathbf{x}} = \mathbf{A}\mathbf{x} + \mathbf{B}u,$$

   where
   $$\mathbf{A} = \begin{bmatrix} 1 & 1 \\ 1 & -1 \end{bmatrix}, \quad \mathbf{B} = \begin{bmatrix} 0 \\ 1 \end{bmatrix}.$$

   a. Show that the pair $(\mathbf{A}, \mathbf{B})$ is stabilizable.
   b. Determine the state feedback gain matrix $\mathbf{K}$ such that the closed-loop system with the feedback control $u = -\mathbf{K}x$ has the closed-loop poles (eigenvalues of $\mathbf{A} - \mathbf{B}\mathbf{K}$) at $\lambda = -1$ and $\lambda = -2$.

3.5 PID Controllers

Control mechanisms most widely used in control engineering are proportional-integral-derivative (PID) feedback controllers. Consider the mass-spring system

$$m\ddot{y} + ky = u(t). \tag{3.24}$$

Given the reference position r, we want to regulate y to r. To achieve this, we try the proportional controller

$$u = -K_1(y - r),\tag{3.25}$$

where K_1 is a positive constant, called the *control gain*. This means that the force applied by the damper is proportional to the difference $y - r$. Substituting this controller into (3.24), we obtain

$$m\ddot{y} + ky = -K_1(y - r).$$

Solving it gives

$$y = c_1 \cos\left(\frac{k + K_1}{m}t\right) + c_2 \sin\left(\frac{k + K_1}{m}t\right) + \frac{rK_1}{k + K_1},$$

where c_1 and c_2 are any constants. Hence $y(t)$ does not converge to r as t tends to infinity.

To improve the proportional controller, we look at the energy of the system

$$E(t) = \frac{1}{2}\left[m(\dot{y})^2 + k(y - r)^2\right].\tag{3.26}$$

Using the equation (3.24), we deduce that

$$\frac{dE}{dt} = \dot{y}(m\ddot{y} + k(y - r)) = \dot{y}(u - kr).$$

This motivate us to take

$$u = kr - K_2\frac{dy}{dt}\tag{3.27}$$

such that $\frac{dE}{dt} = -K_2(\dot{y})^2 \leq 0$, where K_2 is a positive constant. Controller (3.27) is called a *derivative control*. Substituting (3.27) into (3.24) gives

$$m\ddot{y} + ky = kr - K_2\frac{dy}{dt}.$$

Taking $K_2^2 < 4mk$ and solving the equation, we obtain

$$y = e^{-K_2 t/(2m)}\left[c_1 \cos\left(\frac{\sqrt{4mk - K_2^2}}{2m}t\right)\right.$$
$$\left. + c_2 \sin\left(\frac{\sqrt{4mk - K_2^2}}{2m}t\right)\right] + r,$$

where c_1 and c_2 are any constants. Thus, $y(t)$ converges to r, $\dot{y}(t)$ converges to 0, and $u(t)$ converges to kr. The limit kr is called a steady state of control.

Since the system parameter k (the stiffness of the spring) in controller (3.27) may not be estimated accurately, the controller may not be robust to the perturbation of the parameter. To design a controller that does not involve system parameters, we use an integral control to replace the control steady state kr as follows:

$$u = -K_2 \frac{d}{dt}(y-r) - K_3 \int_0^t (y(s) - r)ds, \tag{3.28}$$

where K_3 is a positive constant. This controller is called the *integral-derivative (ID) feedback controller*. The integral accumulates the error $y - r(t)$ and converges to the control steady state kr. Substituting (3.28) into (3.24) gives

$$m\ddot{y} + ky = -K_2 \frac{d}{dt}(y-r) - K_3 \int_0^t (y(s) - r)ds.$$

Differentiating the equation gives that

$$m\dddot{y} + K_2\ddot{y} + k\dot{y} + K_3 y = K_3 r. \tag{3.29}$$

This equation has a particular solution $y_p = r$. Let y_c be the general solution of the associate homogeneous equation

$$m\dddot{y} + K_2\ddot{y} + k\dot{y} + K_3 y = 0.$$

Then

$$y = y_c + r.$$

The characteristic polynomial of the homogeneous equation is

$$m\lambda^3 + K_2\lambda^2 + k\lambda + K_3 = 0.$$

The Routh's table for the equation is

$$
\begin{array}{ccc}
\lambda^3: & m & k \\
\lambda^2: & K_2 & K_3 \\
\lambda: & \frac{K_2k - mK_3}{K_2} & 0 \\
\lambda^0: & K_3 &
\end{array}
$$

By Routh-Hurwitz's stability criterion (Theorem 3.2), it follows that if $K_2>0, K_3>0$, and $mK_3 < K_2k$, then all zeros of the characteristic polynomial have negative real parts. Hence $y_c(t)$ converges to zero as $t \to \infty$ and then $y(t)$ converges to r.

In practice, we usually include a proportional controller to improve a performance. Combining these controllers together, we obtain the proportional-integral-derivative (PID) controller

$$u = -K_1(y-r) - K_2 \frac{d}{dt}(y-r) - K_3 \int_0^t (y(s) - r)ds. \tag{3.30}$$

Exercises 3.5

1. Consider the mass-spring system

$$m\ddot{y} + ky = u(t),$$

and the periodic reference output

$$r(t) = \sin \omega t,$$

where ω is a constant. For $\omega = \sqrt{k/m}$, derive conditions on gains K_1, K_2 such that the following integral-derivative controller

$$u(t) = -K_1(y - r) - K_2 \int_0^t (y(s) - r(s)) ds$$

tracks the reference r, that is, $y(t) - r(t)$ converges to 0 as $t \to \infty$. If $\omega \neq \sqrt{k/m}$, does the controller track the reference r?

2. Consider the third-order system

$$\frac{d^3 y}{dt^3} + y = u(t).$$

Design a derivative controller to stabilize the equilibrium 0.

3.6 Integral State Feedback Control

Let us first look at the mass-spring system

$$\begin{bmatrix} \dot{x}_1 \\ \dot{x}_2 \end{bmatrix} = \begin{bmatrix} 0 & 1 \\ -\frac{k}{m} & 0 \end{bmatrix} \begin{bmatrix} x_1 \\ x_2 \end{bmatrix} + \begin{bmatrix} 0 \\ \frac{1}{m} \end{bmatrix} u(t), \tag{3.31}$$

$$y = \begin{bmatrix} 1 & 0 \end{bmatrix} \begin{bmatrix} x_1 \\ x_2 \end{bmatrix}. \tag{3.32}$$

Suppose that the reference output r is not zero. Then the system has the equilibrium $\bar{x}_1 = r, \bar{x}_2 = 0, \bar{y} = r, \bar{u} = kr$. To change the non-zero reference output to the zero reference output, we introduce new variables:

$$w_1 = x_1 - \bar{x}_1, w_2 = x_2, z = y - r, v = u - kr.$$

It then follows from (3.31) and (3.32) that

$$\begin{bmatrix} \dot{w}_1 \\ \dot{w}_2 \end{bmatrix} = \begin{bmatrix} 0 & 1 \\ -\frac{k}{m} & 0 \end{bmatrix} \begin{bmatrix} w_1 \\ w_2 \end{bmatrix} + \begin{bmatrix} 0 \\ \frac{1}{m} \end{bmatrix} v(t),$$

$$z = \begin{bmatrix} 1 & 0 \end{bmatrix} \begin{bmatrix} w_1 \\ w_2 \end{bmatrix}.$$

From Example 3.8 we know that the state feedback control

$$v = -\mathbf{K}\mathbf{w} = -k_2 w_2$$

regulates z to 0, where $k_2 > 0$. Then the state feedback control

$$u = kr + v = kr - k_2 x_2$$

regulates y to r. So if $r \neq 0$, the control contains the system parameter k. In practice, the estimate of a system parameter is approximate. Hence such a control containing system parameters may not be robust against estimate errors of parameters. To see this, suppose that $\bar{u}$ is calculated by using an approximate value k_0 of k. Then $\bar{u} = k_0 r$. It therefore follows that the equilibrium point of the closed-loop system is given by $\bar{x}_2 = 0$ and

$$-\frac{k}{m}\bar{x}_1 + \frac{k_0 r}{m} = 0.$$

Solving it for $\bar{x}_1$ gives

$$\bar{x}_1 = \frac{k_0 r}{k} \neq r,$$

and then the desired state r is not achieved. In the state equation we used the exact value k, not k_0, because the mass-spring system is a fixed internal system and the states x_1 and x_2 are generated by the system. In the controller we used k_0 because the control signal is generated by the external controller that uses the estimate k_0.

Consider the control system

$$\dot{\mathbf{x}} = \mathbf{A}\mathbf{x} + \mathbf{B}\mathbf{u}, \quad \mathbf{x}(0) = \mathbf{x}_0, \tag{3.33}$$

$$\mathbf{y} = \mathbf{C}\mathbf{x}, \tag{3.34}$$

where $\mathbf{A}, \mathbf{B}, \mathbf{C}$ are $n \times n, m \times n$, and $m \times n$ matrices, respectively. To design a new feedback control that does not contain system parameters, we need to add the integrator

$$\dot{\mathbf{z}} = \mathbf{y} - \mathbf{r} \tag{3.35}$$

to the above system. This integrator accumulates the regulation errors $\mathbf{y} - \mathbf{r}$ and plays a role of the steady state $\bar{u} = kr$ of control. Before we design a feedback control for the augmented system (3.33)-(3.35), we first present the following lemma that will play an important role in the design of integral control.

Lemma 3.4. *Let* $\mathbf{A}, \mathbf{B}, \mathbf{C}$ *be* $n \times n, n \times m$ *and* $m \times n$ *matrices, respectively. If* $(\mathbf{A}, \mathbf{B})$ *is controllable and*

$$\text{rank} \begin{bmatrix} \mathbf{A} & \mathbf{B} \\ \mathbf{C} & \mathbf{0} \end{bmatrix} = n+m, \tag{3.36}$$

then

$$\left(\begin{bmatrix} \mathbf{A} & \mathbf{0} \\ \mathbf{C} & \mathbf{0} \end{bmatrix}, \begin{bmatrix} \mathbf{B} \\ \mathbf{0} \end{bmatrix} \right)$$

is also controllable. Thus the regulator poles of

$$\begin{bmatrix} \mathbf{A} & \mathbf{0} \\ \mathbf{C} & \mathbf{0} \end{bmatrix} - \begin{bmatrix} \mathbf{B} \\ \mathbf{0} \end{bmatrix} \begin{bmatrix} \mathbf{K}_1 & \mathbf{K}_2 \end{bmatrix}$$

can be placed at any locations by choosing the matrices $\mathbf{K}_1, \mathbf{K}_2$.

Proof. By PBH test (Theorem 3.5), it suffices to prove

$$\text{rank} \begin{bmatrix} s\mathbf{I} - \begin{bmatrix} \mathbf{A} & \mathbf{0} \\ \mathbf{C} & \mathbf{0} \end{bmatrix} & \begin{bmatrix} \mathbf{B} \\ \mathbf{0} \end{bmatrix} \end{bmatrix} = m+n.$$

If $s = 0$, this is true because of (3.36). If $s \neq 0$, we also have

$$\text{rank} \begin{bmatrix} s\mathbf{I} - \begin{bmatrix} \mathbf{A} & \mathbf{0} \\ \mathbf{C} & \mathbf{0} \end{bmatrix} & \begin{bmatrix} \mathbf{B} \\ \mathbf{0} \end{bmatrix} \end{bmatrix} = \text{rank} \begin{bmatrix} s\mathbf{I} - \mathbf{A} & \mathbf{B} & \mathbf{0} \\ \mathbf{C} & \mathbf{0} & s\mathbf{I} \end{bmatrix}$$
$$= m + \text{rank} [s\mathbf{I} - \mathbf{A} \quad \mathbf{B}]$$
$$= m+n,$$

because $(\mathbf{A}, \mathbf{B})$ is controllable.

Theorem 3.10. *Let* $\mathbf{A}, \mathbf{B}, \mathbf{C}$ *be* $n \times n, n \times m$, *and* $m \times n$ *matrices, respectively. If* $(\mathbf{A}, \mathbf{B})$ *is controllable and (3.36) holds, then there exist the gain matrices* $\mathbf{K}_1, \mathbf{K}_2$ *such that the integral state feedback control*

$$\mathbf{u} = -\mathbf{K}_1 \mathbf{x} - \mathbf{K}_2 \mathbf{z}, \tag{3.37}$$
$$\dot{\mathbf{z}} = \mathbf{e} = \mathbf{y} - \mathbf{r} \tag{3.38}$$

regulates the output $\mathbf{y}$ *of the control system (3.33)-(3.34) to* $\mathbf{r}$.

Proof. By Lemma 3.4, there exist matrices $\mathbf{K}_1, \mathbf{K}_2$ such that

$$\begin{bmatrix} \mathbf{A} & \mathbf{0} \\ \mathbf{C} & \mathbf{0} \end{bmatrix} - \begin{bmatrix} \mathbf{B} \\ \mathbf{0} \end{bmatrix} \begin{bmatrix} \mathbf{K}_1 & \mathbf{K}_2 \end{bmatrix}$$

is Hurwitz. By (3.36), we deduce that the steady state system

$$0 = \mathbf{A}\bar{\mathbf{x}} + \mathbf{B}\bar{\mathbf{u}}, \tag{3.39}$$
$$\mathbf{r} = \mathbf{C}\bar{\mathbf{x}} \tag{3.40}$$

has a unique solution. We introduce the integrator

$$\dot{\mathbf{z}} = \mathbf{y} - \mathbf{r} = \mathbf{C}(\mathbf{x} - \bar{\mathbf{x}}).$$

We then consider a feedback control law of the form

$$\mathbf{u} = -\mathbf{K}_1\mathbf{x} - \mathbf{K}_2\mathbf{z}. \tag{3.41}$$

Note that $\mathbf{K}_2$ is nonsingular since

$$\begin{bmatrix} \mathbf{A} & 0 \\ \mathbf{C} & 0 \end{bmatrix} - \begin{bmatrix} \mathbf{B} \\ 0 \end{bmatrix} \begin{bmatrix} \mathbf{K}_1 & \mathbf{K}_2 \end{bmatrix}$$

is Hurwitz and then nonsingular. In fact, if $\mathbf{K}_2$ is singular, then the rank of $\mathbf{K}_2$ is less than m, and so the rank of $\mathbf{BK}_2$ is less than m. It therefore follows that the rank of

$$\begin{bmatrix} \mathbf{A} & 0 \\ \mathbf{C} & 0 \end{bmatrix} - \begin{bmatrix} \mathbf{B} \\ 0 \end{bmatrix} \begin{bmatrix} \mathbf{K}_1 & \mathbf{K}_2 \end{bmatrix} = \begin{bmatrix} \mathbf{A} - \mathbf{BK}_1 & -\mathbf{BK}_2 \\ \mathbf{C} & 0 \end{bmatrix}$$

is less than $m + n$. This contradicts with the nonsingularity of the matrix. Thus the equation

$$\bar{\mathbf{u}} = -\mathbf{K}_1\bar{\mathbf{x}} - \mathbf{K}_2\bar{\mathbf{z}} \tag{3.42}$$

has a unique solution $\bar{\mathbf{z}}$. Subtracting (3.39) from (3.33) gives

$$\begin{aligned} \frac{d}{dt} \begin{bmatrix} \mathbf{x} - \bar{\mathbf{x}} \\ \mathbf{z} - \bar{\mathbf{z}} \end{bmatrix} &= \begin{bmatrix} \mathbf{A} & 0 \\ \mathbf{C} & 0 \end{bmatrix} \begin{bmatrix} \mathbf{x} - \bar{\mathbf{x}} \\ \mathbf{z} - \bar{\mathbf{z}} \end{bmatrix} \\ &\quad - \begin{bmatrix} \mathbf{B} \\ 0 \end{bmatrix} \begin{bmatrix} \mathbf{K}_1 & \mathbf{K}_2 \end{bmatrix} \begin{bmatrix} \mathbf{x} - \bar{\mathbf{x}} \\ \mathbf{z} - \bar{\mathbf{z}} \end{bmatrix}. \end{aligned} \tag{3.43}$$

This system is exponentially stable. Then $\mathbf{x}(t)$ converges to its equilibrium $\bar{\mathbf{x}}$. This implies that the controlled output $\mathbf{y}(t) = \mathbf{C}\mathbf{x}(t)$ of the control system (3.33)-(3.34) converges to $\mathbf{r} = \mathbf{C}\bar{\mathbf{x}}$ as t tends to infinity. Therefore, we have proved the theorem.

The equation (3.42) is the key to removing the control steady state $\bar{\mathbf{u}}$ from the control laws designed in previous sections to make the new control law here robust. In fact, if we did not introduce the integrator, we would not have the term $\mathbf{K}_2\bar{\mathbf{z}}$ and then we could not find a matrix $\mathbf{K}_1$ such that $\bar{\mathbf{u}} = -\mathbf{K}_1\bar{\mathbf{x}}$. Therefore we could not obtain the stable system (3.43). Instead, we would have

$$\frac{d}{dt}(\mathbf{x} - \bar{\mathbf{x}}) = \mathbf{A}(\mathbf{x} - \bar{\mathbf{x}}) + \mathbf{B}(\mathbf{u} - \bar{\mathbf{u}}). \tag{3.44}$$

The feedback control that stabilizes this system is

$$\mathbf{u} - \bar{\mathbf{u}} = -\mathbf{K}_1(\mathbf{x} - \bar{\mathbf{x}}).$$

and then

$$\mathbf{u} = -\mathbf{K}_1\mathbf{x} + \bar{\mathbf{u}} + \mathbf{K}_1\bar{\mathbf{x}}.$$

This control law contains the control steady state $\bar{\mathbf{u}}$, which usually depends on system parameters as shown in the mass-spring-damper system, and then it is not robust.

Example 3.11. We now design a robust integral controller for the mass-spring-damper system. It is clear that the matrix

$$\begin{bmatrix} \mathbf{A} & \mathbf{B} \\ \mathbf{C} & 0 \end{bmatrix} = \begin{bmatrix} 0 & 1 & 0 \\ -\frac{k}{m} & 0 & \frac{1}{m} \\ 1 & 0 & 0 \end{bmatrix}$$

has rank 3. Let $\mathbf{K}_1 = [k_1, k_2]$ and $\mathbf{K}_2 = k_3$. A calculation gives

$$\left| \lambda \mathbf{I} - \begin{bmatrix} \mathbf{A} & 0 \\ \mathbf{C} & 0 \end{bmatrix} + \begin{bmatrix} \mathbf{B} \\ 0 \end{bmatrix} [\mathbf{K}_1 \ \mathbf{K}_2] \right| = \lambda^3 + \frac{k_2}{m}\lambda^2 + (k + k_1)\lambda + \frac{k_3}{m}.$$

Using Routh-Hurwitz's criterion, we find that if

$$k_1 > \frac{k_3 - kk_2}{k_2}, \quad k_2 > 0, \quad k_3 > 0,$$

then all eigenvalues have negative real parts. Therefore, the integral feedback control

$$u = -k_1 y - k_2 \dot{y} - k_3 z, \tag{3.45}$$
$$\dot{z} = e = y - r \tag{3.46}$$

regulates the output y to the desired value r. Since this controller does not contain any system parameters, it is robust.

To see that this integral control (3.45) is the same as the PID control proposed in Section 3.5, we take the initial condition of z to be $z(0) = -\frac{k_1 r}{k_3}$. Integrating equation (3.46) and then plugging z into (3.45), we obtain

$$\begin{aligned} u &= -k_1 y - k_2 \dot{y} - k_3 \left(-\frac{k_1 r}{k_3} + \int_0^t (y(s) - r)ds \right) \\ &= -k_1 (y - r) - k_2 \frac{d}{dt}(y - r) - k_3 \int_0^t (y(s) - r)ds, \end{aligned}$$

which is the same as the PID control proposed in Section 3.5.

Exercises 3.6

1. Consider the system

$$\begin{bmatrix} \dot{x}_1 \\ \dot{x}_2 \end{bmatrix} = \begin{bmatrix} 0 & -2 \\ 1 & -1 \end{bmatrix} \begin{bmatrix} x_1 \\ x_2 \end{bmatrix} + \begin{bmatrix} 2 \\ 0 \end{bmatrix} u,$$

$$y = \begin{bmatrix} 1 & 1 \end{bmatrix} \begin{bmatrix} x_1 \\ x_2 \end{bmatrix}.$$

Design an integral feedback control to regulate the output y to a reference output 1.

2. Consider the mass-spring-damper system. If the output is the velocity $y = x_2$, then $\mathbf{C} = [0, 1]$ and the matrix

$$\begin{bmatrix} \mathbf{A} & \mathbf{B} \\ \mathbf{C} & 0 \end{bmatrix} = \begin{bmatrix} 0 & 1 & 0 \\ -\frac{k}{m} & 0 & \frac{1}{m} \\ 0 & 1 & 0 \end{bmatrix}$$

has the rank 2.

a. Can we design a state feedback control to regulate the output y to a nonzero reference output r?

b. Can we design an integral feedback control to regulate the output y to a nonzero reference output r?

3.7 Output Feedback Control

In the state feedback control, we assumed that all state variables are available for feedback. However, in practice, this is not always true. We then need to estimate the state variables using only input and output measurements. Such an estimation scheme is commonly called a *state observer*.

3.7.1 Observer-based Output Feedback Control

Consider the control system

$$\dot{\mathbf{x}} = \mathbf{A}\mathbf{x} + \mathbf{B}\mathbf{u}, \tag{3.47}$$

$$\mathbf{y} = \mathbf{C}\mathbf{x}, \tag{3.48}$$

$$\mathbf{x}(0) = \mathbf{x}_0. \tag{3.49}$$

The estimate $\tilde{\mathbf{x}}$ of $\mathbf{x}$ can be generated by injecting the output into the system as follows

$$\dot{\tilde{\mathbf{x}}} = \mathbf{A}\tilde{\mathbf{x}} + \mathbf{B}\mathbf{u} + \mathbf{K}_e(\mathbf{y} - \mathbf{C}\tilde{\mathbf{x}}). \tag{3.50}$$

This output injection system is called the state observer, known as the Luenberger observer. The mathematical model of the observer is basically the same as the control system except that the estimation error, the difference between the measured output $\mathbf{y}$ and the estimated output $\mathbf{C}\tilde{\mathbf{x}}$, is injected to compensate for inaccuracies in matrices $\mathbf{A}, \mathbf{B}$ and the initial state $\mathbf{x}_0$. The matrix $\mathbf{K}_e$ is the output injection matrix (also called an observer gain matrix). We then use the estimate $\tilde{\mathbf{x}}$ for feedback and introduce the observer-based output feedback control

$$\mathbf{u} = -\mathbf{K}\tilde{\mathbf{x}}, \tag{3.51}$$

which leads to an observer-based output feedback control system

$$\dot{\mathbf{x}} = \mathbf{A}\mathbf{x} - \mathbf{B}\mathbf{K}\tilde{\mathbf{x}}, \tag{3.52}$$

$$\mathbf{y} = \mathbf{C}\mathbf{x}, \tag{3.53}$$

$$\dot{\tilde{\mathbf{x}}} = \mathbf{A}\tilde{\mathbf{x}} - \mathbf{B}\mathbf{K}\tilde{\mathbf{x}} + \mathbf{K}_e(\mathbf{y} - \mathbf{C}\tilde{\mathbf{x}}). \tag{3.54}$$

Theorem 3.11. *If the pair* $(\mathbf{A}, \mathbf{B})$ *is controllable and the pair* $(\mathbf{A}, \mathbf{C})$ *is observable, then for any given closed-loop poles* $S_\lambda = \{\lambda_1, \cdots, \lambda_n\}$ *and any observer poles* $S_\mu = \{\mu_1, \cdots, \mu_n\}$ *with the property that if* $\lambda \in S_\lambda (\mu \in S_\mu)$ *then* $\bar{\lambda} \in S_\lambda (\bar{\mu} \in S_\mu)$, *there exist matrices* $\mathbf{K}$ *and* $\mathbf{K}_e$ *such that the eigenvalues of* $\mathbf{A} - \mathbf{B}\mathbf{K}$ *is equal to* $\lambda_1, \cdots, \lambda_n$ *and the eigenvalues of* $\mathbf{A} - \mathbf{K}_e\mathbf{C}$ *is equal to* $\mu_1, \cdots, \mu_n$. *If these poles are chosen such that they have negative real parts, then the observer-based output feedback control (3.51)-(3.54) regulates the output* $\mathbf{y}$ *of the system (3.52)-(3.54) to zero.*

Proof. We introduce the error vector

$$\mathbf{e} = \mathbf{x} - \tilde{\mathbf{x}}.$$

Subtracting equation (3.54) from equation (3.52) gives

$$\dot{\mathbf{e}} = \mathbf{A}\mathbf{e} - \mathbf{K}_e(\mathbf{y} - \mathbf{C}\tilde{\mathbf{x}}) = \mathbf{A}\mathbf{e} - \mathbf{K}_e(\mathbf{C}\mathbf{x} - \mathbf{C}\tilde{\mathbf{x}}) = (\mathbf{A} - \mathbf{K}_e\mathbf{C})\mathbf{e}.$$

Also the state equation (3.52) can be written as

$$\dot{\mathbf{x}} = (\mathbf{A} - \mathbf{B}\mathbf{K})\mathbf{x} + \mathbf{B}\mathbf{K}\mathbf{e}.$$

Combining these two equations, we obtain

$$\begin{bmatrix} \dot{\mathbf{x}} \\ \dot{\mathbf{e}} \end{bmatrix} = \begin{bmatrix} \mathbf{A} - \mathbf{B}\mathbf{K} & \mathbf{B}\mathbf{K} \\ \mathbf{0} & \mathbf{A} - \mathbf{K}_e\mathbf{C} \end{bmatrix} \begin{bmatrix} \mathbf{x} \\ \mathbf{e} \end{bmatrix}. \tag{3.55}$$

Since $(\mathbf{A}, \mathbf{B})$ is controllable, there exists a gain matrix $\mathbf{K}$ such that the eigenvalues of $\mathbf{A} - \mathbf{B}\mathbf{K}$ are equal to $\lambda_1, \cdots, \lambda_n$. Moreover, since the pair $(\mathbf{A}, \mathbf{C})$ is observable, it follows from Theorem 3.7 that the pair $(\mathbf{A}^*, \mathbf{C}^*)$ is controllable. Then there exists a matrix $\mathbf{G}$ such that the eigenvalues of $\mathbf{A}^* - \mathbf{C}^*\mathbf{G}$ are equal to $\mu_1, \cdots, \mu_n$. But $\mathbf{A} - \mathbf{G}^*\mathbf{C}$ has the same eigenvalues as $\mathbf{A}^* - \mathbf{C}^*\mathbf{G}$. Let $\mathbf{K}_e = \mathbf{G}^*$. Then $\mathbf{A} - \mathbf{K}_e\mathbf{C}$ has eigenvalues $\mu_1, \cdots, \mu_n$. Since

$$\det\left(\lambda\mathbf{I} - \begin{bmatrix} \mathbf{A} - \mathbf{B}\mathbf{K} & \mathbf{B}\mathbf{K} \\ \mathbf{0} & \mathbf{A} - \mathbf{K}_e\mathbf{C} \end{bmatrix}\right)$$

$$= \det(\lambda\mathbf{I} - [\mathbf{A} - \mathbf{B}\mathbf{K}])\det(\lambda\mathbf{I} - [\mathbf{A} - \mathbf{K}_e\mathbf{C}]),$$

the matrix

$$\begin{bmatrix} \mathbf{A} - \mathbf{BK} & \mathbf{BK} \\ \mathbf{0} & \mathbf{A} - \mathbf{K}_e\mathbf{C} \end{bmatrix}$$

has eigenvalues $\lambda_1, \cdots, \lambda_n; \mu_1, \cdots, \mu_n$. If these eigenvalues have negative real parts, then $\mathbf{x}(t)$ converges to zero exponentially and then so $\mathbf{y}(t) = \mathbf{Cx}(t)$ does.

Example 3.12. Consider the system

$$\dot{\mathbf{x}} = \mathbf{Ax} + \mathbf{B}u,$$
$$y = \mathbf{Cx},$$

where

$$\mathbf{A} = \begin{bmatrix} 0 & 1 \\ 0 & -2 \end{bmatrix}, \quad \mathbf{B} = \begin{bmatrix} 0 \\ 4 \end{bmatrix}, \quad \mathbf{C} = \begin{bmatrix} 1 & 0 \end{bmatrix}.$$

Design an output feedback control by observer approach such that the desired closed-loop poles are located at

$$\lambda = -2 + 2\sqrt{3}i, -2 - 2\sqrt{3}i$$

and the desired observer poles are located at

$$\lambda = -8, -8.$$

The feedback gain matrix $\mathbf{K} = [k_1, k_2]$ should be selected such that

$$\det(\lambda\mathbf{I} - \mathbf{A} + \mathbf{BK}) = (\lambda + 2 - 2\sqrt{3}i)(\lambda + 2 + 2\sqrt{3}i),$$

and then

$$\lambda^2 + (2 + 4k_2)\lambda + 4k_1 = \lambda^2 + 4\lambda + 16.$$

Comparing the coefficients gives

$$k_1 = 4, \quad k_2 = 0.5.$$

The observer gain matrix $\mathbf{K}_e = [k_1^e, k_2^e]^*$ should be selected such that

$$\det(\lambda\mathbf{I} - \mathbf{A} + \mathbf{K}_e\mathbf{C}) = (\lambda + 8)^2,$$

and then

$$\lambda^2 + (2 + k_1^e)\lambda + 2k_1^e + k_2^e = \lambda^2 + 16\lambda + 64.$$

Comparing the coefficients gives

$$k_1^e = 14, \quad k_2^e = 36.$$

Then the output feedback control is

$$u = -4\tilde{x}_1 - 0.5\tilde{x}_2,$$

$$\begin{bmatrix} \dot{\tilde{x}}_1 \\ \dot{\tilde{x}}_2 \end{bmatrix} = \begin{bmatrix} 0 & 1 \\ 0 & -2 \end{bmatrix} \begin{bmatrix} \tilde{x}_1 \\ \tilde{x}_2 \end{bmatrix} - \begin{bmatrix} 0 \\ 4 \end{bmatrix} [4 \ 0.5] \begin{bmatrix} \tilde{x}_1 \\ \tilde{x}_2 \end{bmatrix} + \begin{bmatrix} 14 \\ 36 \end{bmatrix} (x_1 - \tilde{x}_1)$$

$$= \begin{bmatrix} 0 & 1 \\ -16 & -4 \end{bmatrix} \begin{bmatrix} \tilde{x}_1 \\ \tilde{x}_2 \end{bmatrix} + \begin{bmatrix} 14 \\ 36 \end{bmatrix} (x_1 - \tilde{x}_1).$$

3.7.2 Integral Output Feedback Control

As in the case of state feedback control, if the reference output is not zero, the above designed output feedback control for the system (3.47)-(3.49) may contain system parameters and then is not robust. To see this, we suppose that the reference output **r** is not zero and the steady state system

$$0 = \mathbf{A}\bar{\mathbf{x}} + \mathbf{B}\bar{\mathbf{u}}, \tag{3.56}$$

$$\mathbf{r} = \mathbf{C}\bar{\mathbf{x}} \tag{3.57}$$

has a solution (solutions of this system may not be unique). Subtracting (3.56) from (3.47) gives

$$\frac{d}{dt}(\mathbf{x} - \bar{\mathbf{x}}) = \mathbf{A}(\mathbf{x} - \bar{\mathbf{x}}) + \mathbf{B}(\mathbf{u} - \bar{\mathbf{u}}). \tag{3.58}$$

Let the estimate $\tilde{\mathbf{x}}$ of $\mathbf{x}$ be generated by the observer (3.50) and consider the output feedback control of the form

$$\mathbf{u} - \bar{\mathbf{u}} = -\mathbf{K}_1(\tilde{\mathbf{x}} - \bar{\mathbf{x}}).$$

It then follows from (3.58) that

$$\begin{aligned} \frac{d}{dt}(\mathbf{x} - \bar{\mathbf{x}}) &= \mathbf{A}(\mathbf{x} - \bar{\mathbf{x}}) - \mathbf{B}\mathbf{K}_1(\tilde{\mathbf{x}} - \bar{\mathbf{x}}) \\ &= \mathbf{A}(\mathbf{x} - \bar{\mathbf{x}}) - \mathbf{B}\mathbf{K}_1(\tilde{\mathbf{x}} - \mathbf{x} + \mathbf{x} - \bar{\mathbf{x}}) \\ &= (\mathbf{A} - \mathbf{B}\mathbf{K}_1)(\mathbf{x} - \bar{\mathbf{x}}) + \mathbf{B}\mathbf{K}_1\mathbf{e}, \end{aligned} \tag{3.59}$$

where $\mathbf{e} = \mathbf{x} - \tilde{\mathbf{x}}$. Subtracting (3.50) from (3.47) gives

$$\dot{\mathbf{e}} = (\mathbf{A} - \mathbf{K}_e\mathbf{C})\mathbf{e}. \tag{3.60}$$

If the pair $(\mathbf{A}, \mathbf{B})$ is controllable and the pair $(\mathbf{A}, \mathbf{C})$ is observable, Theorem 3.11 shows that there exist matrices $\mathbf{K}_1$ and $\mathbf{K}_e$ such that the system (3.59)-(3.60) is exponentially stable. Now the feedback control

$$\mathbf{u} = -\mathbf{K}_1\tilde{\mathbf{x}} + \bar{\mathbf{u}} + \mathbf{K}_1\bar{\mathbf{x}}.$$

contains the control steady state $\bar{\mathbf{u}}$, which usually depends on system parameters as shown in the mass-spring-damper system. Hence it is not robust.

To design a robust feedback control, as in the case of integral state feedback control, we need to add the integrator

$$\dot{\mathbf{z}} = \mathbf{y} - \bar{\mathbf{y}} = \mathbf{C}(\mathbf{x} - \bar{\mathbf{x}}) \tag{3.61}$$

to the above system, where $\bar{\mathbf{y}} = \mathbf{C}\bar{\mathbf{x}}$.

Theorem 3.12. *Let* $\mathbf{A}, \mathbf{B}, \mathbf{C}$ *be* $n \times n, n \times m$, *and* $m \times n$ *matrices, respectively. Let* $\mathbf{r}$ *be the reference output such that the steady state system*

$$0 = \mathbf{A}\bar{\mathbf{x}} + \mathbf{B}\bar{\mathbf{u}}, \tag{3.62}$$

$$\mathbf{r} = \mathbf{C}\bar{\mathbf{x}} \tag{3.63}$$

has a solution. Denote $\bar{\mathbf{y}} = \mathbf{C}\bar{\mathbf{x}}$. *If* $(\mathbf{A}, \mathbf{B})$ *is controllable,* $(\mathbf{A}, \mathbf{C})$ *is observable, and*

$$\mathrm{rank} \begin{bmatrix} \mathbf{A} & \mathbf{B} \\ \mathbf{C} & \mathbf{0} \end{bmatrix} = n + m, \tag{3.64}$$

then there exist the matrices $\mathbf{K}_1, \mathbf{K}_2, \mathbf{K}_e$ *such that the integral output feedback control*

$$\mathbf{u} = -\mathbf{K}_1\tilde{\mathbf{x}} - \mathbf{K}_2\mathbf{z}, \tag{3.65}$$

$$\dot{\mathbf{z}} = \mathbf{e} = \mathbf{y} - \bar{\mathbf{y}} \tag{3.66}$$

regulates the controlled output $\mathbf{y}$ *of the control system* (3.47)-(3.50) *to* $\mathbf{r}$.

Proof. By Lemma 3.4, there exist matrices $\mathbf{K}_1, \mathbf{K}_2$ such that

$$\begin{bmatrix} \mathbf{A} & \mathbf{0} \\ \mathbf{C} & \mathbf{0} \end{bmatrix} - \begin{bmatrix} \mathbf{B} \\ \mathbf{0} \end{bmatrix} \begin{bmatrix} \mathbf{K}_1 & \mathbf{K}_2 \end{bmatrix}$$

is Hurwitz. Consider an output feedback control of the form

$$\mathbf{u} = -\mathbf{K}_1\tilde{\mathbf{x}} - \mathbf{K}_2\mathbf{z}. \tag{3.67}$$

As in the proof of Theorem 3.10, the equation

$$\bar{\mathbf{u}} = -\mathbf{K}_1\bar{\mathbf{x}} - \mathbf{K}_2\bar{\mathbf{z}} \tag{3.68}$$

has a unique solution $\bar{\mathbf{z}}$. Subtracting (3.68) from (3.67) gives

$$\mathbf{u} - \bar{\mathbf{u}} = -\mathbf{K}_1(\tilde{\mathbf{x}} - \bar{\mathbf{x}}) - \mathbf{K}_2(\mathbf{z} - \bar{\mathbf{z}}). \tag{3.69}$$

Subtracting (3.62) from (3.47) and using (3.69), we obtain

$$\frac{d}{dt}(\mathbf{x} - \bar{\mathbf{x}}) = \mathbf{A}(\mathbf{x} - \bar{\mathbf{x}}) - \mathbf{BK}_1(\tilde{\mathbf{x}} - \bar{\mathbf{x}}) - \mathbf{BK}_2(\mathbf{z} - \bar{\mathbf{z}})$$

$$= \mathbf{A}(\mathbf{x} - \bar{\mathbf{x}}) - \mathbf{BK}_1(\tilde{\mathbf{x}} - \mathbf{x} + \mathbf{x} - \bar{\mathbf{x}}) - \mathbf{BK}_2(\mathbf{z} - \bar{\mathbf{z}})$$

$$= (\mathbf{A} - \mathbf{BK}_1)(\mathbf{x} - \bar{\mathbf{x}}) + \mathbf{BK}_1\mathbf{e} - \mathbf{BK}_2(\mathbf{z} - \bar{\mathbf{z}}). \qquad (3.70)$$

It then follows from (3.60), (3.61), and (3.70) that

$$\frac{d}{dt}\begin{bmatrix} \mathbf{x} - \bar{\mathbf{x}} \\ \mathbf{z} - \bar{\mathbf{z}} \\ \mathbf{e} \end{bmatrix} = \begin{bmatrix} \mathbf{A} - \mathbf{BK}_1 & -\mathbf{BK}_2 & \mathbf{BK}_1 \\ \mathbf{C} & 0 & 0 \\ 0 & 0 & \mathbf{A} - \mathbf{K}_e\mathbf{C} \end{bmatrix} \begin{bmatrix} \mathbf{x} - \bar{\mathbf{x}} \\ \mathbf{z} - \bar{\mathbf{z}} \\ \mathbf{e} \end{bmatrix}. \qquad (3.71)$$

Because $(\mathbf{A}, \mathbf{C})$ is observable, there exists a gain matrix $\mathbf{K}_e$ such that $\mathbf{A} - \mathbf{K}_e\mathbf{C}$ is Hurwitz. Since

$$\det\left(\lambda\mathbf{I} - \begin{bmatrix} \mathbf{A} - \mathbf{BK}_1 & -\mathbf{BK}_2 & \mathbf{BK}_1 \\ \mathbf{C} & 0 & 0 \\ 0 & 0 & \mathbf{A} - \mathbf{K}_e\mathbf{C} \end{bmatrix}\right)$$

$$= \det\left(\lambda\mathbf{I} - \begin{bmatrix} \mathbf{A} - \mathbf{BK}_1 & -\mathbf{BK}_2 \\ \mathbf{C} & 0 \end{bmatrix}\right)\det(\lambda\mathbf{I} - [\mathbf{A} - \mathbf{K}_e\mathbf{C}]),$$

the system (3.71) is exponentially stable. Then $\mathbf{x}(t)$ converges to its equilibrium $\bar{\mathbf{x}}$. This implies that the output $\mathbf{y}(t) = \mathbf{C}\mathbf{x}(t)$ of the control system (3.47)-(3.49) converges to $\mathbf{r} = \mathbf{C}\bar{\mathbf{x}}$ as t tends to infinity. Therefore we have proved the theorem.

Example 3.13. We now design a robust integral output controller for the mass-spring-damper system

$$\begin{bmatrix} \dot{x}_1 \\ \dot{x}_2 \end{bmatrix} = \begin{bmatrix} 0 & 1 \\ -\frac{k}{m} & 0 \end{bmatrix}\begin{bmatrix} x_1 \\ x_2 \end{bmatrix} + \begin{bmatrix} 0 \\ \frac{1}{m} \end{bmatrix}u(t),$$

$$y = y = \begin{bmatrix} 1 & 0 \end{bmatrix}\begin{bmatrix} x_1 \\ x_2 \end{bmatrix}.$$

It is clear that the matrix

$$\begin{bmatrix} \mathbf{A} & \mathbf{B} \\ \mathbf{C} & 0 \end{bmatrix} = \begin{bmatrix} 0 & 1 & 0 \\ -\frac{k}{m} & 0 & \frac{1}{m} \\ 1 & 0 & 0 \end{bmatrix}$$

has the full rank 3. Let $\mathbf{K}_1 = [k_1, k_2]$ and $\mathbf{K}_2 = k_3$. A calculation gives

$$\left|\lambda\mathbf{I} - \begin{bmatrix} \mathbf{A} & 0 \\ \mathbf{C} & 0 \end{bmatrix} + \begin{bmatrix} \mathbf{B} \\ 0 \end{bmatrix}[\mathbf{K}_1 \ \mathbf{K}_2]\right| = \lambda^3 + \frac{k_2}{m}\lambda^2 + (k + k_1)\lambda + \frac{k_3}{m}.$$

Using Routh-Hurwitz's criterion, we find that if

$$k_1 > \frac{k_3 - kk_2}{k_2}, \quad k_2 > 0, \quad k_3 > 0,$$

then all eigenvalues have negative real parts. Let $\mathbf{K}_e = [k_1^e, k_2^e]^*$. Then

$$|\lambda\mathbf{I} - \mathbf{A} + \mathbf{K}_e\mathbf{C}| = \lambda^2 + k_1^e\lambda + \frac{k}{m} + k_2^e.$$

So if $k_1^e > 0$ and $k_2^e > \frac{k}{m}$, the eigenvalues have negative real parts. Therefore, the integral output feedback control

$$
\begin{aligned}
u &= -k_1\tilde{x}_1 - k_2\tilde{x}_2 - k_3 z, \\
\dot{\tilde{x}}_1 &= \tilde{x}_2 + k_1^e(x_1 - \tilde{x}_1), \\
\dot{\tilde{x}}_2 &= -\frac{k+k_1}{m}\tilde{x}_1 - \frac{k_2}{m}\tilde{x}_2 + k_2^e(x_1 - \tilde{x}_1), \\
\dot{z} &= x_1 - r
\end{aligned}
$$

regulates the output $y = x_1$ to the desired value r.

Exercises 3.7

1. Consider the system

$$
\begin{aligned}
\dot{\mathbf{x}} &= \mathbf{A}\mathbf{x} + \mathbf{B}u, \\
y &= \mathbf{C}\mathbf{x},
\end{aligned}
$$

where

$$
\mathbf{A} = \begin{bmatrix} 0 & 1 & 0 \\ 0 & 0 & 1 \\ -5 & -6 & 0 \end{bmatrix}, \quad \mathbf{B} = \begin{bmatrix} 0 \\ 0 \\ 1 \end{bmatrix}, \quad \mathbf{C} = \begin{bmatrix} 1 & 0 & 0 \end{bmatrix}.
$$

Design a state observer, assuming that the desired poles for the observer are located at

$$\lambda = -10, -10, -15.$$

2. Consider the system

$$
\begin{aligned}
\dot{\mathbf{x}} &= \mathbf{A}\mathbf{x} + \mathbf{B}u, \\
y &= \mathbf{C}\mathbf{x},
\end{aligned}
$$

where

$$
\mathbf{A} = \begin{bmatrix} 0 & 1 & 0 \\ 0 & 0 & 1 \\ -6 & -11 & -6 \end{bmatrix}, \quad \mathbf{B} = \begin{bmatrix} 0 \\ 0 \\ 1 \end{bmatrix}, \quad \mathbf{C} = \begin{bmatrix} 1 & 0 & 0 \end{bmatrix}.
$$

Design an output feedback control by observer approach such that the desired closed-loop poles are located at

$$\lambda = -1 + i, -1 - i, -5$$

and the desired observer poles are located at

$$\lambda = -6, -6, -6.$$

3. Consider the mass-spring-damper system. If the measured output $y = x_2$, then $\mathbf{C} = [0, 1]$ and the matrix

$$\begin{bmatrix} \mathbf{A} & \mathbf{B} \\ \mathbf{C} & \mathbf{0} \end{bmatrix} = \begin{bmatrix} 0 & 1 & 0 \\ -\frac{k}{m} & 0 & \frac{1}{m} \\ 0 & 1 & 0 \end{bmatrix}$$

has the rank 2. Discuss whether we can design an integral output feedback control to regulate the output y to a reference output r.

3.8 Optimal Control on Finite-time Intervals

Consider the control problem

$$\dot{\mathbf{x}} = \mathbf{A}\mathbf{x} + \mathbf{B}\mathbf{u}(t), \quad \mathbf{x}(0) = \mathbf{x}_0, \tag{3.72}$$

$$\mathbf{y} = \mathbf{C}\mathbf{x}. \tag{3.73}$$

If $(\mathbf{A}, \mathbf{B})$ is controllable, then there exist infinite feedback gain matrices $\mathbf{K}$ to stabilize the system. Which one is best? To answer the question, we first need to set up a criterion. Thus we define the quadratic performance criterion

$$\begin{aligned} J(\mathbf{u}, T; \mathbf{x}_0) &= \int_0^T \left(|\mathbf{C}\mathbf{x}(t)|^2 + |\mathbf{D}\mathbf{u}(t)|^2 \right) dt + \mathbf{x}(T)^* \mathbf{G}\mathbf{x}(T) \\ &= \int_0^T \left(\mathbf{x}(t)^* \mathbf{C}^* \mathbf{C}\mathbf{x}(t) + \mathbf{u}(t)^* \mathbf{D}^* \mathbf{D}\mathbf{u}(t) \right) dt + \mathbf{x}(T)^* \mathbf{G}\mathbf{x}(T) \\ &= \int_0^T \left(\mathbf{x}(t)^* \mathbf{Q}\mathbf{x}(t) + \mathbf{u}(t)^* \mathbf{R}\mathbf{u}(t) \right) dt + \mathbf{x}(T)^* \mathbf{G}\mathbf{x}(T), \end{aligned} \tag{3.74}$$

where $\mathbf{G}$ is a nonnegative matrix and

$$\mathbf{Q} = \mathbf{C}^* \mathbf{C}, \quad \mathbf{R} = \mathbf{D}^* \mathbf{D}.$$

The optimal control problem is to minimize the functional J over $L^2((0, T), \mathbb{R}^m)$:

$$\min_{\mathbf{u} \in L^2((0,T), \mathbb{R}^m)} J(\mathbf{u}, T; \mathbf{x}_0).$$

The function $\mathbf{u}_{opt}$ such that

$$J(\mathbf{u}_{opt}, T; \mathbf{x}_0) = \min_{\mathbf{u} \in L^2((0,T), \mathbb{R}^m)} J(\mathbf{u}, T; \mathbf{x}_0) \tag{3.75}$$

is called an optimal control. The first term $|\mathbf{C}\mathbf{x}(t)|^2$ in the criterion means that the optimal control drives the output $\mathbf{C}\mathbf{x}(t)$ as close to the desired output 0 as possible in the average sense (same meaning for the last term $\mathbf{x}(T)^*\mathbf{G}\mathbf{x}(T)$) and the second term $|\mathbf{D}u(t)|^2$ means that the control effort is minimized in the average sense.

Example 3.14. Consider the control system

$$\dot{\mathbf{x}} = \mathbf{A}\mathbf{x} + \mathbf{B}u,$$
$$\mathbf{y} = \mathbf{C}\mathbf{x},$$

where

$$\mathbf{A} = \begin{bmatrix} 0 & 1 \\ 0 & 0 \end{bmatrix}, \quad \mathbf{B} = \begin{bmatrix} 0 \\ 1 \end{bmatrix}, \quad \mathbf{C} = \begin{bmatrix} 1 & 0 \\ 0 & 1 \end{bmatrix}.$$

The performance criterion is defined by

$$J(u,T;\mathbf{x}_0) = \int_0^T \left(|\mathbf{C}\mathbf{x}(t)|^2 + |u(t)|^2 \right) dt$$
$$= \int_0^T \left(|x_1(t)|^2 + |x_2(t)|^2 + |u(t)|^2 \right) dt,$$

where $\mathbf{D} = [1]$.

3.8.1 Existence and Uniqueness

To solve the optimal control problem (3.75), we need the concept of weak convergence in a Hilbert space.

Definition 3.7. Let H be a Hilbert space. A sequence $\{x_n\} \subset H$ converges weakly to $x \in H$ if

$$\lim_{n \to \infty} (x_n - x, y) = 0 \quad \text{for any } y \in H.$$

We denote $x_n \rightharpoonup x$.

Consider the sequence $\{\sin n\pi x\} \subset L^2(0,1)$. By Theorem 2.5, any function $f \in L^2(0,1)$ has the expansion

$$f(x) = \sum_{n=1}^{\infty} c_n \sin n\pi x,$$

where

$$c_n = 2 \int_0^1 f(x) \sin n\pi x \, dx.$$

It therefore follows that

$$\|f\|_{L^2}^2 = \frac{1}{2} \sum_{n=1}^{\infty} c_n^2,$$

and then

$$\lim_{n\to\infty} \int_0^1 f(x)\sin n\pi x\, dx = \lim_{n\to\infty} c_n/2 = 0.$$

Thus $\sin n\pi x$ converges weakly to 0. Since $\int_0^1 |\sin n\pi x|^2 dx = 1/2$, $\sin n\pi x$ does not converge strongly to 0. So the weak convergence does not imply the strong convergence. But the strong convergence does imply the weak convergence.

Theorem 3.13. *If a sequence $\{x_n\} \subset H$ converges strongly to $x \in H$, then it converges weakly to $x \in H$.*

Proof. For any $y \in H$, it follows from Cauchy-Schwarz inequality (2.7) that

$$|(x_n - x, y)| \le \|x_n - x\|\|y\|,$$

which implies that $\{x_n\}$ converges weakly to $x \in H$.

The following well-known compactness theorem plays an important role in dealing with minimization problems.

Theorem 3.14. *Let H be a Hilbert space. If a sequence $\{x_n\} \subset H$ is bounded, then there exists a subsequence $\{x_{n_i}\}$ such that $\{x_{n_i}\}$ weakly converges to $x \in H$ as $i \to \infty$.*

The proof of this theorem is difficult and referred to [34] or [105]. We use the sequence $\{\sin n\pi x\} \subset L^2(0,1)$ to illustrate the theorem. Since $\int_0^1 |\sin n\pi x|^2 dx = 1/2$, the sequence is bounded. In the above, we have proved that $\sin n\pi x$ converges weakly to 0, but does not converge strongly to 0.

Lemma 3.5. *Let*

$$\dot{\mathbf{x}}_n = \mathbf{A}\mathbf{x}_n + \mathbf{B}\mathbf{u}_n(t),$$
$$\mathbf{x}_n(0) = \mathbf{x}_0,$$

and

$$\dot{\mathbf{x}} = \mathbf{A}\mathbf{x} + \mathbf{B}\mathbf{u}(t),$$
$$\mathbf{x}(0) = \mathbf{x}_0.$$

If

$$\lim_{n\to\infty} \int_0^T \mathbf{u}_n(t) \cdot \mathbf{v}(t)\, dt = \int_0^T \mathbf{u}(t) \cdot \mathbf{v}(t)\, dt, \quad \forall \mathbf{v} \in L^2(0,T;\mathbb{R}^m),$$

then

$$\lim_{n\to\infty} \int_0^T \mathbf{x}_n(t) \cdot \mathbf{y}(t)\, dt = \int_0^T \mathbf{x}(t) \cdot \mathbf{y}(t)\, dt, \quad \forall \mathbf{y} \in L^2(0,T;\mathbb{R}^n).$$

Proof. A direct calculation gives

$$\lim_{n \to \infty} \int_0^T \mathbf{x}_n(t) \cdot \mathbf{y}(t) dt$$

$$= \lim_{n \to \infty} \int_0^T \left[e^{\mathbf{A}t} \mathbf{x}_0 + \int_0^t e^{\mathbf{A}(t-s)} \mathbf{B} \mathbf{u}_n(s) ds \right] \cdot \mathbf{y}(t) dt$$

$$= \int_0^T e^{\mathbf{A}t} \mathbf{x}_0 \cdot \mathbf{y}(t) dt + \lim_{n \to \infty} \int_0^T \left[\int_s^T e^{\mathbf{A}(t-s)} \mathbf{B} \mathbf{u}_n(s) \right] \cdot \mathbf{y}(t) dt ds$$

$$= \int_0^T e^{\mathbf{A}t} \mathbf{x}_0 \cdot \mathbf{y}(t) dt + \lim_{n \to \infty} \int_0^T \left[\int_s^T \mathbf{B}^* e^{\mathbf{A}^*(t-s)} \mathbf{y}(t) \right] \cdot \mathbf{u}_n(s) dt ds$$

$$= \int_0^T e^{\mathbf{A}t} \mathbf{x}_0 \cdot \mathbf{y}(t) dt + \int_0^T \left[\int_s^T \mathbf{B}^* e^{\mathbf{A}^*(t-s)} \mathbf{y}(t) \right] \cdot \mathbf{u}(s) dt ds$$

$$= \int_0^T \left[e^{\mathbf{A}t} \mathbf{x}_0 + \int_0^t e^{\mathbf{A}(t-s)} \mathbf{B} \mathbf{u}(s) ds \right] \cdot \mathbf{y}(t) dt$$

$$= \int_0^T \mathbf{x}(t) \cdot \mathbf{y}(t) dt.$$

Consider

$$x_n'(t) = \bar{x}_n(t) + \frac{n}{n+1}, \quad x_n(0) = x_0.$$

Then the solution

$$x_n(t) = x_0 e^{-t} + \frac{n}{n+1}(1 - e^{-t})$$

converges to $x(t) = x_0 e^{-t} + (1 - e^{-t})$, which is the solution of

$$x'(t) = \bar{x}(t) + 1, \quad x_n(0) = x_0.$$

We now prove that there is a unique optimal control of the problem (3.75).

Theorem 3.15. *Assume that* $\mathbf{R}$ *is positive definite. For every initial condition, there is a unique optimal control such that*

$$J(\mathbf{u}_{opt}, T; \mathbf{x}_0) = \min_{\mathbf{u} \in L^2((0,T), \mathbb{R}^m)} J(\mathbf{u}, T; \mathbf{x}_0).$$

Proof. Existence. Let $\mathbf{u}_n \in L^2((0,T), \mathbb{R}^m)$ be a minimizing sequence such that

$$\lim_{n \to \infty} J(\mathbf{u}_n, T; \mathbf{x}_0) = \min_{\mathbf{u} \in L^2((0,T), \mathbb{R}^m)} J(\mathbf{u}, T; \mathbf{x}_0).$$

Since $\mathbf{R}$ is positive definite, there exists a $C > 0$ such that

$$\|\mathbf{u}_n\|_{L^2}^2 \le CJ(\mathbf{u}_n, T; \mathbf{x}_0) \le M.$$

Then there is a subsequence, still denoted by itself, such that

$$\lim_{n \to \infty} \int_0^T \mathbf{u}_n(t) \cdot \mathbf{v}(t) dt = \int_0^T \mathbf{u}_{opt}(t) \cdot \mathbf{v}(t) dt, \quad \forall \mathbf{v} \in L^2(0, T; \mathbb{R}^m),$$

and

$$\lim_{n \to \infty} \int_0^T \mathbf{x}_n(t) \cdot \mathbf{z}(t) dt = \int_0^T \mathbf{x}_{opt}(t) \cdot \mathbf{z}(t) dt, \quad \forall \mathbf{z} \in L^2(0, T; \mathbb{R}^n),$$

where $\mathbf{x}_n$ and $\mathbf{x}_{opt}$ are solutions of

$$\dot{\mathbf{x}}_n = \mathbf{A}\mathbf{x}_n + \mathbf{B}\mathbf{u}_n(t),$$
$$\mathbf{x}_n(0) = \mathbf{x}_0,$$

and

$$\dot{\mathbf{x}}_{opt} = \mathbf{A}\mathbf{x}_{opt} + \mathbf{B}\mathbf{u}_{opt}(t),$$
$$\mathbf{x}_{opt}(0) = \mathbf{x}_0.$$

A direct calculation gives

$$
\begin{aligned}
J(\mathbf{u}_n, T; \mathbf{x}_0) &= \int_0^T \left(\mathbf{x}_n(t)^* \mathbf{Q} \mathbf{x}_n(t) + \mathbf{u}_n(t)^* \mathbf{R} \mathbf{u}_n(t) \right) dt + \mathbf{x}_n(T)^* \mathbf{G} \mathbf{x}_n(T) \\
&= \int_0^T \left([\mathbf{x}_n(t) - \mathbf{x}_{opt}(t) + \mathbf{x}_{opt}(t)]^* \mathbf{Q} [\mathbf{x}_n(t) - \mathbf{x}_{opt}(t) + \mathbf{x}_{opt}(t)] \right. \\
&\quad \left. + [\mathbf{u}_n(t) - \mathbf{u}_{opt}(t) + \mathbf{u}_{opt}(t)]^* \mathbf{R} [\mathbf{u}_n(t) - \mathbf{u}_{opt}(t) + \mathbf{u}_{opt}(t)] \right) dt \\
&\quad + [\mathbf{x}_n(T) - \mathbf{x}_{opt}(T) + \mathbf{x}_{opt}(T)]^* \mathbf{G} [\mathbf{x}_n(T) - \mathbf{x}_{opt}(T) + \mathbf{x}_{opt}(T)] \\
&= J(\mathbf{u}_n - \mathbf{u}_{opt}, T; 0) + J(\mathbf{u}_{opt}, T; \mathbf{x}_0) \\
&\quad + 2 \int_0^T \left([\mathbf{x}_n(t) - \mathbf{x}_{opt}(t)]^* \mathbf{Q} \mathbf{x}_{opt}(t) + [\mathbf{u}_n(t) - \mathbf{u}_{opt}(t)]^* \mathbf{R} \mathbf{u}_{opt}(t) \right) dt \\
&\quad + 2 [\mathbf{x}_n(T) - \mathbf{x}_{opt}(T)]^* \mathbf{G} \mathbf{x}_{opt}(T) \\
&\geq J(\mathbf{u}_{opt}, T; \mathbf{x}_0) \\
&\quad + 2 \int_0^T \left([\mathbf{x}_n(t) - \mathbf{x}_{opt}(t)]^* \mathbf{Q} \mathbf{x}_{opt}(t) + [\mathbf{u}_n(t) - \mathbf{u}_{opt}(t)]^* \mathbf{R} \mathbf{u}_{opt}(t) \right) dt \\
&\quad + 2 \int_0^T [e^{A(t-r)} B(\mathbf{u}_n(r) - \mathbf{u}_{opt}(r))]^* \mathbf{G} \mathbf{x}_{opt}(T) dr \\
&= J(\mathbf{u}_{opt}, T; \mathbf{x}_0) \\
&\quad + 2 \int_0^T \left([\mathbf{x}_n(t) - \mathbf{x}_{opt}(t)]^* \mathbf{Q} \mathbf{x}_{opt}(t) + [\mathbf{u}_n(t) - \mathbf{u}_{opt}(t)]^* \mathbf{R} \mathbf{u}_{opt}(t) \right) dt \\
&\quad + 2 \int_0^T [\mathbf{u}_n(r) - \mathbf{u}_{opt}(r)]^* [e^{A(t-r)} B]^* \mathbf{G} \mathbf{x}_{opt}(T) dr.
\end{aligned}
$$

It then follows that

$$\min_{\mathbf{u} \in L^2((0,T), \mathbb{R}^m)} J(\mathbf{u}, T; \mathbf{x}_0)$$
$$= \lim_{n \to \infty} J(\mathbf{u}_n, T; \mathbf{x}_0)$$

$$\geq J(\mathbf{u}_{opt}, T; \mathbf{x}_0)$$
$$+2 \lim_{n \to \infty} \int_0^T \left([\mathbf{x}_n(t) - \mathbf{x}_{opt}(t)]^* \mathbf{Q} \mathbf{x}_{opt}(t) + [\mathbf{u}_n(t) - \mathbf{u}_{opt}(t)]^* \mathbf{R} \mathbf{u}_{opt}(t)\right) dt$$
$$+ \lim_{n \to \infty} 2 \int_0^T [\mathbf{u}_n(r) - \mathbf{u}_{opt}(r)]^* [e^{A(t-r)} B]^* \mathbf{G} \mathbf{x}_{opt}(T) dr$$
$$= J(\mathbf{u}_{opt}, T; \mathbf{x}_0)$$
$$\geq \min_{\mathbf{u} \in L^2((0,T), \mathbb{R}^m)} J(\mathbf{u}, T; \mathbf{x}_0).$$

Uniqueness. Assume that there are two optimal controls $\mathbf{u}_1, \mathbf{u}_2$. Using the inequality

$$\left(\frac{a+b}{2}\right)^2 < \frac{1}{2}(a^2 + b^2), \quad \forall a \neq b,$$

we can show that

$$J((\mathbf{u}_1 + \mathbf{u}_2)/2, T; \mathbf{x}_0) < \frac{1}{2} J(\mathbf{u}_1, T; \mathbf{x}_0) + \frac{1}{2} J(\mathbf{u}_2, T; \mathbf{x}_0)$$
$$= \min_{\mathbf{u} \in L^2((0,T), \mathbb{R}^m)} J(\mathbf{u}, T; \mathbf{x}_0). \tag{3.76}$$

This is a contradiction.

A functional satisfying (3.76) is said to be convex.

3.8.2 Optimality System

For a function $f(x)$, minimum occurs at its critical number c, which satisfies

$$f'(c) = 0.$$

This is also true for the optimal control, but we need to generalize the derivative concept.

Definition 3.8. If there exists $\mathbf{u}_d \in L^2((0,T), \mathbb{R}^m)$ such that

$$\int_0^T \mathbf{u}_d(t) \cdot \mathbf{h}(t) dt = \lim_{\lambda \to 0^+} \frac{J(\mathbf{u} + \lambda \mathbf{h}, T; \mathbf{x}_0) - J(\mathbf{u}, T; \mathbf{x}_0)}{\lambda} \tag{3.77}$$

for $\mathbf{h} \in L^2((0,T), \mathbb{R}^m)$, we say J is Gâteaux-differentiable at $\mathbf{u}$ and call $\mathbf{u}_d$ the Gâteaux-differential of J at $\mathbf{u}$ and denote it by $J'(\mathbf{u}, T; \mathbf{x}_0)$.

Let $\mathbf{x}(t; \mathbf{x}_0, \mathbf{u})$ denote the solution of the control system (3.72)-(3.73) with the control $\mathbf{u}$ and the initial condition $\mathbf{x}_0$. Since the equation is linear, we can readily see that

$$\mathbf{x}(t; \mathbf{x}_0, \mathbf{u} + \lambda \mathbf{h}) = \mathbf{x}(t; \mathbf{x}_0, \mathbf{u}) + \lambda \mathbf{x}(t; 0, \mathbf{h}). \tag{3.78}$$

Theorem 3.16. *The Gâteaux-differential of J defined by (3.74) at* $\mathbf{u}$ *is given by*

$$\int_0^T J'(\mathbf{u},T;\mathbf{x}_0)(t) \cdot \mathbf{h}(t)dt = 2\int_0^T \left(\mathbf{x}(t;\mathbf{x}_0,\mathbf{u})^*\mathbf{Q}\mathbf{x}(t;0,\mathbf{h}) + \mathbf{u}(t)^*\mathbf{R}\mathbf{h}(t)\right)dt$$
$$+2\mathbf{x}(T;\mathbf{x}_0,\mathbf{u})^*\mathbf{G}\mathbf{x}(T;0,\mathbf{h}),$$

where $\mathbf{x}(t;0,\mathbf{h})$ *is the solution of*

$$\dot{\mathbf{x}} = \mathbf{A}\mathbf{x} + \mathbf{B}\mathbf{h}(t), \quad \mathbf{x}(0) = 0.$$

The proof of the theorem is left as an exercise. The following is the necessary condition for an optimal control.

Theorem 3.17. *If* $\mathbf{u}_{opt}$ *is the optimal control, then*

$$0 = \int_0^T \left(\mathbf{x}(t;\mathbf{x}_0,\mathbf{u}_{opt})^*\mathbf{Q}\mathbf{x}(t;0,\mathbf{h}) + \mathbf{u}_{opt}^*(t)\mathbf{R}\mathbf{h}(t)\right)dt$$
$$+\mathbf{x}(T;\mathbf{x}_0,\mathbf{u}_{opt})^*\mathbf{G}\mathbf{x}(T;0,\mathbf{h}) \qquad (3.79)$$

for all $\mathbf{h} \in L^2((0,T),\mathbb{R}^m)$, *where* $\mathbf{x}(t;0,\mathbf{h})$ *is the solution of*

$$\dot{\mathbf{x}} = \mathbf{A}\mathbf{x} + \mathbf{B}\mathbf{h}(t), \quad \mathbf{x}(0) = 0. \qquad (3.80)$$

Proof. Since J attains minimum at $\mathbf{u}_{opt}$, we have

$$\int_0^T J'(\mathbf{u}_{opt},T;\mathbf{x}_0)(t) \cdot \mathbf{h}(t)dt = \lim_{\lambda \to 0^+} \frac{J(\mathbf{u}_{opt}+\lambda\mathbf{h},T;\mathbf{x}_0) - J(\mathbf{u}_{opt},T;\mathbf{x}_0)}{\lambda} \geq 0$$

for all $\mathbf{h} \in L^2((0,T),\mathbb{R}^m)$. Replacing $\mathbf{h}$ by $-\mathbf{h}$, we obtain

$$\int_0^T J'(\mathbf{u}_{opt},T;\mathbf{x}_0)(t) \cdot (-\mathbf{h}(t))dt \geq 0.$$

Combining these two inequalities gives

$$\int_0^T J'(\mathbf{u}_{opt},T;\mathbf{x}_0)(t) \cdot \mathbf{h}(t)dt = 0$$

for all $\mathbf{h} \in L^2((0,T),\mathbb{R}^m)$.

From the above necessary condition theorem, we can derive an optimality (also called Hamiltonian) system.

Theorem 3.18. *The optimal control* $\mathbf{u}_{opt}$ *satisfies the following system*

$$\dot{\mathbf{x}} = \mathbf{A}\mathbf{x} + \mathbf{B}\mathbf{u}_{opt}, \qquad (3.81)$$
$$\dot{\mathbf{y}} = -\mathbf{A}^*\mathbf{y} + \mathbf{Q}\mathbf{x}, \qquad (3.82)$$
$$\mathbf{u}_{opt} = \mathbf{R}^{-1}\mathbf{B}^*\mathbf{y}, \qquad (3.83)$$

$$\mathbf{x}(0) = \mathbf{x}_0, \tag{3.84}$$

$$\mathbf{y}(T) = -\mathbf{G}\mathbf{x}(T). \tag{3.85}$$

Proof. Multiplying (3.82) by the solution of (3.80), we obtain

$$\mathbf{x}^*(t;0,\mathbf{h})\dot{\mathbf{y}} = -\mathbf{x}^*(t;0,\mathbf{h})\mathbf{A}^*\mathbf{y} + \mathbf{x}^*(t;0,\mathbf{h})\mathbf{Q}\mathbf{x}(t;\mathbf{x}_0,\mathbf{u}_{opt}). \tag{3.86}$$

Multiplying (3.80) by the solution of (3.82), we obtain

$$\mathbf{y}^*\dot{\mathbf{x}}(t;0,\mathbf{h}) = \mathbf{y}^*\mathbf{A}\mathbf{x}(t;0,\mathbf{h}) + \mathbf{y}^*\mathbf{B}\mathbf{h}. \tag{3.87}$$

Adding equations (3.86) and (3.87) and integrating from 0 to T, we obtain

$$-\mathbf{x}^*(t;0,\mathbf{h})\mathbf{G}\mathbf{x}(T;\mathbf{x}_0,\mathbf{u}_{opt}) = \mathbf{x}^*(t;0,\mathbf{h})\mathbf{y}(T) \quad \text{(use (3.85))}$$
$$= \int_0^T [\mathbf{x}^*(t;\mathbf{x}_0,\mathbf{u}_{opt})\mathbf{Q}\mathbf{x}(t;\mathbf{x}_0,\mathbf{u}_{opt}) + \mathbf{y}^*(t)\mathbf{B}\mathbf{h}(t)]dt.$$

Substituting this equation into (3.79), we obtain

$$\int_0^T [\mathbf{u}_{opt}^*(t)\mathbf{R}\mathbf{h}(t) - \mathbf{y}^*(t)\mathbf{B}\mathbf{h}(t)]dt$$

for all $\mathbf{h} \in L^2((0,T),\mathbb{R}^m)$. Thus we have

$$\mathbf{u}_{opt} = \mathbf{R}^{-1}\mathbf{B}^*\mathbf{y}.$$

Solving the equation (3.82) for $\mathbf{y}$ gives

$$\mathbf{y}(t) = -e^{\mathbf{A}^*(T-t)}\mathbf{G}\mathbf{x}(T) - \int_t^T e^{\mathbf{A}^*(s-t)}\mathbf{Q}\mathbf{x}(s)ds.$$

Define

$$\mathbf{P}(t)\mathbf{x}(t) = -\mathbf{y}(t) = e^{\mathbf{A}^*(T-t)}\mathbf{G}\mathbf{x}(T) + \int_t^T e^{\mathbf{A}^*(s-t)}\mathbf{Q}\mathbf{x}(s)ds.$$

Then we obtain the optimal feedback control

$$\mathbf{u}_{opt} = -\mathbf{R}^{-1}\mathbf{B}^*\mathbf{P}(t)\mathbf{x}(t). \tag{3.88}$$

Theorem 3.19. *The matrix* $\mathbf{P}(t)$ *has the following properties:*

(1) $\mathbf{P}(t)$ *solves the differential Riccati equation*

$$\mathbf{P}' = -\mathbf{P}\mathbf{A} - \mathbf{A}^*\mathbf{P} - \mathbf{Q} + \mathbf{P}\mathbf{B}\mathbf{R}^{-1}\mathbf{B}^*\mathbf{P}, \tag{3.89}$$

$$\mathbf{P}(T) = \mathbf{G}. \tag{3.90}$$

(2) $\mathbf{P}(t)$ *is symmetric.*

(3) The following identity holds:

$$J(\mathbf{u}, T; \mathbf{x}_0) = \mathbf{x}_0^* \mathbf{P}(0)\mathbf{x}_0 + \int_0^T [\mathbf{u} + \mathbf{R}^{-1}\mathbf{B}^*\mathbf{P}(t)\mathbf{x}(t)]^* \mathbf{R}[\mathbf{u} + \mathbf{R}^{-1}\mathbf{B}^*\mathbf{P}(t)\mathbf{x}(t)]dt.$$

(3.91)

(4) If $\mathbf{u}_{opt}$ is the optimal control, then

$$J(\mathbf{u}_{opt}, T; \mathbf{x}_0) = \min_{\mathbf{u} \in L^2((0,T), \mathbb{R}^m)} J(\mathbf{u}, T; \mathbf{x}_0) = \mathbf{x}_0^* \mathbf{P}(0)\mathbf{x}_0.$$

(3.92)

Hence $\mathbf{P}(0)$ is nonnegative.

Proof. Plugging $\mathbf{y}(t) = -\mathbf{P}(t)\mathbf{x}(t)$ into the optimality system gives the differential Riccati equation. From the differential equation we can see that $\mathbf{P}(t)$ is symmetric.
    Define

$$V(t) = \mathbf{x}(t)^* \mathbf{P}(t)\mathbf{x}(t).$$

Then

$$\begin{aligned}
\dot{V}(t) &= \mathbf{x}'(t)^* \mathbf{P}(t)\mathbf{x}(t) + \mathbf{x}(t)^* \mathbf{P}'(t)\mathbf{x}(t) + \mathbf{x}(t)^* \mathbf{P}(t)\mathbf{x}'(t) \\
&= [\mathbf{x}(t)^* \mathbf{A}^* + \mathbf{u}^*\mathbf{B}^*]\mathbf{P}(t)\mathbf{x}(t) + \mathbf{x}(t)^* \mathbf{P}'(t)\mathbf{x}(t) + \mathbf{x}(t)^* \mathbf{P}(t)[\mathbf{A}\mathbf{x}(t) + \mathbf{B}\mathbf{u}] \\
&= \mathbf{x}(t)^*[\mathbf{A}^*\mathbf{P}(t) + \mathbf{P}'(t) + \mathbf{P}(t)\mathbf{A}]\mathbf{x}(t) + \mathbf{u}^*\mathbf{B}^*\mathbf{P}(t)\mathbf{x}(t) + \mathbf{x}(t)^* \mathbf{P}(t)\mathbf{B}\mathbf{u} \\
&= \mathbf{x}(t)^*[-\mathbf{Q} + \mathbf{P}\mathbf{B}\mathbf{R}^{-1}\mathbf{B}^*\mathbf{P}]\mathbf{x}(t) + \mathbf{u}^*\mathbf{B}^*\mathbf{P}(t)\mathbf{x}(t) + \mathbf{x}(t)^* \mathbf{P}(t)\mathbf{B}\mathbf{u} \\
&= -\mathbf{x}(t)^*\mathbf{Q}\mathbf{x}(t) - \mathbf{u}(t)^*\mathbf{R}\mathbf{u}(t) \\
&\quad + \mathbf{u}(t)^*\mathbf{R}\mathbf{u}(t) + \mathbf{u}^*\mathbf{B}^*\mathbf{P}(t)\mathbf{x}(t) + \mathbf{x}(t)^* \mathbf{P}(t)\mathbf{B}\mathbf{u} + \mathbf{x}(t)^*\mathbf{P}\mathbf{B}\mathbf{R}^{-1}\mathbf{B}^*\mathbf{P}\mathbf{x}(t) \\
&= -\mathbf{x}(t)^*\mathbf{Q}\mathbf{x}(t) - \mathbf{u}(t)^*\mathbf{R}\mathbf{u}(t) \\
&\quad + [\mathbf{u} + \mathbf{R}^{-1}\mathbf{B}^*\mathbf{P}(t)\mathbf{x}(t)]^*\mathbf{R}[\mathbf{u} + \mathbf{R}^{-1}\mathbf{B}^*\mathbf{P}(t)\mathbf{x}(t)]
\end{aligned}$$

and

$$\dot{V}(t) + \mathbf{x}(t)^*\mathbf{Q}\mathbf{x}(t) + \mathbf{u}(t)^*\mathbf{R}\mathbf{u}(t) = [\mathbf{u} + \mathbf{R}^{-1}\mathbf{B}^*\mathbf{P}(t)\mathbf{x}(t)]^*\mathbf{R}[\mathbf{u} + \mathbf{R}^{-1}\mathbf{B}^*\mathbf{P}(t)\mathbf{x}(t)].$$

Integrating from 0 to T, we obtain

$$J(\mathbf{u}, T; \mathbf{x}_0) = \mathbf{x}_0^* \mathbf{P}(0)\mathbf{x}_0 + \int_0^T [\mathbf{u} + \mathbf{R}^{-1}\mathbf{B}^*\mathbf{P}(t)\mathbf{x}(t)]^* \mathbf{R}[\mathbf{u} + \mathbf{R}^{-1}\mathbf{B}^*\mathbf{P}(t)\mathbf{x}(t)]dt.$$

Taking

$$\mathbf{u} = \mathbf{u}_{opt} = \mathbf{R}^{-1}\mathbf{B}^*\mathbf{P}(t)\mathbf{x}(t),$$

we obtain

$$J(\mathbf{u}_{opt}, T; \mathbf{x}_0) = \min_{\mathbf{u} \in L^2((0,T), \mathbb{R}^m)} J(\mathbf{u}, T; \mathbf{x}_0) = \mathbf{x}_0^* \mathbf{P}(0)\mathbf{x}_0.$$

Example 3.15. Consider

$$x'(t) = x(t) + u(t), \quad x(0) = x_0,$$

with the performance criterion

$$J(u,x_0) = \int_0^T [3x^2(t) + u^2(t)]dt.$$

Here $G = 0, Q = 3, R = 1$. The differential Riccati equation is

$$p'(t) + 2p(t) - p^2(t) + 3 = 0, \quad p(T) = 0,$$

which has the solution

$$p(t) = \frac{3(1 - e^{-4(T-t)})}{1 + 3e^{-4(T-t)}}.$$

We then have the optimal control

$$u(t) = -p(t)x(t).$$

Example 3.16. Consider the system

$$\dot{\mathbf{x}} = \mathbf{A}\mathbf{x} + \mathbf{B}u,$$
$$\mathbf{y} = \mathbf{C}\mathbf{x},$$

where

$$\mathbf{A} = \begin{bmatrix} 0 & 1 \\ 0 & 0 \end{bmatrix}, \quad \mathbf{B} = \begin{bmatrix} 0 \\ 1 \end{bmatrix}, \quad \mathbf{C} = \begin{bmatrix} 1 & 0 \\ 0 & 1 \end{bmatrix}.$$

The performance criterion is defined by

$$J(u,T;\mathbf{x}_0) = \int_0^T \left(|x_1(t)|^2 + |x_2(t)|^2 + |u(t)|^2 \right) dt.$$

Thus $\mathbf{D} = [1]$. The corresponding differential Riccati equation is

$$-\begin{bmatrix} 0 & 0 \\ 1 & 0 \end{bmatrix}\begin{bmatrix} p_{11} & p_{12} \\ p_{12} & p_{22} \end{bmatrix} - \begin{bmatrix} p_{11} & p_{12} \\ p_{12} & p_{22} \end{bmatrix}\begin{bmatrix} 0 & 1 \\ 0 & 0 \end{bmatrix}$$
$$+ \begin{bmatrix} p_{11} & p_{12} \\ p_{12} & p_{22} \end{bmatrix}\begin{bmatrix} 0 \\ 1 \end{bmatrix}[1][0 \ 1]\begin{bmatrix} p_{11} & p_{12} \\ p_{12} & p_{22} \end{bmatrix} - \begin{bmatrix} 1 & 0 \\ 0 & 1 \end{bmatrix} = \begin{bmatrix} p'_{11} & p'_{12} \\ p'_{12} & p'_{22} \end{bmatrix}.$$

This equation gives three equations:

$$p'_{11} = -1 + p_{12}^2,$$
$$p'_{12} = -p_{11} + p_{12}p_{22},$$
$$p'_{22} = -1 - 2p_{12} + p_{22}^2.$$

It is very difficult or impossible to solve it analytically! This example demonstrates that it is usually difficult to solve the differential Riccati equation (3.89) explicitly.

Exercises 3.8

1. Prove Theorem 3.16.
2. Prove the inequality (3.76).
3. Consider

$$x'(t) = 4x(t) + u(t), \quad x(0) = x_0,$$

with the performance criterion

$$J(u, x_0) = \int_0^T [x^2(t) + u^2(t)] dt.$$

Write down the corresponding differential Riccati equation and then derive the optimal feedback controller.

3.9 Optimal State Feedback Controllers

Consider the control problem

$$\dot{\mathbf{x}} = \mathbf{A}\mathbf{x} + \mathbf{B}\mathbf{u}(t), \quad \mathbf{x}(0) = \mathbf{x}_0, \tag{3.93}$$

$$\mathbf{y} = \mathbf{C}\mathbf{x}. \tag{3.94}$$

We define the quadratic performance criterion

$$\begin{aligned} J(\mathbf{u}; \mathbf{x}_0) &= \int_0^\infty \left(|\mathbf{C}\mathbf{x}(t)|^2 + |\mathbf{D}\mathbf{u}(t)|^2 \right) dt \\ &= \int_0^\infty \left(\mathbf{x}(t)^* \mathbf{C}^* \mathbf{C}\mathbf{x}(t) + \mathbf{u}(t)^* \mathbf{D}^* \mathbf{D}\mathbf{u}(t) \right) dt \\ &= \int_0^\infty \left(\mathbf{x}(t)^* \mathbf{Q}\mathbf{x}(t) + \mathbf{u}(t)^* \mathbf{R}\mathbf{u}(t) \right) dt, \end{aligned} \tag{3.95}$$

where

$$\mathbf{Q} = \mathbf{C}^* \mathbf{C}, \quad \mathbf{R} = \mathbf{D}^* \mathbf{D}.$$

The optimal control problem is to minimize the functional J over $L^2((0, \infty), \mathbb{R}^m)$:

$$\min_{\mathbf{u} \in L^2((0, \infty), \mathbb{R}^m)} J(\mathbf{u}; \mathbf{x}_0).$$

The function $\mathbf{u}_{opt}$ such that

$$J(\mathbf{u}_{opt}; \mathbf{x}_0) = \min_{\mathbf{u} \in L^2((0, \infty), \mathbb{R}^m)} J(\mathbf{u}; \mathbf{x}_0) \tag{3.96}$$

is called an optimal control.

Consider the system

$$\dot{\mathbf{x}} = \mathbf{A}\mathbf{x} + \mathbf{B}u,$$

where

$$\mathbf{A} = \begin{bmatrix} 0 & 1 \\ 0 & 0 \end{bmatrix}, \quad \mathbf{B} = \begin{bmatrix} 1 \\ 0 \end{bmatrix}.$$

The solution is

$$x_1 = c_1 + \int_0^t (c_2 + u(s))ds, \quad x_2 = c_2.$$

Then

$$J(u, \mathbf{x}_0) = \int_0^\infty (x_1^2(t) + x_2^2(t))dt = \int_0^\infty \left[\left(c_1 + \int_0^t (c_2 + u(s))ds \right)^2 + c_2^2 \right] dt = \infty$$

for any u. Thus we must impose a condition on the system to ensure that the performance criterion is finite for some controls.

Definition 3.9. The optimal control problem (3.96) is well posed if for every initial condition $\mathbf{x}_0$, there is a control $\mathbf{u} \in L^2((0, \infty), \mathbb{R}^m)$ such that $J(\mathbf{u}; \mathbf{x}_0) < \infty$.

Consider the system

$$\dot{\mathbf{x}} = \mathbf{A}\mathbf{x} + \mathbf{B}u,$$

where

$$\mathbf{A} = \begin{bmatrix} 0 & 1 \\ 0 & 0 \end{bmatrix}, \quad \mathbf{B} = \begin{bmatrix} 0 \\ 1 \end{bmatrix}.$$

For any initial condition $\mathbf{x}_0 = (c_1, c_2)$, we can construct

$$u = -c_2 e^{-t} - (c_1 + c_2) \frac{d}{dt} (te^{-t}) \in L^2((0, \infty), \mathbb{R})$$

such that the solution

$$x_1 = [c_1 + (c_1 + c_2)t]e^{-t}, \quad x_2 = [c_2 - (c_1 + c_2)t]e^{-t}$$

is in $L^2((0, \infty), \mathbb{R}^2)$. Note that this system is controllable. In general, the optimal control problem of a controllable system is always well posed.

Theorem 3.20. If $(\mathbf{A}, \mathbf{B})$ is controllable, then the optimal control problem (3.96) is well posed.

Proof. Let $T > 0$. For every initial condition $\mathbf{x}_0$, there is a control $\mathbf{u}_T \in L^2((0, T), \mathbb{R}^m)$ such that the solution of

$$\dot{\mathbf{x}}_T = \mathbf{A}\mathbf{x}_T + \mathbf{B}\mathbf{u}_T(t),$$
$$\mathbf{x}_T(0) = \mathbf{x}_0$$

satisfies that

$$\mathbf{x}_T(T) = 0.$$

Define

$$\mathbf{u}(t) = \begin{cases} \mathbf{u}_T(t), & 0 \le t \le T, \\ 0, & t > T. \end{cases}$$

Then the solution of

$$\dot{\mathbf{x}} = \mathbf{A}\mathbf{x} + \mathbf{B}\mathbf{u}(t),$$
$$\mathbf{x}(0) = \mathbf{x}_0$$

is given by

$$\mathbf{x}(t) = \begin{cases} \mathbf{x}_T(t), & 0 \le t \le T, \\ 0, & t > T, \end{cases}$$

and so $\mathbf{x} \in L^2((0,\infty),\mathbb{R}^n)$.

It can be expected that the optimal control problem (3.96) over an infinite time interval would be the limit of the optimal control problem (3.75) over a finite time interval as $T \to \infty$ in some sense. Thus we further study the differential Riccati equation and investigate the asymptotic behavior of its solution as $t \to \infty$.

Lemma 3.6. *The differential Riccati equation*

$$\Pi' = \Pi\mathbf{A} + \mathbf{A}^*\Pi + \mathbf{Q} - \Pi\mathbf{B}\mathbf{R}^{-1}\mathbf{B}^*\Pi, \tag{3.97}$$
$$\Pi(0) = \mathbf{G}. \tag{3.98}$$

has a unique continuous symmetric nonnegative solution satisfying

$$\min_{\mathbf{u}\in L^2((0,t),\mathbb{R}^m)} J(\mathbf{u},t;\mathbf{x}_0) = \mathbf{x}_0^*\Pi(t)\mathbf{x}_0. \tag{3.99}$$

If $\mathbf{G} = 0$, then Π is increasing, that is

$$\mathbf{x}_0^*\Pi(t_1)\mathbf{x}_0 \le \mathbf{x}_0^*\Pi(t_2)\mathbf{x}_0$$

for all $\mathbf{x}_0$ and $t_1 \le t_2$.

Proof. Define

$$\Pi_T(t) = \mathbf{P}(T - t),$$

where $\mathbf{P}$ is the solution of the differential Riccati equation (3.89). It is clear that $\Pi_T(t)$ is a solution of (3.97) and the equation (3.99) follows from Theorem 3.19. Uniqueness of the solution can be established by standard results in ordinary differential equations.

To show that Π_T is independent of T, we take any $T_1 \le T_2$. Since the solution is unique, we have

$$\Pi_{T_1}(t) = \Pi_{T_2}(t), \quad 0 \le t \le T_1.$$

Now assume that $\mathbf{G} = 0$. Then, for any $\mathbf{x}_0$ and $t_1 \le t_2$, we have

$$\mathbf{x}_0^* \Pi(t_1)\mathbf{x}_0 = \min_{\mathbf{u} \in L^2((0,t_1),\mathbb{R}^m)} J(\mathbf{u}, t_1; \mathbf{x}_0)$$

$$\leq \min_{\mathbf{u} \in L^2((0,t_2),\mathbb{R}^m)} J(\mathbf{u}, t_2; \mathbf{x}_0)$$

$$= \mathbf{x}_0^* \Pi(t_2)\mathbf{x}_0.$$

Consider

$$x'(t) = x(t) + u(t), \quad x(0) = x_0$$

with the performance criterion

$$J(u, x_0) = \int_0^\infty [3x^2(t) + u^2(t)]dt.$$

Here $G = 0, Q = 3, R = 1$. The differential Riccati equation is

$$p'(t) - 2p(t) + p^2(t) - 3 = 0, \quad p(0) = 0,$$

which has a unique positive increasing solution

$$p(t) = \frac{3(1 - e^{-4t})}{1 + 3e^{-4t}}.$$

The limit of the differential Riccati equation leads to algebraic Riccati equations.

Lemma 3.7. *If the optimal control problem (3.96) is well posed, then the algebraic Riccati equation*

$$\mathbf{PA} + \mathbf{A}^*\mathbf{P} + \mathbf{Q} - \mathbf{PBR}^{-1}\mathbf{B}^*\mathbf{P} = 0$$

has a symmetric nonnegative solution given by

$$\mathbf{P} = \lim_{t \to \infty} \Pi(t),$$

where Π is the solution of the differential Riccati equation (3.97).

Proof. Assume $\bar{\mathbf{u}} \in L^2((0,\infty), \mathbb{R}^m)$ such that $J(\bar{\mathbf{u}}; \mathbf{x}_0) < \infty$. Since $\mathbf{x}_0^* \Pi(t)\mathbf{x}_0$ is increasing and

$$\mathbf{x}_0^* \Pi(t)\mathbf{x}_0 = \min_{\mathbf{u} \in L^2((0,t),\mathbb{R}^m)} J(\mathbf{u}, t; \mathbf{x}_0) \leq J(\bar{\mathbf{u}}; \mathbf{x}_0) < \infty$$

the limit

$$\lim_{t \to \infty} \mathbf{x}_0^* \Pi(t)\mathbf{x}_0$$

exists. Taking limit in the differential Riccati equations, we can show that the limit is the required solution of the algebraic Riccati equation.

Example 3.17. Consider the system

$$\dot{\mathbf{x}} = \mathbf{A}\mathbf{x} + \mathbf{B}u, \tag{3.100}$$

$$\mathbf{y} = \mathbf{C}\mathbf{x}, \tag{3.101}$$

where

$$\mathbf{A} = \begin{bmatrix} 0 & 1 \\ 0 & 0 \end{bmatrix}, \quad \mathbf{B} = \begin{bmatrix} 0 \\ 1 \end{bmatrix}, \quad \mathbf{C} = \begin{bmatrix} 1 & 0 \\ 0 & 1 \end{bmatrix}.$$

The performance criterion is defined by

$$\begin{aligned} J(u, \mathbf{x}_0) &= \int_0^\infty \left(|\mathbf{C}\mathbf{x}(t)|^2 + |u(t)|^2 \right) dt \\ &= \int_0^\infty \left(|x_1(t)|^2 + |x_2(t)|^2 + |u(t)|^2 \right) dt. \end{aligned}$$

Thus $\mathbf{D} = [1]$. The corresponding Riccati equation is

$$\begin{bmatrix} 0 & 0 \\ 1 & 0 \end{bmatrix} \begin{bmatrix} p_{11} & p_{12} \\ p_{12} & p_{22} \end{bmatrix} + \begin{bmatrix} p_{11} & p_{12} \\ p_{12} & p_{22} \end{bmatrix} \begin{bmatrix} 0 & 1 \\ 0 & 0 \end{bmatrix}$$
$$- \begin{bmatrix} p_{11} & p_{12} \\ p_{12} & p_{22} \end{bmatrix} \begin{bmatrix} 0 \\ 1 \end{bmatrix} [1] [0 \ 1] \begin{bmatrix} p_{11} & p_{12} \\ p_{12} & p_{22} \end{bmatrix} + \begin{bmatrix} 1 & 0 \\ 0 & 1 \end{bmatrix} = \begin{bmatrix} 0 & 0 \\ 0 & 0 \end{bmatrix}.$$

This equation gives three equations:

$$1 - p_{12}^2 = 0,$$
$$p_{11} - p_{12}p_{22} = 0,$$
$$1 + 2p_{12} - p_{22}^2 = 0.$$

Solving the system, we obtain the positive definite symmetric matrix

$$\mathbf{P} = \begin{bmatrix} \sqrt{3} & 1 \\ 1 & \sqrt{3} \end{bmatrix}.$$

Note that this Riccati equation has other complex solutions.

As in the proof of the identity (3.91), we can prove the following identity.

Lemma 3.8. *Let* $\mathbf{P}$ *be a symmetric solution of the algebraic Riccati equation*

$$\mathbf{P}\mathbf{A} + \mathbf{A}^*\mathbf{P} + \mathbf{Q} - \mathbf{P}\mathbf{B}\mathbf{R}^{-1}\mathbf{B}^*\mathbf{P} = 0.$$

Then

$$\int_0^t \left(\mathbf{x}(s)^*\mathbf{Q}\mathbf{x}(s) + \mathbf{u}(s)^*\mathbf{R}\mathbf{u}(s) \right) dt \tag{3.102}$$
$$= \mathbf{x}_0^*\mathbf{P}\mathbf{x}_0 - \mathbf{x}(t)^*\mathbf{P}\mathbf{x}(t) + \int_0^t [\mathbf{u}(s) + \mathbf{R}^{-1}\mathbf{B}^*\mathbf{P}\mathbf{x}(s)]^*\mathbf{R}[\mathbf{u}(s) + \mathbf{R}^{-1}\mathbf{B}^*\mathbf{P}\mathbf{x}(s)]ds.$$

We are now ready to derive the optimal state feedback controller.

Theorem 3.21. *If the optimal control problem (3.96) is well posed, then there is a unique optimal control*

$$\mathbf{u}_{opt}(t) = -\mathbf{R}^{-1}\mathbf{B}^*\mathbf{P}\mathbf{x}(t) \in L^2((0,\infty),\mathbb{R}^m)$$

such that

$$J(\mathbf{u}_{opt};\mathbf{x}_0) = \min_{\mathbf{u}\in L^2((0,\infty),\mathbb{R}^m)} J(\mathbf{u};\mathbf{x}_0) = \mathbf{x}_0^*\mathbf{P}\mathbf{x}_0,$$

where $\mathbf{P}$ *is the minimal nonnegative solution of the algebraic Riccati equation. By minimal we mean that any other symmetric solution* $\mathbf{P}_1$ *satisfies*

$$\mathbf{x}_0^*\mathbf{P}\mathbf{x}_0 \le \mathbf{x}_0^*\mathbf{P}_1\mathbf{x}_0, \quad \forall \mathbf{x}_0 \in \mathbb{R}^n.$$

Proof. Existence. Assume $\bar{\mathbf{u}} \in L^2((0,\infty),\mathbb{R}^m)$ such that $J(\bar{\mathbf{u}};\mathbf{x}_0) < \infty$. For any $t \ge 0$, we have

$$\begin{aligned}
\infty &> J(\bar{\mathbf{u}};\mathbf{x}_0) \\
&\ge \min_{\mathbf{u}\in L^2((0,\infty),\mathbb{R}^m)} J(\mathbf{u};\mathbf{x}_0) \\
&\ge \min_{\mathbf{u}\in L^2((0,\infty),\mathbb{R}^m)} J(\mathbf{u},t;\mathbf{x}_0) \\
&= \min_{\mathbf{u}\in L^2((0,t),\mathbb{R}^m)} J(\mathbf{u},t;\mathbf{x}_0) \\
&= \mathbf{x}_0^*\Pi(t)\mathbf{x}_0.
\end{aligned}$$

Taking limit gives

$$\min_{\mathbf{u}\in L^2((0,\infty),\mathbb{R}^m)} J(\mathbf{u};\mathbf{x}_0) \ge \mathbf{x}_0^*\mathbf{P}\mathbf{x}_0. \tag{3.103}$$

On the other hand, it follows from (3.102) that

$$\begin{aligned}
J(\mathbf{u},t;\mathbf{x}_0) &= \int_0^t (\mathbf{x}(s)^*\mathbf{Q}\mathbf{x}(s) + \mathbf{u}(s)^*\mathbf{R}\mathbf{u}(s))\,ds \\
&= \mathbf{x}_0^*\mathbf{P}\mathbf{x}_0 - \mathbf{x}(t)^*\mathbf{P}\mathbf{x}(t) \\
&\quad + \int_0^t [\mathbf{u}(s) + \mathbf{R}^{-1}\mathbf{B}^*\mathbf{P}\mathbf{x}(s)]^*\mathbf{R}[\mathbf{u}(s) + \mathbf{R}^{-1}\mathbf{B}^*\mathbf{P}\mathbf{x}(s)]\,ds \\
&\le \mathbf{x}_0^*\mathbf{P}\mathbf{x}_0 + \int_0^t [\mathbf{u}(s) + \mathbf{R}^{-1}\mathbf{B}^*\mathbf{P}\mathbf{x}(s)]^*\mathbf{R}[\mathbf{u}(s) + \mathbf{R}^{-1}\mathbf{B}^*\mathbf{P}\mathbf{x}(s)]\,ds.
\end{aligned}$$

Taking

$$\mathbf{u}(s) = \mathbf{u}_{opt}(s) = -\mathbf{R}^{-1}\mathbf{B}^*\mathbf{P}\mathbf{x}(s)$$

we obtain

$$J(\mathbf{u}_{opt},t;\mathbf{x}_0) \le \mathbf{x}_0^*\mathbf{P}\mathbf{x}_0,$$

and then

$$J(\mathbf{u}_{opt};\mathbf{x}_0) \le \mathbf{x}_0^*\mathbf{P}\mathbf{x}_0. \tag{3.104}$$

Now we show that

$$\mathbf{u}_{opt} \in L^2((0,\infty),\mathbb{R}^m).$$

Since $\mathbf{R}$ is positive definite, then there exists a $C > 0$ such that

$$\int_0^\infty |\mathbf{u}_{opt}(t)|^2 dt \leq C \int_0^\infty \mathbf{u}_{opt}(t)^* \mathbf{R} \mathbf{u}_{opt}(t) dt$$
$$\leq C J(\mathbf{u}_{opt};\mathbf{x}_0)$$
$$< \infty.$$

Combing (3.103) and (3.104) gives

$$J(\mathbf{u}_{opt};\mathbf{x}_0) = \mathbf{x}_0^* \mathbf{P} \mathbf{x}_0 = \min_{\mathbf{u} \in L^2((0,\infty),\mathbb{R}^m)} J(\mathbf{u};\mathbf{x}_0).$$

Uniqueness. The uniqueness follows from the convexity:

$$J((\mathbf{u}_1 + \mathbf{u}_2)/2;\mathbf{x}_0) < \frac{1}{2} J(\mathbf{u}_1;\mathbf{x}_0) + \frac{1}{2} J(\mathbf{u}_2;\mathbf{x}_0).$$

$\mathbf{P}$ is minimal. Let $\mathbf{P}_1$ be another solution. Then

$$J(\mathbf{u},t;\mathbf{x}_0) = \int_0^t (\mathbf{x}(s)^* \mathbf{Q} \mathbf{x}(s) + \mathbf{u}(s)^* \mathbf{R} \mathbf{u}(s))\, ds$$
$$= \mathbf{x}_0^* \mathbf{P}_1 \mathbf{x}_0 - \mathbf{x}(t)^* \mathbf{P}_1 \mathbf{x}(t)$$
$$+ \int_0^t [\mathbf{u}(s) + \mathbf{R}^{-1} \mathbf{B}^* \mathbf{P}_1 \mathbf{x}(s)]^* \mathbf{R}[\mathbf{u}(s) + \mathbf{R}^{-1} \mathbf{B}^* \mathbf{P}_1 \mathbf{x}(s)] ds$$
$$\leq \mathbf{x}_0^* \mathbf{P}_1 \mathbf{x}_0$$
$$+ \int_0^t [\mathbf{u}(s) + \mathbf{R}^{-1} \mathbf{B}^* \mathbf{P}_1 \mathbf{x}(s)]^* \mathbf{R}[\mathbf{u}(s) + \mathbf{R}^{-1} \mathbf{B}^* \mathbf{P}_1 \mathbf{x}(s)] ds.$$

Taking

$$\mathbf{u}(s) = \mathbf{u}_1(s) = -\mathbf{R}^{-1} \mathbf{B}^* \mathbf{P}_1 \mathbf{x}(s)$$

we obtain

$$J(\mathbf{u}_1,t;\mathbf{x}_0) \leq \mathbf{x}_0^* \mathbf{P}_1 \mathbf{x}_0,$$

and so

$$J(\mathbf{u}_1;\mathbf{x}_0) \leq \mathbf{x}_0^* \mathbf{P}_1 \mathbf{x}_0.$$

Thus

$$\mathbf{x}_0^* \mathbf{P} \mathbf{x}_0 = \min_{\mathbf{u} \in L^2((0,\infty),\mathbb{R}^m)} J(\mathbf{u};\mathbf{x}_0) \leq J(\mathbf{u}_1;\mathbf{x}_0) \leq \mathbf{x}_0^* \mathbf{P}_1 \mathbf{x}_0.$$

The next question is whether the optimal feedback controller stabilizes the system. Before answering this question, we first look at an example.

Example 3.18. In Example 3.17, we have obtained the positive definite symmetric matrix

$$\mathbf{P} = \begin{bmatrix} \sqrt{3} & 1 \\ 1 & \sqrt{3} \end{bmatrix}.$$

We then obtain the optimal feedback gain matrix for the system (3.100)-(3.101)

$$\mathbf{K}_{opt} = \mathbf{R}^{-1}\mathbf{B}^*\mathbf{P} = [1][0\ 1]\begin{bmatrix} \sqrt{3} & 1 \\ 1 & \sqrt{3} \end{bmatrix} = \begin{bmatrix} 1 & \sqrt{3} \end{bmatrix}.$$

Thus the optimal state feedback controller of (3.100)-(3.101) is

$$u = -\mathbf{K}_{opt}\mathbf{x} = -x_1 - \sqrt{3}x_2.$$

The matrix

$$\mathbf{A} - \mathbf{B}\mathbf{K}_{opt} = \begin{bmatrix} 0 & 1 \\ -1 & -\sqrt{3} \end{bmatrix}$$

is Hurwitz. In this case, one can check that $(\mathbf{A}, \mathbf{C})$ is observable.

If we change $\mathbf{C}$ to

$$\mathbf{C} = \begin{bmatrix} 0 & 0 \\ 0 & 1 \end{bmatrix},$$

the performance criterion is

$$\begin{aligned} J(u, \mathbf{x}_0) &= \int_0^\infty \left(|\mathbf{C}\mathbf{x}(t)|^2 + |u(t)|^2 \right) dt \\ &= \int_0^\infty \left(|x_2(t)|^2 + |u(t)|^2 \right) dt. \end{aligned}$$

The corresponding Riccati equation is

$$\begin{bmatrix} 0 & 0 \\ 1 & 0 \end{bmatrix}\begin{bmatrix} p_{11} & p_{12} \\ p_{12} & p_{22} \end{bmatrix} + \begin{bmatrix} p_{11} & p_{12} \\ p_{12} & p_{22} \end{bmatrix}\begin{bmatrix} 0 & 1 \\ 0 & 0 \end{bmatrix}$$
$$- \begin{bmatrix} p_{11} & p_{12} \\ p_{12} & p_{22} \end{bmatrix}\begin{bmatrix} 0 \\ 1 \end{bmatrix}[1][0\ 1]\begin{bmatrix} p_{11} & p_{12} \\ p_{12} & p_{22} \end{bmatrix} + \begin{bmatrix} 0 & 0 \\ 0 & 1 \end{bmatrix} = \begin{bmatrix} 0 & 0 \\ 0 & 0 \end{bmatrix}.$$

This equation gives three equations:

$$p_{12}^2 = 0,$$
$$p_{11} - p_{12}p_{22} = 0,$$
$$1 + 2p_{12} - p_{22}^2 = 0.$$

Solving the system, we obtain the semi-definite symmetric matrix

$$\mathbf{P} = \begin{bmatrix} 0 & 0 \\ 0 & 1 \end{bmatrix},$$

and then the optimal feedback gain matrix

$$\mathbf{K}_{opt} = \mathbf{R}^{-1}\mathbf{B}^*\mathbf{P} = [1][0\ 1]\begin{bmatrix} 0 & 0 \\ 0 & 1 \end{bmatrix} = [0\ 1].$$

Thus the optimal state feedback control is

$$u = -\mathbf{K}_{opt}\mathbf{x} = -x_2.$$

The matrix

$$\mathbf{A} - \mathbf{B}\mathbf{K}_{opt} = \begin{bmatrix} 0 & 1 \\ 0 & -1 \end{bmatrix}$$

is not Hurwitz. This happened because x_1 was not penalized or not detected by the performance criterion. In fact, we can check that $(\mathbf{A}, \mathbf{C})$ is not detectable.

This example demonstrates that certain conditions such as detectability are needed to be imposed on $(\mathbf{A}, \mathbf{C})$ to ensure that the optimal feedback controller stabilizes the system.

Lemma 3.9. *The function* $\exp(\mathbf{A}t)\mathbf{x}_0 \in L^2((0,\infty),\mathbb{R}^n)$ *for every* $\mathbf{x}_0 \in \mathbb{R}^n$ *if only if* $\mathbf{A}$ *is Hurwitz.*

Proof. If $\mathbf{A}$ is Hurwitz, then

$$\|\exp(\mathbf{A}t)\mathbf{x}_0\| \le Me^{-at}\|\mathbf{x}_0\|$$

for some $M, a > 0$. So $\exp(\mathbf{A}t)\mathbf{x}_0 \in L^2((0,\infty),\mathbb{R}^n)$.

If $\mathbf{A}$ is not Hurwitz, then there is an eigenvalue s_0 with $\mathrm{Re}(s_0) \ge 0$ and corresponding eigenvector $\mathbf{x}_{s_0}$. Then

$$\exp(\mathbf{A}t)\mathbf{x}_{s_0} = \exp(s_0 t)\mathbf{x}_{s_0} \notin L^2((0,\infty),\mathbb{R}^n).$$

Theorem 3.22. *If the optimal control problem (3.96) is well posed and the pair* $(\mathbf{A}, \mathbf{C})$ *is detectable, then the optimal feedback*

$$\mathbf{K}_{opt} = -\mathbf{R}^{-1}\mathbf{B}^*\mathbf{P}$$

stabilizes $(\mathbf{A}, \mathbf{B})$, *that is,* $\mathbf{A} - \mathbf{B}\mathbf{K}_{opt}$ *is Hurwitz.*

Proof. Since

$$\int_0^\infty \left(\|\mathbf{C}\mathbf{x}(s)\|^2 + \mathbf{u}(s)^*\mathbf{R}\mathbf{u}_{opt}(s) \right) ds = J(\mathbf{u}_{opt}; \mathbf{x}_0) < \infty,$$

and $\mathbf{R}$ is positive definite, we have

$$\mathbf{C}\mathbf{x}(s) \in L^2((0,\infty),\mathbb{R}^n), \quad \mathbf{u}_{opt} \in L^2((0,\infty),\mathbb{R}^m).$$

Since the system $(\mathbf{A}, \mathbf{C})$ is detectable, there is a $\mathbf{F}$ such that $\mathbf{A} - \mathbf{F}\mathbf{C}$ is Hurwitz. Writing

$$\dot{\mathbf{x}} = \mathbf{A}\mathbf{x} - \mathbf{B}\,\mathbf{K}_{opt}\mathbf{x} = \mathbf{A}\mathbf{x} + \mathbf{B}\mathbf{u}_{opt} = (\mathbf{A} - \mathbf{F}\mathbf{C})\mathbf{x} + \mathbf{F}\mathbf{C}\mathbf{x} + \mathbf{B}\mathbf{u}_{opt},$$

we obtain

$$\mathbf{x}(t) = e^{(\mathbf{A}-\mathbf{FC})t}\mathbf{x}_0 + \int_0^t e^{(\mathbf{A}-\mathbf{FC})(t-s)}[\mathbf{FC}\mathbf{x}(s) + \mathbf{B}\mathbf{u}_{opt}(s)]\,ds,$$

which implies that

$$\mathbf{x} \in L^2((0,\infty),\mathbb{R}^n).$$

So $\mathbf{A} - \mathbf{BK}_{opt}$ is Hurwitz.

We next show that the optimal feedback matrix $\mathbf{K}_{opt}$ minimizes the performance criterion among all gain matrices. Let $\mathscr{K}$ denote the set of all $m \times n$ gain matrices. Consider a state feedback control

$$\mathbf{u}(t) = -\mathbf{K}\mathbf{x}(t). \tag{3.105}$$

We define the performance criterion

$$\begin{aligned}
J(\mathbf{K},\mathbf{x}_0) &= \int_0^\infty \left(|\mathbf{C}\mathbf{x}(t)|^2 + |\mathbf{D}\mathbf{u}(t)|^2\right) dt \\
&= \int_0^\infty \left(\mathbf{x}(t)^*\mathbf{C}^*\mathbf{C}\mathbf{x}(t) + \mathbf{x}(t)^*\mathbf{K}^*\mathbf{D}^*\mathbf{D}\mathbf{K}\mathbf{x}(t)\right) dt \\
&= \int_0^\infty \left(\mathbf{x}(t)^*\mathbf{Q}\mathbf{x}(t) + \mathbf{x}(t)^*\mathbf{K}^*\mathbf{R}\mathbf{K}\mathbf{x}(t)\right) dt,
\end{aligned}$$

where

$$\mathbf{Q} = \mathbf{C}^*\mathbf{C}, \quad \mathbf{R} = \mathbf{D}^*\mathbf{D}.$$

Theorem 3.23. *Suppose that the optimal control problem (3.96) is well posed and let*

$$\mathbf{K}_{opt} = -\mathbf{R}^{-1}\mathbf{B}^*\mathbf{P}$$

be the optimal feedback matrix. Then

$$J(\mathbf{K}_{opt},\mathbf{x}_0) = \min_{\mathbf{K}\in\mathscr{K}} J(\mathbf{K},\mathbf{x}_0). \tag{3.106}$$

Proof. We first show that

$$\min_{\mathbf{u}\in L^2((0,\infty),\mathbb{R}^m)} J(\mathbf{u};\mathbf{x}_0) \leq \min_{\mathbf{K}\in\mathscr{K}} J(\mathbf{K},\mathbf{x}_0).$$

In fact, if $\mathbf{u}(t) = -\mathbf{K}\mathbf{x}(t) \notin L^2((0,\infty),\mathbb{R}^m)$, then $J(\mathbf{K},\mathbf{x}_0) = \infty$. If $\mathbf{u}(t) = -\mathbf{K}\mathbf{x}(t) \in L^2((0,\infty),\mathbb{R}^m)$, then

$$\min_{\mathbf{u}\in L^2((0,\infty),\mathbb{R}^m)} J(\mathbf{u};\mathbf{x}_0) \leq J(\mathbf{K},\mathbf{x}_0).$$

It therefore follows that

$$J(\mathbf{K}_{opt},\mathbf{x}_0) = \min_{\mathbf{u}\in L^2((0,\infty),\mathbb{R}^m)} J(\mathbf{u};\mathbf{x}_0) \leq \min_{\mathbf{K}\in\mathscr{K}} J(\mathbf{K},\mathbf{x}_0) \leq J(\mathbf{K}_{opt},\mathbf{x}_0).$$

Exercises 3.9

1. Prove the identity (3.102).
2. Consider the system

$$\dot{\mathbf{x}} = \mathbf{A}\mathbf{x} + \mathbf{B}u,$$
$$\mathbf{y} = \mathbf{C}\mathbf{x},$$

where

$$\mathbf{A} = \begin{bmatrix} 0 & 0 \\ 1 & 0 \end{bmatrix}, \quad \mathbf{B} = \begin{bmatrix} 1 \\ 0 \end{bmatrix}, \quad \mathbf{C} = \begin{bmatrix} 1 & 0 \\ 0 & 1 \end{bmatrix}.$$

The performance criterion is defined by

$$J(u, \mathbf{x}_0) = \int_0^\infty \left(|\mathbf{C}\mathbf{x}(t)|^2 + |u(t)|^2 \right) dt$$
$$= \int_0^\infty \left(|x_1(t)|^2 + |x_2(t)|^2 + |u(t)|^2 \right) dt.$$

   a. Write down the corresponding algebraic Riccati equation.
   b. Derive the optimal feedback matrix and show that it stabilizes the system.

3. Consider the control problem

$$\dot{\mathbf{x}} = \mathbf{A}\mathbf{x} + \mathbf{B}u(t),$$
$$\mathbf{y} = \mathbf{C}\mathbf{x},$$
$$\mathbf{x}(0) = \mathbf{x}_0$$

with a state feedback control

$$\mathbf{u}(t) = -\mathbf{K}\mathbf{x}(t).$$

Suppose that the pair $(\mathbf{A}, \mathbf{B})$ is stabilizable and define

$$\mathcal{K} = \{\mathbf{K} : \mathbf{A} - \mathbf{B}\mathbf{K} \text{ is Hurwitz}\}.$$

Let $\mathbf{D}$ be a nonsingular matrix and define the quadratic performance criterion

$$J(\mathbf{K}) = \int_0^\infty \left(|\mathbf{C}\mathbf{x}(t)|^2 + |\mathbf{D}\mathbf{u}(t)|^2 \right) dt$$
$$= \int_0^\infty \left(\mathbf{x}(t)^* \mathbf{C}^* \mathbf{C}\mathbf{x}(t) + \mathbf{x}(t)^* \mathbf{K}^* \mathbf{D}^* \mathbf{D}\mathbf{K}\mathbf{x}(t) \right) dt$$
$$= \int_0^\infty \left(\mathbf{x}(t)^* \mathbf{Q}\mathbf{x}(t) + \mathbf{x}(t)^* \mathbf{K}^* \mathbf{R}\mathbf{K}\mathbf{x}(t) \right) dt,$$

where

$$\mathbf{Q} = \mathbf{C}^* \mathbf{C}, \quad \mathbf{R} = \mathbf{D}^* \mathbf{D}.$$

Let $\mathbf{P}$ be a real symmetric nonnegative solution of the algebraic Riccati equation

$$\mathbf{A}^*\mathbf{P} + \mathbf{P}\mathbf{A} + \mathbf{Q} - \mathbf{P}\mathbf{B}\mathbf{R}^{-1}\mathbf{B}^*\mathbf{P} = \mathbf{0}.$$

Consider a quadratic Lyapunov function candidate

$$V(\mathbf{x}) = \mathbf{x}^*\mathbf{P}\mathbf{x}.$$

a. Show that for any gain matrix $\mathbf{K}$

$$\dot{V} = -\mathbf{x}^*\mathbf{Q}\mathbf{x} - \mathbf{x}^*\mathbf{K}^*\mathbf{D}^*\mathbf{D}\mathbf{K}\mathbf{x}$$
$$+ \mathbf{x}^*[(\mathbf{D}^*)^{-1}\mathbf{B}^*\mathbf{P} - \mathbf{D}\mathbf{K}]^*[(\mathbf{D}^*)^{-1}\mathbf{B}^*\mathbf{P} - \mathbf{D}\mathbf{K}]\mathbf{x}.$$

b. Show that for any $\mathbf{K} \in \mathcal{K}$

$$J(\mathbf{K}) = \mathbf{x}_0^*\mathbf{P}\mathbf{x}_0 + \int_0^\infty \mathbf{x}^*(t)[(\mathbf{D}^*)^{-1}\mathbf{B}^*\mathbf{P} - \mathbf{D}\mathbf{K}]^*[(\mathbf{D}^*)^{-1}\mathbf{B}^*\mathbf{P} - \mathbf{D}\mathbf{K}]\mathbf{x}(t)\, dt$$
$$\geq \mathbf{x}_0^*\mathbf{P}\mathbf{x}_0.$$

c. Let

$$\mathbf{K}_{opt} = \mathbf{D}^{-1}(\mathbf{D}^*)^{-1}\mathbf{B}^*\mathbf{P} = \mathbf{R}^{-1}\mathbf{B}^*\mathbf{P}.$$

Show that

$$J(\mathbf{K}_{opt}) \leq \mathbf{x}_0^*\mathbf{P}\mathbf{x}_0 \leq \min_{\mathbf{K} \in \mathcal{K}} J(\mathbf{K}).$$

3.10 Asymptotic Tracking and Disturbance Rejection

Consider the control system subject to a disturbance $\mathbf{d}$

$$\dot{\mathbf{x}} = \mathbf{A}\mathbf{x} + \mathbf{B}\mathbf{u} + \mathbf{E}_d\mathbf{d}, \tag{3.107}$$
$$\mathbf{y} = \mathbf{C}\mathbf{x} + \mathbf{D}\mathbf{u} + \mathbf{F}_d\mathbf{d}. \tag{3.108}$$

The reference $\mathbf{r}(t)$ and the disturbance $\mathbf{d}(t)$ are assumed to be generated by the exosystem

$$\dot{\mathbf{r}} = \mathbf{A}_{1r}\mathbf{r}, \tag{3.109}$$
$$\dot{\mathbf{d}} = \mathbf{A}_{1d}\mathbf{d}. \tag{3.110}$$

The constant reference $\mathbf{r}$ can be generated by the system with $\mathbf{A}_{1r} = 0$ and the sinusoidal disturbance $d = M\sin(\omega t)$ can be generated with

$$\mathbf{A}_{1d} = \begin{bmatrix} 0 & \omega \\ -\omega & 0 \end{bmatrix}.$$

The tracking error is given by

$$e = Cx + Du + F_d d - r.$$

Define

$$v = \begin{bmatrix} r \\ d \end{bmatrix}, \quad A_1 = \begin{bmatrix} A_{1r} & 0 \\ 0 & A_{1d} \end{bmatrix}, \quad E = [0 \; E_d], \quad F = [-I \; F_d].$$

Combining the control system (3.107)-(3.108) with the exosystem (3.109)-(3.110), we obtain

$$\dot{x} = Ax + Bu + Ev, \tag{3.111}$$
$$\dot{v} = A_1 v, \tag{3.112}$$
$$e = Cx + Du + Fv. \tag{3.113}$$

The control problem is to design a controller u to track the reference r:

$$\lim_{t \to \infty} e(t) = 0.$$

This problem is called *asymptotic tracking and disturbance rejection*.

Consider the static state feedback control

$$u = -K_1 x + K_2 v. \tag{3.114}$$

The two matrices K_1 and K_2 are called the *feedback gain* and the *feedforward gain*, respectively. Substituting (3.114) into (3.111)-(3.113) gives

$$\dot{x} = (A - BK_1)x + (BK_2 + E)v, \tag{3.115}$$
$$\dot{v} = A_1 v, \tag{3.116}$$
$$e = (C - DK_1)x + (DK_2 + F)v. \tag{3.117}$$

To eliminate v from the equation (3.115), we introduce the transformation

$$x = w + Xv,$$

where the matrix X is to be determined. Substituting this transformation into (3.115)-(3.117) gives

$$\dot{w} + XA_1 v = (A - BK_1)w + ((A - BK_1)X + BK_2 + E)v, \tag{3.118}$$
$$e = (C - DK_1)w + ((C - DK_1)X + DK_2 + F)v. \tag{3.119}$$

This leads to the matrix equations:

$$XA_1 = (A - BK_1)X + BK_2 + E, \tag{3.120}$$
$$0 = (C - DK_1)X + DK_2 + F. \tag{3.121}$$

The equation (3.120) for X is called the Sylvester equation. Let

$$\mathbf{U} = \mathbf{K}_2 - \mathbf{K}_1 \mathbf{X}.$$

Then the above matrix equations are changed to

$$\mathbf{X}\mathbf{A}_1 = \mathbf{A}\mathbf{X} + \mathbf{B}\mathbf{U} + \mathbf{E}, \tag{3.122}$$

$$0 = \mathbf{C}\mathbf{X} + \mathbf{D}\mathbf{U} + \mathbf{F}. \tag{3.123}$$

These equations are called the *regulator equations*.

Lemma 3.10. *Let $\mathbf{A}$ be an $n \times n$ matrix and $\mathbf{C}$ be a $p \times n$ matrix. For any matrices $\mathbf{E}$ and $\mathbf{F}$, the regulator equations (3.122)-(3.123) are solvable if and only if*

$$\text{rank} \begin{bmatrix} \mathbf{A} - \lambda \mathbf{I} & \mathbf{B} \\ \mathbf{C} & \mathbf{D} \end{bmatrix} = n + p \tag{3.124}$$

for all $\lambda \in \sigma(\mathbf{A}_1)$, where $\sigma(\mathbf{A}_1)$ denotes the spectrum of $\mathbf{A}_1$.

The proof of this lemma needs advanced linear algebra and thus is referred to [43, p.9, Theorem 1.9].

Theorem 3.24. *Suppose that $(\mathbf{A}, \mathbf{B})$ is stabilizable and the following holds:*

$$\text{rank} \begin{bmatrix} \mathbf{A} - \lambda \mathbf{I} & \mathbf{B} \\ \mathbf{C} & \mathbf{D} \end{bmatrix} = n + p$$

for all $\lambda \in \sigma(\mathbf{A}_1)$. Then there exist $\mathbf{K}_1$ and $\mathbf{K}_2$ such that the controller (3.114) drives the output to track the reference:

$$\lim_{t \to \infty} \mathbf{e}(t) = \lim_{t \to \infty} (\mathbf{y}(t) - \mathbf{r}(t)) = 0.$$

Proof. By Lemma 3.10, the regulator equations (3.122)-(3.123) have a solution $\mathbf{X}, \mathbf{U}$. Since $(\mathbf{A}, \mathbf{B})$ is stabilizable, there exists $\mathbf{K}_1$ such that $\mathbf{A} - \mathbf{K}_1 \mathbf{B}$ is Hurwitz. Set

$$\mathbf{K}_2 = \mathbf{U} + \mathbf{K}_1 \mathbf{X}.$$

Then the equations (3.118)-(3.119) become

$$\dot{\mathbf{w}} = (\mathbf{A} - \mathbf{B}\mathbf{K}_1)\mathbf{w},$$

$$\mathbf{e} = (\mathbf{C} - \mathbf{D}\mathbf{K}_1)\mathbf{w}.$$

It then follows that

$$\lim_{t \to \infty} \mathbf{e}(t) = \lim_{t \to \infty} (\mathbf{C} - \mathbf{D}\mathbf{K}_1)\mathbf{w}(t) = 0.$$

We now consider the Luenberger-observer based dynamic output feedback control

$$u = -Kz, \tag{3.125}$$

$$\dot{z} = \begin{bmatrix} A & E \\ 0 & A_1 \end{bmatrix} z + \begin{bmatrix} B \\ 0 \end{bmatrix} u + L (e - [C \ F]z - Du)$$

$$= G_1 z + G_2 e, \tag{3.126}$$

where the gain matrices K and L are to be designed and

$$G_1 = \begin{bmatrix} A & E \\ 0 & A_1 \end{bmatrix} - \begin{bmatrix} B \\ 0 \end{bmatrix} K - L([C \ F] - DK), \quad G_2 = L.$$

Substituting (3.125) and (3.126) into (3.111)-(3.113) gives

$$\dot{x} = Ax - BKz + Ev, \tag{3.127}$$

$$\dot{z} = G_2 Cx + (G_1 - G_2 DK)z + G_2 Fv, \tag{3.128}$$

$$\dot{v} = A_1 v, \tag{3.129}$$

$$e = Cx - DKz + Fv. \tag{3.130}$$

To eliminate v from the equation (3.127), we introduce the transformation

$$x = w + Xv, \quad z = q + Zv,$$

where matrices X and Z are to be determined. Substituting this transformation into (3.127)-(3.130) gives

$$\dot{w} + XA_1 v = Aw - BKq + (AX - BKZ + E)v, \tag{3.131}$$

$$\dot{q} + ZA_1 v = G_2 Cw + (G_1 - G_2 DK)q$$

$$+ (G_2 CX + (G_1 - G_2 DK)Z + G_2 F)v, \tag{3.132}$$

$$e = Cw - DKq + (CX - DKZ + F)v. \tag{3.133}$$

This leads to the matrix equations:

$$XA_1 = AX - BKZ + E, \tag{3.134}$$

$$0 = CX - DKZ + F, \tag{3.135}$$

$$ZA_1 = G_1 Z. \tag{3.136}$$

Let

$$U = -KZ.$$

Then the above matrix equations are changed to

$$XA_1 = AX + BU + E, \tag{3.137}$$

$$0 = CX + DU + F, \tag{3.138}$$

$$ZA_1 = G_1 Z. \tag{3.139}$$

By Lemma 3.10, the regulator equations (3.137)-(3.138) have a solution if the rank condition (3.124) is satisfied. The equation (3.139) has a trivial solution 0. To find a nonzero solution, we rewrite the regulator equations (3.134)-(3.135) as the following matrix form:

$$
\mathbf{XA}_1 = [\mathbf{A}\ \mathbf{E}] \begin{bmatrix} \mathbf{X} \\ \mathbf{I} \end{bmatrix} - \mathbf{BKZ}
$$

$$
= [\mathbf{A}\ \mathbf{E}] \begin{bmatrix} \mathbf{X} \\ \mathbf{I} \end{bmatrix} - \mathbf{BKZ} + \mathbf{L0}
$$

$$
= [\mathbf{A}\ \mathbf{E}] \begin{bmatrix} \mathbf{X} \\ \mathbf{I} \end{bmatrix} - \mathbf{BKZ} + \mathbf{L}(\mathbf{CX} - \mathbf{DKZ} + \mathbf{F})
$$

$$
= [\mathbf{A}\ \mathbf{E}] \begin{bmatrix} \mathbf{X} \\ \mathbf{I} \end{bmatrix} - \mathbf{BKZ} + \mathbf{L}\left([\mathbf{C}\ \mathbf{F}] \begin{bmatrix} \mathbf{X} \\ \mathbf{I} \end{bmatrix} - \mathbf{DKZ} \right)
$$

$$
\mathbf{A}_1 = [\mathbf{0}\ \mathbf{A}_1] \begin{bmatrix} \mathbf{X} \\ \mathbf{I} \end{bmatrix}.
$$

If we let

$$
\mathbf{Z} = \begin{bmatrix} \mathbf{X} \\ \mathbf{I} \end{bmatrix},
$$

we obtain

$$
\mathbf{ZA}_1 = \left(\begin{bmatrix} \mathbf{A} & \mathbf{E} \\ \mathbf{0} & \mathbf{A}_1 \end{bmatrix} - \begin{bmatrix} \mathbf{B} \\ \mathbf{0} \end{bmatrix} \mathbf{K} - \mathbf{L}([\mathbf{C}\ \mathbf{F}] - \mathbf{DK}) \right) \mathbf{Z} = \mathbf{G}_1 \mathbf{Z}.
$$

Thus $\mathbf{Z} = \begin{bmatrix} \mathbf{X} \\ \mathbf{I} \end{bmatrix}$ is a solution of the equation (3.139).

Theorem 3.25. *Suppose that* $(\mathbf{A}, \mathbf{B})$ *is stabilizable, the pair*

$$
\left([\mathbf{C}\ \mathbf{F}], \begin{bmatrix} \mathbf{A} & \mathbf{E} \\ \mathbf{0} & \mathbf{A}_1 \end{bmatrix} \right)
$$

is detectable, and the following holds:

$$
\mathrm{rank} \begin{bmatrix} \mathbf{A} - \lambda \mathbf{I} & \mathbf{B} \\ \mathbf{C} & \mathbf{D} \end{bmatrix} = n + p
$$

for all $\lambda \in \sigma(\mathbf{A}_1)$. *Then there exist* $\mathbf{K}$ *and* $\mathbf{L}$ *such that the controller (3.125)-(3.126) drives the output to track the reference:*

$$
\lim_{t \to \infty} \mathbf{e}(t) = \lim_{t \to \infty} (\mathbf{y}(t) - \mathbf{r}(t)) = 0.
$$

Proof. By Lemma 3.10, the regulator equations (3.137)-(3.138) have a solution $\mathbf{X}, \mathbf{U}$. Since $(\mathbf{A}, \mathbf{B})$ is stabilizable, there exists $\mathbf{K}_1$ such that $\mathbf{A} - \mathbf{K}_1 \mathbf{B}$ is Hurwitz. Define $\mathbf{K}_2 = -\mathbf{U} - \mathbf{K}_1 \mathbf{X}$. Then $\mathbf{U} = -\mathbf{KZ}$, where $\mathbf{Z} = \begin{bmatrix} \mathbf{X} \\ \mathbf{I} \end{bmatrix}$ satisfies the equation

(3.139). Then the equations (3.131)-(3.133) become

$$\dot{\mathbf{w}} = \mathbf{Aw} - \mathbf{BKq},$$
$$\dot{\mathbf{q}} = \mathbf{G}_2\mathbf{Cw} + (\mathbf{G}_1 - \mathbf{G}_2\mathbf{DK})\mathbf{q},$$
$$\mathbf{e} = \mathbf{Cw} - \mathbf{DKq}.$$

Let

$$\mathbf{L} = \begin{bmatrix} \mathbf{L}_1 \\ \mathbf{L}_2 \end{bmatrix}.$$

Then the coefficient matrix can be written as

$$\begin{bmatrix} \mathbf{A} & -\mathbf{BK} \\ \mathbf{G}_2\mathbf{C} \ \mathbf{G}_1 - \mathbf{G}_2\mathbf{DK} \end{bmatrix} = \begin{bmatrix} \mathbf{A} & -\mathbf{BK}_1 & -\mathbf{BK}_2 \\ \mathbf{L}_1\mathbf{C} \ \mathbf{A} - \mathbf{BK}_1 - \mathbf{L}_1\mathbf{C} \ \mathbf{E} - \mathbf{BK}_2 - \mathbf{L}_1\mathbf{F} \\ \mathbf{L}_2\mathbf{C} & -\mathbf{L}_2\mathbf{C} & \mathbf{A}_1 - \mathbf{L}_2\mathbf{F} \end{bmatrix}.$$

Subtracting the first row from the second row and adding the second column to the first column, we find that the coefficient matrix is equivalent to

$$\begin{bmatrix} \mathbf{A} - \mathbf{BK}_1 & -\mathbf{BK}_1 & -\mathbf{BK}_2 \\ 0 & \mathbf{A} - \mathbf{L}_1\mathbf{C} & \mathbf{E} - \mathbf{L}_1\mathbf{F} \\ 0 & -\mathbf{L}_2\mathbf{C} & \mathbf{A}_1 - \mathbf{L}_2\mathbf{F} \end{bmatrix}.$$

Thus the eigenvalues of the coefficient matrix are the union of the eigenvalues of $\mathbf{A} - \mathbf{BK}_1$ and the eigenvalues of

$$\begin{bmatrix} \mathbf{A} - \mathbf{L}_1\mathbf{C} & \mathbf{E} - \mathbf{L}_1\mathbf{F} \\ -\mathbf{L}_2\mathbf{C} & \mathbf{A}_1 - \mathbf{L}_2\mathbf{F} \end{bmatrix}.$$

Because the pair

$$\left([\mathbf{C} \ \mathbf{F}], \begin{bmatrix} \mathbf{A} & \mathbf{E} \\ 0 & \mathbf{A}_1 \end{bmatrix} \right)$$

is detectable, there exists a matrix $\mathbf{L}$ such that the above matrix is Hurwitz. Thus we have shown that the coefficient matrix is Hurwitz and then

$$\lim_{t \to \infty} \mathbf{e}(t) = \lim_{t \to \infty} (\mathbf{Cw}(t) - \mathbf{DKq}(t)) = 0.$$

Exercises 3.10

1. Consider the control system subject to a disturbance

$$\dot{x} = x + u + d_1,$$
$$\dot{r} = r,$$
$$\dot{d}_1 = d_2,$$
$$\dot{d}_2 = -d_1,$$
$$y = x.$$

Solve the regulator equations (3.122)-(3.123) and then design a static state feedback control

$$u = k_1 x + k_2 r + k_3 d_1 + k_4 d_2$$

such that the output y tracks the reference r.

2. Let $\mathbf{A} = [a_{ij}]$ be an $m \times q$ matrix and $\mathbf{B} = [b_{ij}]$ be a $p \times n$ matrix. The Kronecker product of $\mathbf{A}$ and $\mathbf{B}$, denoted by $\mathbf{A} \otimes \mathbf{B}$, is defined by

$$\mathbf{A} \otimes \mathbf{B} = \begin{bmatrix} a_{11}\mathbf{B} & \cdots & a_{1q}\mathbf{B} \\ \vdots & \vdots & \vdots \\ a_{m1}\mathbf{B} & \cdots & a_{mq}\mathbf{B} \end{bmatrix}.$$

Show that for any matrices $\mathbf{A}, \mathbf{B}, \mathbf{C}$, and $\mathbf{D}$ of appropriate dimensions

$$(\mathbf{A} \otimes \mathbf{B})(\mathbf{C} \otimes \mathbf{D}) = (\mathbf{AC}) \otimes (\mathbf{BD}),$$
$$(\mathbf{A} + \mathbf{B}) \otimes (\mathbf{C} + \mathbf{D}) = \mathbf{A} \otimes \mathbf{C} + \mathbf{A} \otimes \mathbf{D} + \mathbf{B} \otimes \mathbf{C} + \mathbf{B} \otimes \mathbf{D}.$$

3. For an $n \times m$ matrix $\mathbf{X}$, define

$$\mathrm{vec}(\mathbf{X}) = \begin{bmatrix} X_1 \\ \vdots \\ X_m \end{bmatrix},$$

where X_i is the ith column of $\mathbf{X}$ for $i = 1, \cdots, m$. Prove that

$$\mathrm{vec}(\mathbf{BXA}) = (\mathbf{A}^* \otimes \mathbf{B})\mathrm{vec}(\mathbf{X}).$$

4. Prove that the Sylvester equation

$$\mathbf{XA} - \mathbf{BX} = \mathbf{Q}$$

has a unique solution if and only if $\mathbf{A}$ and $\mathbf{B}$ have no eigenvalues in common.

3.11 References and Notes

This chapter is based on the references [38, 45, 86, 90, 104]. Section 3.2 is adopted from [45], Section 3.3 from [86], Section 3.4 from [104], Sections 3.6 and 3.7 from [38, 45, 90], Sections 3.8 and 3.9 from [86], and Section 3.10 from [43]. Most of exercise problems are adopted from [86, 90]. For classical control theory, which is based on frequency-domain analysis, we refer to [38, 86, 90]; for nonlinear controls, we refer to [10, 43, 45]; for more discussions about PID trajectory tracking control for mechanical systems, we refer to [19].

Chapter 4
Linear Reaction-Convection-Diffusion Equation

In this chapter, we discuss the control problem of the linear reaction-convection-diffusion equation

$$\frac{\partial u}{\partial t} = \mu \nabla^2 u + \nabla \cdot (u\mathbf{v}) + au. \tag{4.1}$$

Depending on a particular real problem, u can represent a temperature or the concentration of a chemical species. The constant $\mu > 0$ is the diffusivity of the temperature or the species, the vector $\mathbf{v}(\mathbf{x}) = (v_1(\mathbf{x}), \cdots, v_n(\mathbf{x}))$ is the velocity field of a fluid flow, and $a(\mathbf{x})$ is a reaction rate. ∇^2 is defined by

$$\nabla^2 u = \frac{\partial^2 u}{\partial x_1^2} + \cdots + \frac{\partial^2 u}{\partial x_n^2},$$

which is called the Laplace operator. The gradient of u is denoted by

$$\nabla u = \left(\frac{\partial u}{\partial x_1}, \cdots, \frac{\partial u}{\partial x_n} \right).$$

$\nabla \cdot (u\mathbf{v}) = \mathrm{div}(u\mathbf{v}) = \sum_{i=1}^{n} \frac{\partial (uv_i)}{\partial x_i}$ denotes the divergence of the vector $u\mathbf{v}$.

We briefly mention that the equation (4.1) is a quite simplified model for real problems. As demonstrated in the equation (1.6), the reaction kinetics is usually nonlinear. Thus the linear term au is a linear approximation of the reaction kinetics originating in real problems. In a chemical reaction, there is usually more than one chemical. Competitive-consecutive reactions are the simplest model system and are described by $A + B \rightarrow P$, $B + P \rightarrow W$, where A and B are the initial reactants, P is the desired product, and W is a by-product (waste). The dynamics of the competitive-consecutive reactions can be modeled by the system of reaction-convection-diffusion equations [87]

$$\frac{\partial A}{\partial t} + (\mathbf{v} \cdot \nabla)A = \mu \nabla^2 A - k_1 AB,$$

W. Liu, *Elementary Feedback Stabilization of the Linear Reaction-Convection-Diffusion Equation and the Wave Equation*, Mathématiques et Applications 66, DOI 10.1007/978-3-642-04613-1_4, © Springer-Verlag Berlin Heidelberg 2010

$$\frac{\partial B}{\partial t} + (\mathbf{v} \cdot \nabla)B = \mu \nabla^2 B - k_1 AB - k_2 BP,$$

$$\frac{\partial P}{\partial t} + (\mathbf{v} \cdot \nabla)P = \mu \nabla^2 P + k_1 AB - k_2 BP,$$

$$\frac{\partial W}{\partial t} + (\mathbf{v} \cdot \nabla)W = \mu \nabla^2 W + k_2 BP,$$

where k_1, k_2 are the reaction rate constants. In addition, in real problems, the velocity field $\mathbf{v}$ should satisfy the incompressible Navier-Stokes equations (see, e.g., [37])

$$\frac{\partial \mathbf{v}}{\partial t} + (\mathbf{v} \cdot \nabla)\mathbf{v} = -\frac{1}{\rho}\nabla p + \nu \nabla^2 \mathbf{v} + \mathbf{F}(\mathbf{x}, t) \quad \text{in } \Omega,$$

$$\nabla \cdot \mathbf{v} = 0 \quad \text{in } \Omega,$$

where p denotes the pressure, ρ the density of the fluid, ν the kinematic viscosity of the fluid, and $\mathbf{F}(\mathbf{x}, t)$ some external forces.

In what follows, Ω denotes a bounded domain in $\mathbb{R}^n$, Γ denotes its boundary $\partial \Omega$, and $\mathbf{n}$ denotes the unit normal on the boundary pointing the outside of Ω. In this text, for convenience, we will use the subscripts u_t, u_x or $\frac{\partial u}{\partial t}, \frac{\partial u}{\partial x}$ interchangeably to denote the derivatives of u with respect to t, x, respectively.

In this chapter, we focus on the Dirichlet boundary condition

$$u|_\Gamma = f,$$

where f is a given function. However, the theories presented here also hold for other boundary conditions such as the Neumann boundary condition

$$\left.\frac{\partial u}{\partial \mathbf{n}}\right|_\Gamma = f.$$

4.1 Stability

Before we study our control problems, we consider the stability of the boundary initial value problem

$$\frac{\partial u}{\partial t} = \mu \nabla^2 u + \nabla \cdot (u\mathbf{v}) + au \quad \text{in } \Omega, \tag{4.2}$$

$$u|_\Gamma = 0, \tag{4.3}$$

$$u(\mathbf{x}, 0) = u_0(\mathbf{x}), \tag{4.4}$$

where u_0 is an initial condition. The steady state equation of the problem is

$$\mu \nabla^2 \psi + \nabla \cdot (\psi \mathbf{v}) + a\psi = 0 \quad \text{in } \Omega, \tag{4.5}$$

$$\psi|_\Gamma = 0. \tag{4.6}$$

This problem has a trivial steady solution $\psi = 0$. Depending on $\mathbf{v}$ and a, it may have other non-trivial solutions. For example, if $\Omega = (0, \pi)$, $\mathbf{v} = 0$, and $a = \mu$, it has the non-trivial solution $\psi = \sin x$. We are interested in the stability of the equilibrium $\psi = 0$.

Definition 4.1. The equilibrium point $\psi = 0$ of (4.2)-(4.4) is

(1) stable if, for any $\varepsilon > 0$, there exists $\delta = \delta(\varepsilon) > 0$ such that

$$\int_\Omega |u(\mathbf{x},t)|^2 dV < \varepsilon \quad \text{for} \quad \int_\Omega |u_0(\mathbf{x})|^2 dV < \delta, \, t \geq 0.$$

(2) unstable if it is not stable.
(3) asymptotically stable if for every $u_0 \in L^2(\Omega)$

$$\lim_{t \to \infty} \int_\Omega |u(\mathbf{x},t)|^2 dV = 0.$$

(4) exponentially stable if there exist constants $\gamma, C > 0$ such that the solution of (4.2)-(4.4) satisfies

$$\int_\Omega |u(\mathbf{x},t)|^2 dV \leq Ce^{-\gamma t} \int_\Omega |u_0(\mathbf{x})|^2 dV.$$

Whenever the equilibrium is stable, unstable, asymptotically stable, or exponentially stable, we usually say that so is the equation. The number γ is called the *decay rate*. In this definition, the L^2-norm is used and thus the stability is called the L^2-*stability*. Unlike finite dimensional systems, other norms can be used, for example, the H^1 norm

$$\left(\int_\Omega \left(|u(\mathbf{x},t)|^2 + |\nabla u(\mathbf{x},t)|^2 \right) dV \right)^{1/2}.$$

In this case, the stability is called the H^1-*stability*. In this text, for simplicity, we discuss only the L^2-stability. Other stabilities with high regularity, such as H^1- or H^2-stability, are usually difficult to establish.

Analogous to the finite dimensional systems, the stability of the equilibrium point $\psi = 0$ of (4.2)-(4.4) is determined by the eigenvalues of the eigenvalue problem

$$\mu \nabla^2 \psi + \nabla \cdot (\psi \mathbf{v}) + a\psi = \lambda \psi \quad \text{in } \Omega, \tag{4.7}$$
$$\psi = 0 \quad \text{on } \Gamma. \tag{4.8}$$

Readers who are not familiar with semigroup theory may skip this part and jump to the elementary Theorem 4.2. Corresponding to this problem, we define the operator A on $L^2(\Omega)$ by

$$A\psi = \mu \nabla^2 \psi + \nabla \cdot (\psi \mathbf{v}) + a(\mathbf{x})\psi \tag{4.9}$$

with the domain $D(A) = H^2(\Omega) \cap H_0^1(\Omega)$. According to Theorem 2.7, the spectrum $\sigma(A)$ consists of infinite eigenvalues of A.

Theorem 4.1. *For any* $\omega > \sup\{\mathrm{Re}(\lambda) \mid \lambda \in \sigma(A)\}$, *there exists* $M(\gamma) > 0$ *such that the solutions of (4.2)-(4.4) satisfies*

$$\int_{\Omega} |u(\mathbf{x},t)|^2 dV \leq M e^{2\omega t} \int_{\Omega} |u_0(\mathbf{x})|^2 dV. \tag{4.10}$$

Proof. Using the operator A defined by (4.9), the problem (4.2)-(4.4) can be formulated into the abstract Cauchy problem (2.64). By Theorem 2.16, A generates an analytic semigroup e^{At} on $L^2(\Omega)$. Thus by Theorem 2.23, the problem (4.2)-(4.4) has a unique solution given by $u = e^{At}u_0$. Finally, (4.10) follows from Theorem 2.21. $\qquad\blacksquare$

This theorem tells that if $\sup\{\mathrm{Re}(\lambda) \mid \lambda \in \sigma(A)\} < 0$, then the equilibrium 0 of (4.2)-(4.4) is exponentially stable.

In the case where $\mathbf{v} = 0$, we can give an elementary proof of this theorem. In this case, by Theorem 2.5, all the eigenvalues of (4.7)-(4.8) consist of a sequence of real numbers converging to $-\infty$ and all eigenfunctions form a basis in $L^2(\Omega)$. Using the separation of variables, we can obtain the series solutions of (4.2)-(4.4).

Theorem 4.2. *Assume that* $\mathbf{v} = 0$. *Let* $\lambda_1 \geq \lambda_2 \geq \cdots \geq \lambda_n \geq \cdots$ *with* $\lim_{n\to\infty} \lambda_n = -\infty$ *be the eigenvalues of (4.7)-(4.8) and let* $\{\psi_i\}_{i=1}^{\infty}$ *be the corresponding normalized (by normalized we mean that* $\int_{\Omega} |\psi_i|^2 dV = 1$) *orthogonal eigenfunctions. Then the solution of (4.2)-(4.4) is given by*

$$u = \sum_{i=1}^{\infty} c_i e^{\lambda_i t} \psi_i, \tag{4.11}$$

where

$$c_i = \int_{\Omega} u_0 \psi_i dV.$$

Moreover,

$$\int_{\Omega} |u(\mathbf{x},t)|^2 dV \leq e^{2\lambda_1 t} \int_{\Omega} |u_0(\mathbf{x})|^2 dV. \tag{4.12}$$

Proof. Let $u = f(t)\psi(\mathbf{x})$. Substituting u into (4.2)-(4.4), we obtain (noting that $\mathbf{v} = 0$)

$$\frac{f_t}{f} = \frac{\mu \nabla^2 \psi + a(\mathbf{x})\psi}{\psi} = \lambda, \quad \psi|_{\Gamma} = 0.$$

For each eigenvalue λ_i, solving

$$\frac{df}{dt} = \lambda_i f,$$

we obtain $f_i = f_i(0)e^{\lambda_i t}$. Therefore the solution u is given by the series

$$u = \sum_{i=1}^{\infty} f_i(0)e^{\lambda_i t} \psi_i.$$

Setting $t = 0$ in the above equation, we obtain

$$u_0 = \sum_{i=1}^{\infty} f_i(0)\psi_i.$$

Since $\{\psi_i\}_{i=1}^{\infty}$ is an orthonormal basis in $L^2(\Omega)$, multiplying the equation by ψ_i and integrating over Ω, we obtain

$$f_i(0) = \int_{\Omega} u_0 \psi_i dV.$$

Moreover,

$$\begin{aligned}
\int_{\Omega} |u(\mathbf{x},t)|^2 dV &= \sum_{i=1}^{\infty} c_i^2 e^{2\lambda_i t} \int_{\Omega} |\psi_i|^2 dV \\
&= \sum_{i=1}^{\infty} c_i^2 e^{2\lambda_i t} \\
&= e^{2\lambda_1 t} \sum_{i=1}^{\infty} c_i^2 e^{2(\lambda_i - \lambda_1)t} \\
&\leq e^{2\lambda_1 t} \sum_{i=1}^{\infty} c_i^2 \\
&= e^{2\lambda_1 t} \int_{\Omega} |u_0(\mathbf{x})|^2 dV.
\end{aligned}$$

This theorem tells that the equilibrium 0 of (4.2)-(4.4) is exponentially stable if $\lambda_1 < 0$, stable if $\lambda_1 = 0$, and unstable if $\lambda_1 > 0$. The above two theorems do not provide conditions on a and $\mathbf{v}$ under which the equilibrium $\psi = 0$ of (4.2)-(4.4) is exponentially stable. To obtain such conditions, we need Gronwall's inequality.

Lemma 4.1. *(Gronwall's inequality) If*

$$x' \leq g(t)x + h(t) \quad \text{for } t \geq s, \tag{4.13}$$

then

$$x(t) \leq x(s)e^{\int_s^t g(r)dr} + e^{\int_s^t g(r)dr} \int_s^t h(z)e^{-\int_s^z g(r)dr} dz. \tag{4.14}$$

Proof. Multiplying (4.13) by $e^{-\int_s^t g(r)dr}$ gives

$$x' e^{-\int_s^t g(r)dr} - g(t)x e^{-\int_s^t g(r)dr} \leq h(t)e^{-\int_s^t g(r)dr},$$

and then

$$\frac{d}{dt}\left(x e^{-\int_s^t g(r)dr}\right) \leq h(t)e^{-\int_s^t g(r)dr}.$$

Integrating from s to t gives (4.14).

In what follows, we denote by div the divergence of a vector $\mathbf{v}$: $\text{div}(\mathbf{v}) = \frac{\partial v_1}{\partial x_1} + \cdots + \frac{\partial v_n}{\partial x_n}$.

Theorem 4.3. *Assume that the velocity* $\mathbf{v}$ *is differentiable. Denote*

$$v_m = \frac{1}{2} \max_{\mathbf{x} \in \Omega} \{ \mathrm{div}(\mathbf{v}(\mathbf{x}, t)) \}, \quad a_m = \max_{\mathbf{x} \in \Omega} a(\mathbf{x}).$$

Let λ_1 *be the constant in Poincaré's inequality (2.29). If* $a_m + v_m < \mu \lambda_1$, *then the equilibrium* 0 *of (4.2)-(4.4) is exponentially stable:*

$$\int_\Omega |u(t)|^2 dV \le e^{2(a_m + v_m - \mu\lambda_1)t} \int_\Omega |u_0|^2 dV. \tag{4.15}$$

Proof. Multiplying (4.2) by u, integrating over Ω by parts, and using the boundary conditions on u, we deduce that

$$\frac{1}{2}\frac{d}{dt}\int_\Omega |u(t)|^2 dV = \mu \int_\Omega u \nabla^2 u(t) dV + \int_\Omega (\nabla \cdot (u\mathbf{v})) u(t) dV + \int_\Omega a|u(t)|^2 dV$$

$$\text{(use (2.25) and (2.26))}$$

$$= \mu \int_\Gamma u(t) \frac{\partial u}{\partial \mathbf{n}} dS - \mu \int_\Omega |\nabla u(t)|^2 dV + \int_\Gamma u^2(t)\mathbf{n} \cdot v dS$$

$$\quad - \int_\Omega u(t)\mathbf{v} \cdot \nabla u(t) dV + \int_\Omega a|u(t)|^2 dV$$

$$= -\mu \int_\Omega |\nabla u(t)|^2 dV - \frac{1}{2}\int_\Omega \mathbf{v} \cdot \nabla(u^2(t)) dV + \int_\Omega a|u(t)|^2 dV$$

$$= -\mu \int_\Omega |\nabla u(t)|^2 dV - \frac{1}{2}\int_\Gamma u^2(t)\mathbf{n} \cdot v dS$$

$$\quad + \frac{1}{2}\int_\Omega \mathrm{div}(\mathbf{v})(u^2(t) dV + \int_\Omega a|u(t)|^2 dV$$

$$= -\mu \int_\Omega |\nabla u(t)|^2 dV + \frac{1}{2}\int_\Omega \mathrm{div}(\mathbf{v})(u^2(t) dV + \int_\Omega a|u(t)|^2 dV.$$

The application of Poincaré's inequality (2.29) gives

$$\frac{d}{dt}\int_\Omega |u(t)|^2 dV \le 2(a_m + v_m - \mu\lambda_1)\int_\Omega |u(t)|^2 dV.$$

Then (4.15) follows from Gronwall's inequality.

Example 4.1. Consider the initial boundary value problem

$$\frac{\partial u}{\partial t} = \frac{\partial^2 u}{\partial x^2} + au,$$

$$u(0,t) = u(\pi,t) = 0,$$

$$u(x,0) = u_0,$$

where a is a constant. Using the method of separation of variables, we can solve the problem to obtain the series solution

$$u(x,t) = \sum_{n=1}^{\infty} c_n \sin(nx) e^{(a-n^2)t},$$

where the Fourier coefficient c_n is given by

$$c_n = \frac{2}{\pi} \int_0^{\pi} u_0(x) \sin(nx) \, dx.$$

Evidently, if $a < 1$, the equilibrium 0 is exponentially stable; if $a = 1$, it is stable, but not exponentially stable; if $a > 1$, it is unstable because $\sin(x) e^{(a-1)t}$ goes to infinity as $t \to \infty$.

We now extend the definition of stability for the reaction-convection-diffusion equation to an abstract Cauchy problem

$$\frac{du}{dt} = Au, \quad u(0) = u_0, \tag{4.16}$$

where A is the infinitesimal generator of a C_0 semigroup $T(t)$ on a Banach space X and $u_0 \in X$ is an initial condition. The space X is called the *state space* of the problem. By Theorem 2.23, the solution is given by $u(t) = T(t)u_0$. Since $A0 = 0$, the zero element is an equilibrium point. Readers who do not have a background in operator theory may skip this part.

Definition 4.2. The equilibrium point 0 of (4.16) is

(1) stable if, for any $\varepsilon > 0$, there exists $\delta = \delta(\varepsilon) > 0$ such that

$$\|u(t)\| < \varepsilon \quad \text{for } \|u_0\| < \delta, t \geq 0.$$

(2) unstable if it is not stable.
(3) asymptotically stable if $\lim_{t \to \infty} \|u(t)\| = 0$ for each $u_0 \in X$.
(4) exponentially stable if there exist constants $\gamma, C > 0$ such that the solution of (4.16) satisfies

$$\|u(t)\| \leq Ce^{-\gamma t} \|u_0\|.$$

Whenever the equilibrium is stable, unstable, asymptotically stable, or exponentially stable, we usually say that so is the system or the semigroup.

If A is the infinitesimal generator of an analytic semigroup, Theorem 2.21 indicates that the equilibrium is exponentially stable if the spectrum $\sigma(A)$ lies in the left-half complex plane. Unfortunately, this is not always true for general generators. For such an example, we refer to [82, p.114, Example 3.6].

To find conditions on A that ensure the exponential stability, we examine the inequality (2.45). This inequality tells that if the semigroup $T(t)$ is exponentially stable, that is, $\omega < 0$, then the resolvent $R(\lambda, A)$ as an operator-valued function of

λ is bounded on the right-half complex plane $\mathrm{Re}(\lambda) > 0$. In fact, this condition is both necessary and sufficient.

Theorem 4.4. *Let $T(t)$ be a C_0 semigroup with infinitesimal generator A on a Hilbert space H. Then $T(t)$ is exponentially stable if and only if the resolvent $R(\lambda,A)$ as an operator-valued function of λ is bounded on the right-half complex plane* $\mathrm{Re}(\lambda) > 0$.

The proof uses the Laplace transform of operator-valued functions and is beyond the scope of this text. Thus we refer to [24, p.222, Theorem 5.1.5]. This is a nice theoretical result, but the condition is usually difficult to verify in applications to partial differential equations.

The above tests for stability rely on the spectrum of A. Finding out the spectrum itself is a difficult problem. Thus it is valuable to have a test without using the spectrum. For this, we examine the simple exponential function e^{at}. If

$$\int_0^\infty e^{at}\,dt < \infty,$$

then $a < 0$. This is actually true for any semigroups.

Theorem 4.5. *If the semigroup $T(t)$ on a Banach space X satisfies*

$$\int_0^\infty \|T(t)\|^p\,dt \le C < \infty \tag{4.17}$$

for some $p > 0$, then there exist constants $\alpha, \sigma > 0$ such that

$$\|T(t)\| \le \alpha e^{-\sigma t}. \tag{4.18}$$

Proof. From the properties of semigroups (Theorem 2.8) it follows that there exist $K, \kappa >$ such that

$$\|T(t)\| \le K e^{\kappa t}. \tag{4.19}$$

Then for $0 \le t \le 1$, we have that

$$\|T(t)\| \le K e^{\kappa}.$$

For $t \ge 1$, we deduce that

$$\frac{1 - e^{-p\kappa}}{p\kappa}\|T(t)\|^p \le \frac{1 - e^{-p\kappa t}}{p\kappa}\|T(t)\|^p$$

$$= \int_0^t e^{-p\kappa s}\|T(t)\|^p\,ds$$

$$= \int_0^t e^{-p\kappa s}\|T(s)T(t-s)\|^p\,ds$$

$$\le K^p \int_0^t e^{-p\kappa s}e^{p\kappa s}\|T(t-s)\|^p\,ds \quad \text{(use (4.19))}$$

$$\leq K^p \int_0^\infty \|T(t)\|^p dt$$

$$\leq CK^p. \quad \text{(use (4.17))}$$

Thus, for all $t \geq 0$, we obtain

$$\|T(t)\|^p \leq M$$

for some $M > 0$. Using this inequality, we deduce that

$$t\|T(t)\|^p = \int_0^t \|T(t)\|^p ds$$

$$= \int_0^t \|T(s)T(t-s)\|^p ds$$

$$\leq \int_0^t M\|T(t-s)\|^p ds$$

$$\leq M \int_0^\infty \|T(t)\|^p dt$$

$$\leq MC. \quad \text{(use (4.17))}$$

Hence

$$\|T(t)\|^p \leq \frac{MC}{t},$$

and then

$$\|T(2MC)\|^p \leq \frac{1}{2}.$$

Let $t = 2MCn + r$, where n is a non-negative integer and $0 \leq r < 2MC$. It then follows that

$$\|T(t)\|^p = \|T(2MCn + r)\|^p$$

$$= \|T(2MC)T(2MC(n-1) + r)\|^p$$

$$\leq \frac{1}{2}\|T(2MC(n-1) + r)\|^p$$

$$\vdots$$

$$\leq \left(\frac{1}{2}\right)^n \|T(r)\|^p$$

$$\leq K^p \left(\frac{1}{2}\right)^{n+1} e^{p\kappa MC}$$

$$= K^p e^{-(n+1)\ln 2} e^{p\kappa MC}$$

$$= K^p e^{-\frac{\ln 2}{2MC} t} e^{p\kappa MC}.$$

For the finite dimensional system (3.7), the number of the eigenvectors of the matrix $\mathbf{A}$ is finite. For operators defined through differential equations in a function space, such as the operator A in (4.9), the number of eigenvectors is infinite. Thus such a system as (4.2) is called an *infinite dimensional system*.

The extension of control theory of finite dimensional systems to infinite dimensional systems is not trivial. New methods and theories are needed to create to deal with the infinite dimensional control systems. Some of these methods originate from the finite dimensional control systems, but many others are completely new.

Exercises 4.1

1. Consider the initial boundary value problem of the heat equation

$$\frac{\partial u}{\partial t} = \mu \frac{\partial^2 u}{\partial x^2} + au,$$
$$u(0,t) = 0, \quad u(1,t) = 0,$$
$$u(x,0) = u_0(x),$$

   where a is a constant.

   a. Use the method of separation of variables to find its solution.
   b. For what values of the reaction constant a, is the equilibrium 0 unstable?

2. Consider the initial boundary value problem

$$\frac{\partial u}{\partial t} = \mu \frac{\partial^2 u}{\partial x^2} + au,$$
$$u(0,t) = \frac{\partial u}{\partial x}(\pi,t) = 0,$$
$$u(x,0) = u_0(x),$$

   where a is a constant.

   a. Use the method of separation of variables to find its solution.
   b. For what values of the reaction constant a, is the equilibrium 0 unstable?

3. Consider the initial boundary value problem

$$\frac{\partial u}{\partial t} = \mu \frac{\partial^2 u}{\partial x^2},$$
$$u(0,t) = 0, \quad \frac{\partial u}{\partial x}(1,t) = -\frac{1}{2}u(1,t),$$
$$u(x,0) = u_0(x).$$

   Show that

$$\int_0^1 |u(x,t)|^2 dx \le e^{-\sigma t} \int_0^1 |u_0(x)|^2 dx,$$

   where σ is a positive constant.

4. Consider the initial boundary value problem of the Burgers equation

$$\frac{\partial u}{\partial t} = \mu \frac{\partial^2 u}{\partial x^2} + u \frac{\partial u}{\partial x},$$
$$u(0,t) = 0, \quad u(1,t) = 0,$$
$$u(x,0) = u_0(x).$$

Show that

$$\int_0^1 |u(x,t)|^2 dx \le e^{-\sigma t} \int_0^1 |u_0(x)|^2 dx,$$

where σ is a positive constant.

5. Assume that the velocity $\mathbf{v} = (v_1(\mathbf{x},t), v_2(\mathbf{x},t), v_3(\mathbf{x},t))$ over a bounded domain Ω satisfies the conditions

$$\text{div}(\mathbf{v}) = \frac{\partial v_1}{\partial x_1} + \frac{\partial v_2}{\partial x_2} + \frac{\partial v_3}{\partial x_3} = 0, \quad \mathbf{v}|_\Gamma = 0.$$

Show that the equilibrium 0 of the convection-diffusion equation

$$\frac{\partial u}{\partial t} + \mathbf{v} \cdot \nabla u = \mu \nabla^2 u \quad \text{in } \Omega,$$
$$u|_\Gamma = 0,$$
$$u(\mathbf{x},0) = u_0(\mathbf{x})$$

is exponentially L^2-stable. Is it exponentially H^1-stable?

4.2 Interior Feedback Stabilization

Consider the interior control problem

$$\frac{\partial u}{\partial t} = \mu \nabla^2 u + \nabla \cdot (u\mathbf{v}) + au + \mathbf{b}(\mathbf{x}) \cdot \phi(t) \quad \text{in } \Omega, \tag{4.20}$$

$$u|_\Gamma = 0, \tag{4.21}$$

$$u(\mathbf{x},0) = u_0(\mathbf{x}), \tag{4.22}$$

where $\phi(t) = (\phi_1(t), \cdots, \phi_m(t))$ is a control input vector, $\mathbf{b}(\mathbf{x}) = (b_1(\mathbf{x}), \cdots b_m(\mathbf{x}))$ is a given vector which prescribes how a control action is distributed over the domain Ω, and $\mathbf{b}(\mathbf{x}) \cdot \phi(t) = \sum_{i=1}^m b_i(\mathbf{x})\phi_i(t)$. Typical functions b_i are the characteristic functions χ_{ω_i} ($\chi_{\omega_i}(\mathbf{x}) = 1$ if $\mathbf{x} \in \omega_i$ and $\chi_{\omega_i}(\mathbf{x}) = 0$ if $\mathbf{x} \notin \omega_i$), where ω_i ($i = 1, 2, \cdots, m$) are subsets of Ω. In this control problem, the state variable is u.

The output of the system can be proposed in many ways in accordance with specific physical problems. Concentration of a chemical or a room temperature is usually desired to be homogenized. Then the controlled output is the concentration

$$o_c(u) = u(\mathbf{x},t). \tag{4.23}$$

If the average concentration on different subsets ω_i^o ($i = 1, 2, \cdots, l$) can be measured, then the measured output is

$$\mathbf{o}_m(u) = \left(\int_{\omega_1^o} h_1(\mathbf{x}) u(\mathbf{x}, t) dV, \cdots, \int_{\omega_l^o} h_l(\mathbf{x}) u(\mathbf{x}, t) dV \right), \qquad (4.24)$$

where h_i ($i = 1, 2, \cdots, l$) are given weight functions.

This control problem could simulate the regulation of room temperature. The controller $\sum_{i=1}^m b_i(\mathbf{x}) \phi_i(t)$ represents m different heaters that are placed at different locations of the room and the measured output $\mathbf{o}_m$ represents l different thermo-sensors that are placed at different locations.

In Section 4.6, we will show how to formulate this control problem into a control system similar to the finite dimensional system (3.1)-(3.2).

Let the reference output $o_r(\mathbf{x}) = r(\mathbf{x})$. Since $u|_\Gamma = 0$, r should satisfy $r|_\Gamma = 0$. If the controls ϕ_i regulate u to r, then ϕ_i and u will converge to $\bar{\phi}_i$ and r, respectively, and u_t will converge to zero. Thus the control steady states $\bar{\phi}_i$ satisfy

$$\mu \nabla^2 r + \nabla \cdot (r\mathbf{v}) + ar + \sum_{i=1}^m b_i(\mathbf{x}) \bar{\phi}_i = 0.$$

Introducing new variables

$$w = u - r, \quad \psi_i = \phi_i - \bar{\phi}_i, \quad \eta_c = o_c - o_r = u - r = w,$$

and

$$\eta_m = \mathbf{o}_m(u) - \mathbf{o}_m(r) = \left(\int_{\omega_1^o} h_1(\mathbf{x}) w(\mathbf{x}, t) dV, \cdots, \int_{\omega_l^o} h_l(\mathbf{x}) w(\mathbf{x}, t) dV \right),$$

we then transform the control problem to

$$\frac{\partial w}{\partial t} = \mu \nabla^2 w + \nabla \cdot (w\mathbf{v}) + aw + \sum_{i=1}^m b_i(\mathbf{x}) \psi_i(t),$$

$$\eta_c = w(x, t),$$

$$\eta_m = \left(\int_{\omega_1^o} h_1(\mathbf{x}) w(\mathbf{x}, t) dV, \cdots, \int_{\omega_l^o} h_l(\mathbf{x}) w(\mathbf{x}, t) dV \right)$$

with the zero reference output. Therefore, in what follows, we assume that the reference output is zero.

The control problem is to design a feedback controller ϕ to regulate the controlled output o_c to zero, that is, the solution u of the problem (4.20)-(4.22) converges to zero in some function space. Note that zero is the equilibrium of the problem (4.20)-(4.22) with zero steady-state control. Hence, as in the case of finite dimensional systems, the control problem is to stabilize the equilibrium.

4.2.1 State Feedback Controls

We assume that the state u on the whole domain Ω is available for feedback and design a controller of the form

$$\phi = \mathbf{F}(u), \tag{4.25}$$

where $\mathbf{F}$ is a linear operator from $L^2(\Omega)$ to $\mathbb{R}^m$, called a *feedback operator*.

Definition 4.3. If there is a state feedback control $\phi = \mathbf{F}(u)$ such that the closed-loop system (4.20)-(4.22) is exponentially stable, we say that the reaction-convection-diffusion equation is exponentially stabilizable by interior feedback.

We first consider the case where $\mathbf{v} = 0$. In this case, the eigenfunctions of (2.11)-(2.12) are orthogonal and then we can use them to decompose the reaction-diffusion equation into a finite dimensional control system and an infinite dimensional system.

Let $\lambda_1 > \lambda_2 > \cdots > \lambda_n > \cdots$ with $\lim_{n \to \infty} \lambda_n = -\infty$ be the eigenvalues of (2.11)-(2.12) and let $\psi_{i1}, \psi_{i2}, \cdots, \psi_{in_i}$ be n_i normalized (by normalized we mean that $\int_\Omega |\psi_{ij}|^2 dV = 1$) orthogonal eigenfunctions corresponding to the eigenvalue λ_i. Consider the eigenfunction expansion of u

$$u = \sum_{i=1}^{\infty} \sum_{j=1}^{n_i} c_{ij}(t) \psi_{ij}(\mathbf{x}), \tag{4.26}$$

$$u_0 = \sum_{i=1}^{\infty} \sum_{j=1}^{n_i} c_{ij}^0 \psi_{ij}(\mathbf{x}), \tag{4.27}$$

where

$$c_{ij}(t) = \int_\Omega u(\mathbf{x}, t) \psi_{ij}(\mathbf{x}) dV, \tag{4.28}$$

$$c_{ij}^0 = \int_\Omega u_0(\mathbf{x}) \psi_{ij}(\mathbf{x}) dV. \tag{4.29}$$

Substituting (4.26) into (4.20), we obtain (noting that we are assuming $\mathbf{v} = 0$)

$$\sum_{i=1}^{\infty} \sum_{j=1}^{n_i} \frac{dc_{ij}(t)}{dt} \psi_{ij}(\mathbf{x}) = \mu \sum_{i=1}^{\infty} \sum_{j=1}^{n_i} c_{ij}(t) \nabla^2 \psi_{ij}(\mathbf{x}) + \sum_{i=1}^{\infty} \sum_{j=1}^{n_i} c_{ij}(t) a(\mathbf{x}) \psi_{ij}(\mathbf{x})$$

$$+ \sum_{p=1}^{m} b_p(\mathbf{x}) \phi_p(t)$$

$$= \sum_{i=1}^{\infty} \sum_{j=1}^{n_i} c_{ij}(t) \lambda_i \psi_{ij}(\mathbf{x}) + \sum_{p=1}^{m} b_p(\mathbf{x}) \phi_p(t).$$

Multiplying the equation by $\psi_{ij}(\mathbf{x})$ and integrating over Ω, we obtain

$$\frac{dc_{ij}}{dt} = \lambda_i c_{ij} + \sum_{p=1}^{m} b_{ijp} \phi_p(t), \tag{4.30}$$

$$c_{ij}(0) = c_{ij}^0, \quad i = 1, 2, \cdots; j = 1, 2, \cdots, n_i, \tag{4.31}$$

where

$$b_{ijp} = \int_\Omega \psi_{ij}(\mathbf{x}) b_p(\mathbf{x}) dV.$$

Denote

$$\mathbf{c}_i = \begin{bmatrix} c_{i1} \\ c_{i2} \\ \vdots \\ c_{in_i} \end{bmatrix}, \quad \mathbf{c}_f = \begin{bmatrix} \mathbf{c}_1 \\ \mathbf{c}_2 \\ \vdots \\ \mathbf{c}_N \end{bmatrix}, \quad \mathbf{c}_\infty = \begin{bmatrix} \mathbf{c}_{N+1} \\ \mathbf{c}_{N+2} \\ \vdots \end{bmatrix},$$

$$\mathbf{c}_i^0 = \begin{bmatrix} c_{i1}^0 \\ c_{i1}^0 \\ \vdots \\ c_{in_i}^0 \end{bmatrix}, \quad \mathbf{c}_f^0 = \begin{bmatrix} \mathbf{c}_1^0 \\ \mathbf{c}_2^0 \\ \vdots \\ \mathbf{c}_N^0 \end{bmatrix}, \quad \mathbf{c}_\infty^0 = \begin{bmatrix} \mathbf{c}_{N+1}^0 \\ \mathbf{c}_{N+2}^0 \\ \vdots \end{bmatrix},$$

$$\Lambda_i = \begin{bmatrix} \lambda_i & 0 & \cdots & 0 \\ 0 & \lambda_i & \cdots & 0 \\ \vdots & \vdots & \ddots & \vdots \\ 0 & 0 & \cdots & \lambda_i \end{bmatrix} \quad (n_i \times n_i \text{ matrix}),$$

$$\Lambda_f = \begin{bmatrix} \Lambda_1 & 0 & \cdots & 0 \\ 0 & \Lambda_2 & \cdots & 0 \\ \vdots & \vdots & \ddots & \vdots \\ 0 & 0 & \cdots & \Lambda_N \end{bmatrix}, \quad \Lambda_\infty = \begin{bmatrix} \Lambda_{N+1} & 0 & 0 & \cdots \\ 0 & \Lambda_{N+2} & 0 & \cdots \\ 0 & 0 & \Lambda_{N+3} & \cdots \\ \vdots & \vdots & \vdots & \ddots \end{bmatrix},$$

$$\mathbf{B}_i = \begin{bmatrix} b_{i11} & b_{i12} & \cdots & b_{i1m} \\ b_{i21} & b_{i22} & \cdots & b_{i2m} \\ \vdots & \vdots & \ddots & \vdots \\ b_{in_i1} & b_{in_i2} & \cdots & b_{in_im} \end{bmatrix}, \quad \mathbf{B}_f = \begin{bmatrix} \mathbf{B}_1 \\ \mathbf{B}_2 \\ \vdots \\ \mathbf{B}_N \end{bmatrix}, \quad \mathbf{B}_\infty = \begin{bmatrix} \mathbf{B}_{N+1} \\ \mathbf{B}_{N+2} \\ \mathbf{B}_{N+3} \\ \vdots \end{bmatrix},$$

and

$$\phi = \begin{bmatrix} \phi_1 \\ \phi_2 \\ \vdots \\ \phi_m \end{bmatrix}.$$

Then (4.30) can be decomposed into a finite dimensional control system

$$\frac{d\mathbf{c}_f}{dt} = \Lambda_f \mathbf{c}_f + \mathbf{B}_f \phi, \tag{4.32}$$

$$\mathbf{c}_f(0) = \mathbf{c}_f^0, \tag{4.33}$$

and an infinite dimensional control system

$$\frac{d\mathbf{c}_\infty}{dt} = \Lambda_\infty \mathbf{c}_\infty + \mathbf{B}_\infty \phi, \tag{4.34}$$

$$\mathbf{c}_\infty(0) = \mathbf{c}_\infty^0. \tag{4.35}$$

If N is large enough, the eigenvalues in Λ_∞ are negative. So the system (4.34) is exponentially stable. Thus the original control problem (4.20) is reduced to the stabilization of the finite dimensional system (4.32).

Lemma 4.2. *If* $\text{Rank}(\mathbf{B}_i) = n_i$ *for* $i = 1, 2, \cdots, N$, *then the pair* $(\Lambda_f, \mathbf{B}_f)$ *is controllable.*

Proof. By PHB test (Theorem 3.5), it suffices to show that

$$\text{Rank}[\lambda \mathbf{I} - \Lambda_f \ \ \mathbf{B}_f] = \sum_{i=1}^{N} n_i$$

for any $\lambda \in \mathbb{C}$. This is true if $\lambda \neq \lambda_1, \cdots \lambda_N$. Let $\lambda = \lambda_i$, $1 \leq i \leq N$. Since $\text{Rank}(\mathbf{B}_i) = n_i$, there exists an $n_i \times n_i$ sub-matrix $\mathbf{b}_i$ of $\mathbf{B}_i$ such that $\det(\mathbf{b}_i) \neq 0$. Let $\mathbf{P}$ denote the $\sum_{i=1}^{N} n_i \times \sum_{i=1}^{N} n_i$ matrix $\lambda \mathbf{I} - \Lambda_f$ with $\lambda_i \mathbf{I} - \Lambda_i$ replaced by $\mathbf{b}_i$, that is,

$$\mathbf{P} = \begin{bmatrix} \lambda_i \mathbf{I} - \Lambda_1 & 0 & \cdots & 0 & 0 & 0 & \cdots & 0 \\ 0 & \lambda_i \mathbf{I} - \Lambda_2 & \cdots & 0 & 0 & 0 & \cdots & 0 \\ \vdots & \vdots & \ddots & \vdots & \vdots & \vdots & \vdots & \vdots \\ 0 & 0 & \cdots & \lambda_i \mathbf{I} - \Lambda_{i-1} & 0 & 0 & \cdots & 0 \\ 0 & 0 & \cdots & 0 & \mathbf{b}_i & 0 & \cdots & 0 \\ 0 & 0 & \cdots & 0 & 0 & \lambda_i \mathbf{I} - \Lambda_{i+1} & \cdots & 0 \\ 0 & 0 & \cdots & 0 & 0 & 0 & \ddots & 0 \\ 0 & 0 & \cdots & 0 & 0 & 0 & \cdots & \lambda_i \mathbf{I} - \Lambda_N \end{bmatrix}.$$

It is easy to see that

$$\det(\mathbf{P}) = (\lambda_i - \lambda_1)^{n_1} \cdots (\lambda_i - \lambda_{i-1})^{n_{i-1}} \det(\mathbf{b}_i)(\lambda_i - \lambda_{i+1})^{n_{i+1}} \cdots (\lambda_i - \lambda_N)^{n_N} \neq 0$$

and so $\text{Rank}[\lambda \mathbf{I} - \Lambda_f \ \ \mathbf{B}_f] = \text{Rank}(\mathbf{P}) = \sum_{i=1}^{N} n_i$.

Using the design method of feedback control for finite dimensional control systems (Theorem 3.8), we can now derive a state feedback control law for the reaction-diffusion equation (4.20).

Theorem 4.6. *Assume that* $\mathbf{v} = 0$. *Let* $\lambda_1 > \lambda_2 > \cdots > \lambda_n > \cdots$ *with* $\lim_{n \to \infty} \lambda_n = -\infty$ *be the eigenvalues of (2.11)-(2.12) and let* $\psi_{i1}, \psi_{i2}, \cdots, \psi_{in_i}$ *be* n_i *normalized orthogonal eigenfunctions corresponding to the eigenvalue* λ_i. *Let* $\gamma > 0$ *be a given number and* N *an integer such that* $\lambda_N \geq -\gamma > \lambda_{N+1}$. *If* $\text{Rank}(\mathbf{B}_i) = n_i$ *for* $n = 1, 2, \cdots, N$, *there exists a state feedback control*

$$\phi_p(t) = F_p(u(t)) = \sum_{i=1}^{N} \sum_{j=1}^{n_i} K_{p,q(i,j)} \int_{\Omega} \psi_{ij}(\mathbf{x}) u(\mathbf{x},t) dV, \quad p = 1, \cdots, m, \quad (4.36)$$

such that the solution of (4.20) satisfies

$$\int_{\Omega} |u(\mathbf{x},t)|^2 dV \leq M e^{-2\gamma t} \int_{\Omega} |u_0(\mathbf{x})|^2 dV, \quad (4.37)$$

where $q(i,j) = \sum_{r=1}^{i-1} n_r + j$, $\mathbf{K} = (K_{pq})$ is an $m \times \sum_{i=1}^{N} n_i$ feedback gain matrix and M is a positive constant.

Proof. Let $L = \sum_{i=1}^{N} n_i$. For any numbers $\mu_1, \mu_2, \cdots, \mu_L < -\gamma$, it follows from Theorem 3.8 and Lemma 4.2 that there exists a feedback gain matrix $\mathbf{K} = (K_{pq})$ such that the eigenvalues of $\Lambda_f - \mathbf{B}_f \mathbf{K}$ are equal to $\mu_1, \mu_2, \cdots, \mu_L$. This means that the solution of the finite dimensional system (4.32) with the state feedback control

$$\phi = -\mathbf{K} \mathbf{c}_f \quad (4.38)$$

satisfies

$$\sum_{i=1}^{N} \sum_{j=1}^{n_i} c_{ij}^2(t) \leq M \sum_{i=1}^{N} \sum_{j=1}^{n_i} c_{ij}^2(0) e^{-2\gamma t}, \quad (4.39)$$

where M is a positive constant that may change from line to line. It then follows from (4.38) that

$$\sum_{p=1}^{m} \phi_p^2(t) \leq M \sum_{i=1}^{N} \sum_{j=1}^{n_i} c_{ij}^2(t) \leq M \sum_{i=1}^{N} \sum_{j=1}^{n_i} c_{ij}^2(0) e^{-2\gamma t}. \quad (4.40)$$

To estimate $\mathbf{c}_\infty$, we solve (4.34) to obtain

$$c_{ij}(t) = e^{\lambda_i t} c_{ij}(0) + \sum_{p=1}^{m} b_{ijp} \int_0^t e^{\lambda_i(t-s)} \phi_p(s) ds.$$

It then follows from (4.40) that

$$\left(\sum_{i=N+1}^{\infty} \sum_{j=1}^{n_i} c_{ij}^2(t) \right)^{1/2} \leq \left(\sum_{i=N+1}^{\infty} \sum_{j=1}^{n_i} e^{2\lambda_i t} c_{ij}^2(0) \right)^{1/2} \quad (4.41)$$

$$+ \left(\sum_{i=N+1}^{\infty} \sum_{j=1}^{n_i} \left(\sum_{p=1}^{m} b_{ijp} \int_0^t e^{\lambda_i(t-s)} \phi_p(s) ds \right)^2 \right)^{1/2}$$

$$\leq \left(\sum_{i=N+1}^{\infty} \sum_{j=1}^{n_i} e^{2\lambda_i t} c_{ij}^2(0) \right)^{1/2} + M \sum_{i=1}^{N} \sum_{j=1}^{n_i} c_{ij}^2(0) e^{-\gamma t}.$$

Since

$$\int_\Omega |u(\mathbf{x},t)|^2 dV = \sum_{i=1}^\infty \sum_{j=1}^{n_i} c_{ij}^2(t),$$

$$\int_\Omega |u_0(\mathbf{x})|^2 dV = \sum_{i=1}^\infty \sum_{j=1}^{n_i} c_{ij}^2(0),$$

(4.37) follows from (4.39) and (4.41). Since

$$c_{ij}(t) = \int_\Omega \psi_{ij}(\mathbf{x})u(\mathbf{x},t)dV,$$

the state feedback control (4.36) follows from (4.38).

Example 4.2. We set $\mathbf{v} = 0$, $\mu = 0.5, a = 3\mu\pi^2, b(x,y) = 1, u_0(x,y) = 5xy(1-x)(1-y)$ and the domain $\Omega = (0,1) \times (0,1)$. In this case, the eigenvalue problem (2.11)-(2.12) has a positive eigenvalue $\lambda_1 = a - 2\mu\pi^2 = \mu\pi^2$ with the eigenfunction $\psi_{11} = 2\sin\pi x\sin\pi y$. To verify the conditions of Theorem 4.6, we calculate

$$\begin{aligned} b_{111} &= \int_0^1\int_0^1 \psi_{11}(x,y)b(x,y)dxdy \\ &= 2\int_0^1\int_0^1 \sin\pi x\sin\pi ydxdy \\ &= \frac{8}{\pi^2}. \end{aligned}$$

Thus the conditions of Theorem 4.6 are satisfied. The finite dimensional system (4.32) becomes

$$\frac{dc}{dt} = \mu\pi^2 c + \frac{8}{\pi^2}\phi(t).$$

Therefore the feedback control

$$\phi(t) = -Kc = -K\int_0^1\int_0^1 u(x,y,t)\sin\pi x\sin\pi ydxdy, \quad K > \frac{\mu\pi^4}{8} \qquad (4.42)$$

exponentially stabilizes the equation:

$$\begin{aligned} \frac{\partial u}{\partial t} &= \frac{1}{2}\nabla^2 u + \frac{3}{2}\pi^2 u - K\int_0^1\int_0^1 u(x,y,t)\sin\pi x\sin\pi ydxdy, \\ u(0,y) &= u(1,y) = u(x,0) = u(x,1) = 0. \end{aligned}$$

Example 4.3. Point control. In Example 4.2, we take $b(x,y)$ to be the Dirac-delta function

$$b(x,y) = \delta(x-0.5,y-0.5).$$

Such a control acting at one point is called a *point control*. By the definition of the Dirac-delta function, we have

$$\int_0^1 \int_0^1 f(x,y)\delta(x-0.5,y-0.5)dxdy = f(0.5,0.5)$$

for any continuous functions f. Thus

$$\begin{aligned}
b_{111} &= \int_0^1 \int_0^1 \psi_{11}(x,y)\delta(x-0.5,y-0.5)dxdy \\
&= 2\int_0^1 \int_0^1 \sin\pi x \sin\pi y\delta(x-0.5,y-0.5)dxdy \\
&= 2\sin(\pi/2)\sin(\pi/2) \\
&= 2
\end{aligned}$$

and then the conditions of Theorem 4.6 are satisfied. The finite dimensional system (4.32) becomes

$$\frac{dc}{dt} = \mu\pi^2 c + 2\phi(t).$$

Then the feedback control

$$\phi(t) = -Kc = -K\int_0^1 \int_0^1 u(x,y,t)\sin\pi x \sin\pi y dxdy, \ K > \frac{\mu\pi^2}{2} \qquad (4.43)$$

exponentially stabilizes the equation:

$$\frac{\partial u}{\partial t} = \frac{1}{2}\nabla^2 u + \frac{3}{2}\pi^2 u - K\delta(x-0.5,y-0.5)\int_0^1 \int_0^1 u(x,y,t)\sin\pi x \sin\pi y dxdy,$$
$$u(0,y) = u(1,y) = u(x,0) = u(x,1) = 0.$$

We now consider the case where $\mathbf{v} \neq 0$. Readers who are not familiar with semigroup theory can skip this case. By Theorem 2.7, we have the decomposition:

$$L^2(\Omega) = H_N \oplus H_\infty, \qquad (4.44)$$

where H_N is a finite dimensional space containing all eigenfunctions corresponding to the eigenvalues $\lambda_1, \lambda_2, \cdots, \lambda_N$ and H_∞ is an infinite dimensional space containing all eigenfunctions corresponding to the eigenvalues $\lambda_{N+1}, \lambda_{N+2}, \cdots$. Moreover, H_N and H_∞ are invariant under the operator A defined by (4.9), that is, $A(H_N) \subset H_N$ and $A(H_\infty \cap D(A)) \subset H_\infty$. Therefore we can still decompose the reaction-convection-diffusion equation into a finite dimensional control system and an infinite dimensional system.

Usually, the space $L^2(\Omega)$ is decomposed in the way such that H_N contains all eigenfunctions whose eigenvalues may have positive real parts and H_∞ contains all eigenfunctions whose eigenvalues have negative real parts. In this case, H_N is called the *unstable (or slow) manifold* and H_∞ is called the *stable (or fast) manifold*.

We define the restriction A_N of A to H_N and the restriction A_∞ of A to H_∞ by

$$A_N u = Au \quad \text{for every } u \in H_N,$$
$$A_\infty u = Au \quad \text{for every } u \in H_\infty \cap D(A).$$

Theorem 4.7. *Let* $\lambda_1, \lambda_2, \cdots$ *be the eigenvalues of the operator* A *defined by* (4.9) *ordered in their real parts:* $\text{Re}(\lambda_1) > \text{Re}(\lambda_2), \cdots$. *Then the restrictions* A_N, A_∞ *are the infinitesimal generators of analytic semigroups* $e^{A_N t}, e^{A_\infty t}$ *on* H_N, *and* H_∞, *respectively. Moreover, for any* $\gamma > \text{Re}(\lambda_{N+1})$, *there exist constants* $M_1, M_2(\gamma)$ *such that*

$$\|e^{A_N t}\| \leq M_1 e^{\text{Re}(\lambda_1)t}, \tag{4.45}$$
$$\|e^{A_\infty t}\| \leq M_2 e^{\gamma t}. \tag{4.46}$$

This theorem is a consequence of Theorem 2.21. For the detailed proof, we refer to [42, p.30, Theorem 1.5.3].

For the direct composition $L^2(\Omega) = H_N \oplus H_\infty$, we define the projection P_N from $L^2(\Omega)$ onto H_N and the projection P_∞ from $L^2(\Omega)$ onto H_∞ by

$$P_N u = u_N, \quad P_\infty u = u_\infty,$$

where $u = u_N + u_\infty$ with $u_N \in H_N$ and $u_\infty \in H_\infty$. Evidently we have $P_\infty = I - P_N$.

Let $u_{0,N} = P_N u_0$ and $u_{0,\infty} = P_\infty u_0$. Applying the projection P_N to (4.20), we obtain a finite dimensional system on H_N

$$\frac{\partial u_N}{\partial t} = \mu \nabla^2 u_N + \nabla \cdot (u_N \mathbf{v}) + a u_N + \sum_{i=1}^m P_N b_i(\mathbf{x}) \phi_i(t), \tag{4.47}$$

$$u_N = 0 \quad \text{on } \Gamma, \tag{4.48}$$
$$u_N(\mathbf{x}, 0) = u_{0,N}(\mathbf{x}), \tag{4.49}$$

and an infinite dimensional system on H_∞

$$\frac{\partial u_\infty}{\partial t} = \mu \nabla^2 u_\infty + \nabla \cdot (u_\infty \mathbf{v}) + a u_\infty + \sum_{i=1}^m P_\infty b_i(\mathbf{x}) \phi_i(t), \tag{4.50}$$

$$u_\infty = 0 \quad \text{on } \Gamma, \tag{4.51}$$
$$u_\infty(\mathbf{x}, 0) = u_{0,\infty}(\mathbf{x}). \tag{4.52}$$

Consider the eigenfunction expansions

$$u_N = \sum_{i=1}^N \sum_{j=1}^{n_i} c_{ij}(t) \psi_{ij}(\mathbf{x}), \tag{4.53}$$

$$u_{0,N} = \sum_{i=1}^N \sum_{j=1}^{n_i} c_{ij}^0 \psi_{ij}(\mathbf{x}), \tag{4.54}$$

$$P_N b_p = \sum_{i=1}^N \sum_{j=1}^{n_i} b_{ijp} \psi_{ij}(\mathbf{x}), \tag{4.55}$$

where ψ_{ij} are eigenfunctions corresponding the eigenvalue λ_i. Substituting these expressions into (4.47) and using the independence of the eigenfunctions, we obtain

$$\frac{dc_{ij}}{dt} = \lambda_i c_{ij} + \sum_{p=1}^{m} b_{ijp}\phi_p(t), \tag{4.56}$$

$$c_{ij}(0) = c_{ij}^0, \quad i = 1, 2, \cdots, N; j = 1, 2, \cdots, n_i. \tag{4.57}$$

This system is the same as (4.32) except that b_{ijp} is defined by (4.55).

To derive a state feedback control for (4.20), we need to solve the system (4.53) for the coefficients c_{ij}. Thus we multiply (4.53) by ψ_{pq} and integrate over Ω to obtain the system

$$\int_{\Omega} u_N(\mathbf{x})\psi_{pq}(\mathbf{x})dV = \sum_{i=1}^{N}\sum_{j=1}^{n_i} c_{ij}(t) \int_{\Omega} \psi_{ij}(\mathbf{x})\psi_{pq}(\mathbf{x})dV, \tag{4.58}$$

where $p = 1, 2, \cdots, N$ and $q = 1, 2, \cdots, n_i$. This system has the solution

$$\mathbf{c}_f = \Psi^{-1}\mathbf{U}_N, \tag{4.59}$$

where the matrix Ψ and the column vector $\mathbf{U}_N$ is defined by

$$\Psi_{r(p,q),r(i,j)} = \int_{\Omega} \psi_{ij}(\mathbf{x})\psi_{pq}(\mathbf{x})dV,$$
$$\Psi = (\Psi_{r(p,q),r(i,j)}),$$
$$U_{N,r(p,q)} = \int_{\Omega} u_N(\mathbf{x})\psi_{pq}(\mathbf{x})dV,$$
$$\mathbf{U}_N = (U_{N,r(p,q)}).$$

with $n_0 = 0$ and $r(i, j) = \sum_{p=1}^{i} n_{p-1} + j$. Note that Ψ is a $\sum_{i=1}^{N} n_i \times \sum_{i=1}^{N} n_i$ matrix and $\mathbf{U}_N$ is a $\sum_{i=1}^{N} n_i$ dimensional column vector.

Theorem 4.8. *Let $\lambda_1, \lambda_2, \cdots$ be the eigenvalues of the operator A defined by (4.9) ordered in their real parts: $\text{Re}(\lambda_1) > \text{Re}(\lambda_2), \cdots$. Let $\psi_{i1}, \psi_{i2}, \cdots, \psi_{in_i}$ be n_i eigenfunctions corresponding to the eigenvalue λ_i. Let $\gamma > 0$ be a given number and N an integer such that $\text{Re}(\lambda_N) \geq -\gamma > \text{Re}(\lambda_{N+1})$. If $\text{Rank}(\mathbf{B}_i) = n_i$ for $n = 1, 2, \cdots, N$, there exists a state feedback control*

$$\phi = -\mathbf{K}\mathbf{c}_f = -\mathbf{K}\Psi^{-1}\mathbf{U}_N \tag{4.60}$$

such that the solution of (4.20) satisfies

$$\|u(t)\|_{L^2} \leq M\|u_0\|_{L^2}e^{-\gamma t}, \tag{4.61}$$

where $\mathbf{K} = (K_{ij})$ is an $m \times \sum_{i=1}^{N} n_i$ feedback gain matrix and M is a positive constant.

Proof. Let $L = \sum_{i=1}^{N} n_i$. For any numbers $\mu_1, \mu_2, \cdots, \mu_L < -\gamma$, it follows from Theorem 3.8 and Lemma 4.2 that there exists a feedback gain matrix $\mathbf{K} = (K_{ij})$ such that the eigenvalues of $\Lambda_f - \mathbf{B}_f\mathbf{K}$ are equal to $\mu_1, \mu_2, \cdots, \mu_L$. This means that there exists a constant $M_1 > 0$ such that the solution of the finite dimensional system (4.56) with the state feedback control (4.60) satisfies

$$\left(\sum_{i=1}^{N}\sum_{j=1}^{n_i} c_{ij}^2(t)\right)^{1/2} \leq M \left(\sum_{i=1}^{N}\sum_{j=1}^{n_i} c_{ij}^2(0)\right)^{1/2} e^{-\gamma t}. \tag{4.62}$$

Here M denotes a generic constant that may change from line to line. It then follows from (4.60) that

$$\|\phi\| \leq M\|u_N(0)\|_{L^2}e^{-\gamma t}. \tag{4.63}$$

The estimate (4.62) is equivalent to

$$\|u_N(t)\|_{L^2} \leq M\|u_N(0)\|_{L^2}e^{-\gamma t}. \tag{4.64}$$

To estimate u_∞, we can use (2.66) to rewrite the infinite system (4.50)-(4.52) as

$$u_\infty = e^{A_\infty t}u_{0,\infty} + \sum_{p=1}^{m}\int_0^t e^{A_\infty(t-s)}P_\infty b_p\phi_p(s)ds.$$

It then follows from (4.46) and (4.63) that for $-\gamma > -\gamma_1 > \operatorname{Re}(\lambda_{N+1})$

$$
\begin{aligned}
\|u_\infty(t)\|_{L^2} &\leq \|u_\infty(0)\|_{L^2}e^{-\gamma_1 t} + \sum_{p=1}^{m}\int_0^t \|e^{A_\infty(t-s)}P_\infty b_p\phi_p(s)\|_{L^2}ds \\
&\leq \|u_\infty(0)\|_{L^2}e^{-\gamma_1 t} + M\|u_N(0)\|_{L^2}\int_0^t e^{-\gamma_1(t-s)}e^{-\gamma s}ds \\
&= \|u_\infty(0)\|_{L^2}e^{-\gamma_1 t} + \frac{M}{\gamma_1 - \gamma}\|u_N(0)\|_{L^2}\left(e^{-\gamma t} - e^{-\gamma_1 t}\right) \\
&\leq \|u_\infty(0)\|_{L^2}e^{-\gamma_1 t} + \frac{M}{\gamma_1 - \gamma}\|u_N(0)\|_{L^2}e^{-\gamma t} \\
&\leq M\|u(0)\|_{L^2}e^{-\gamma t}.
\end{aligned}
\tag{4.65}
$$

Thus (4.61) follows from (4.64) and (4.65).

There is a difficulty in computing the controller (4.60). To compute it, we need to compute the project operator P_N, but the expression (2.22) for P_N is difficult to evaluate.

4.2.2 Output Feedback Controls

In reality, since only finite sensors are placed in the domain, only the output $\mathbf{o}_m(u)$ is available for feedback. To design an output feedback controller, we need to consider the output injection equation

$$\frac{\partial u}{\partial t} = \mu\nabla^2 u + \nabla\cdot(u\mathbf{v}) + au - \mathbf{K}\cdot\mathbf{o}_m(u) \quad \text{in } \Omega, \tag{4.66}$$

$$u|_\Gamma = 0, \tag{4.67}$$

where $\mathbf{K} = \mathbf{K}(\mathbf{x}) = (K_1(\mathbf{x}), \cdots, K_l(\mathbf{x}))$ is an *output injection vector*.

Definition 4.4. If there is an output injection vector $\mathbf{K}$ such that (4.66)-(4.67) is exponentially stable, we say that the reaction-convection-diffusion equation is exponentially detectable by interior output injection.

We show that the reaction-convection-diffusion is exponentially detectable. We look for the output injection vector $\mathbf{K} = (K_1, \cdots, K_l)$ in $(H_N)^l$. Thus we assume that

$$K_p = \sum_{i=1}^{N}\sum_{j=1}^{n_i} k_{ijp}\psi_{ij}, \quad p = 1,\cdots,l. \tag{4.68}$$

Applying the projection P_N to (4.66), we obtain a finite dimensional output injection system on H_N

$$\frac{\partial u_N}{\partial t} = \mu\nabla^2 u_N + \nabla\cdot(u_N\mathbf{v}) + au_N - \mathbf{K}\cdot\mathbf{o}_m(u), \tag{4.69}$$

$$u_N = 0 \quad \text{on } \Gamma, \tag{4.70}$$

and an infinite dimensional system on H_∞

$$\frac{\partial u_\infty}{\partial t} = \mu\nabla^2 u_\infty + \nabla\cdot(u_\infty\mathbf{v}) + au_\infty, \tag{4.71}$$

$$u_\infty = 0 \quad \text{on } \Gamma. \tag{4.72}$$

Substituting the expression (4.53) into (4.69) and using the independence of the eigenfunctions, we obtain

$$\frac{dc_{ij}}{dt} = \lambda_i c_{ij} - \sum_{p=1}^{l} k_{ijp}\left(\sum_{r=1}^{N}\sum_{s=1}^{n_r} c_{rs}h_{prs} + \int_{\omega_p^o} h_p u_\infty dV\right), \tag{4.73}$$

where $i = 1,2,\cdots,N$, $j = 1,2,\cdots,n_i$, and

$$h_{prs} = \int_{\omega_p^o} h_p\psi_{rs}dV. \tag{4.74}$$

Define

$$\mathbf{K}_i = \begin{bmatrix} k_{i11} & k_{i12} & \cdots & k_{i1l} \\ k_{i21} & k_{i22} & \cdots & k_{i2l} \\ \vdots & \vdots & \ddots & \vdots \\ k_{in_i1} & k_{in_i2} & \cdots & k_{in_il} \end{bmatrix}, \quad \mathbf{K}_f = \begin{bmatrix} \mathbf{K}_1 \\ \mathbf{K}_2 \\ \vdots \\ \mathbf{K}_N \end{bmatrix},$$

and

$$\mathbf{H}_i = \begin{bmatrix} h_{1i1} & h_{1i2} & \cdots & h_{1in_i} \\ h_{2i1} & h_{2i2} & \cdots & h_{2in_i} \\ \vdots & \vdots & \ddots & \vdots \\ h_{li1} & h_{li2} & \cdots & h_{lin_i} \end{bmatrix}, \quad \mathbf{H}_f = [\mathbf{H}_1, \mathbf{H}_2 \cdots, \mathbf{H}_N].$$

Theorem 4.9. *Let* $\lambda_1, \lambda_2, \cdots$ *be the eigenvalues of the operator A defined by (4.9) ordered in their real parts:* $\mathrm{Re}(\lambda_1) > \mathrm{Re}(\lambda_2), \cdots$. *Let* $\psi_{i1}, \psi_{i2}, \cdots, \psi_{in_i}$ *be* n_i *eigenfunctions corresponding to the eigenvalue* λ_i. *Let* $\gamma > 0$ *be a given number and N an integer such that* $\mathrm{Re}(\lambda_N) \geq -\gamma > \mathrm{Re}(\lambda_{N+1})$. *If* $\mathrm{Rank}(\mathbf{H}_i) = n_i$ *for* $n = 1, 2, \cdots, N$, *then there exists an output injection vector* $\mathbf{K}(\mathbf{x})$ *in* $(H_N)^l$ *such that the solution of (4.66)-(4.67) satisfies*

$$\|u(t)\|_{L^2} \leq M \|u_0\|_{L^2} e^{-\gamma t}, \tag{4.75}$$

where M is a positive constant.

Proof. By (4.46), we deduce that

$$\|u_\infty(t)\|_{L^2} \leq M \|u(0)\|_{L^2} e^{-\gamma t}. \tag{4.76}$$

Let $L = \sum_{i=1}^N n_i$. For any numbers $\mu_1, \mu_2, \cdots, \mu_L < -\gamma$, it follows from Theorem 3.8 and Lemma 4.2 that there exists a feedback gain matrix $\mathbf{K}_f$ such that the eigenvalues of $\Lambda_f - \mathbf{K}_f \mathbf{H}_f$ are equal to $\mu_1, \mu_2, \cdots, \mu_L$. It therefore follows from Theorem 2.25 that there exists a constant $M > 0$ such that the solution of the finite dimensional system (4.73) satisfies

$$\left(\sum_{i=1}^N \sum_{j=1}^{n_i} c_{ij}^2(t) \right)^{1/2} \leq M \|u(0)\|_{L^2} \left(\sum_{i=1}^N \sum_{j=1}^{n_i} c_{ij}^2(0) \right)^{1/2} e^{-\gamma t}. \tag{4.77}$$

The estimate (4.77) is equivalent to

$$\|u_N(t)\|_{L^2} \leq M \|u(0)\|_{L^2} e^{-\gamma t}. \tag{4.78}$$

Thus (4.75) follows from (4.76) and (4.78).

To design an output feedback controller, we consider the following state observer

$$\frac{\partial w}{\partial t} = \mu \nabla^2 w + \nabla \cdot (w\mathbf{v}) + aw + \sum_{i=1}^m b_i(\mathbf{x}) \phi_i(t)$$

$$+ \mathbf{K} \cdot (\mathbf{o}_m(u) - \mathbf{o}_m(w)) \quad \text{in } \Omega, \tag{4.79}$$

$$w|_\Gamma = 0, \tag{4.80}$$

$$w(\mathbf{x}, 0) = 0. \tag{4.81}$$

We then use the estimate w for feedback and introduce the observer-based output feedback controller

$$\phi = \mathbf{F}(w) = (F_1(w), \cdots, F_m(w)),$$

which leads to the observer-based output feedback control system

$$\frac{\partial u}{\partial t} = \mu \nabla^2 u + \nabla \cdot (u\mathbf{v}) + au + \sum_{i=1}^{m} b_i(\mathbf{x}) F_i(w) \quad \text{in } \Omega, \tag{4.82}$$

$$\frac{\partial w}{\partial t} = \mu \nabla^2 w + \nabla \cdot (w\mathbf{v}) + aw + \sum_{i=1}^{m} b_i(\mathbf{x}) F_i(w)$$

$$+\mathbf{K} \cdot (\mathbf{o}_m(u) - \mathbf{o}_m(w)) \quad \text{in } \Omega, \tag{4.83}$$

$$u|_\Gamma = w|_\Gamma = 0. \tag{4.84}$$

Introducing the error $e = u - w$, we then transform the above system to

$$\frac{\partial u}{\partial t} = \mu \nabla^2 u + \nabla \cdot (u\mathbf{v}) + au + \sum_{i=1}^{m} b_i(\mathbf{x}) F_i(u) - \sum_{i=1}^{m} b_i(\mathbf{x}) F_i(e), \tag{4.85}$$

$$\frac{\partial e}{\partial t} = \mu \nabla^2 e + \nabla \cdot (e\mathbf{v}) + ae - \mathbf{K} \cdot \mathbf{o}_m(e) \quad \text{in } \Omega, \tag{4.86}$$

$$u|_\Gamma = e|_\Gamma = 0. \tag{4.87}$$

Using this error equation and combing Theorems 2.25, 4.8 and 4.9, we can prove the following theorem.

Theorem 4.10. *Let $\lambda_1, \lambda_2, \cdots$ be the eigenvalues of the operator A defined by (4.9) ordered in their real parts: $\mathrm{Re}(\lambda_1) > \mathrm{Re}(\lambda_2), \cdots$. Let $\psi_{i1}, \psi_{i2}, \cdots, \psi_{in_i}$ be n_i eigenfunctions corresponding to the eigenvalue λ_i. Let $\gamma > 0$ be a given number and N an integer such that $\mathrm{Re}(\lambda_N) \geq -\gamma > \mathrm{Re}(\lambda_{N+1})$. If $\mathrm{Rank}(\mathbf{B}_i) = \mathrm{Rank}(\mathbf{H}_i) = n_i$ for $n = 1, 2, \cdots, N$, then there exists an output injection vector $\mathbf{K}$ in $(H_N)^l$ and an output feedback control*

$$\phi = \mathbf{F}(w) \tag{4.88}$$

such that the solution of (4.82)-(4.84) satisfies

$$\|u(t)\|_{L^2} \leq M \|u_0\|_{L^2} e^{-\gamma t}, \tag{4.89}$$

where M is a positive constant .

Example 4.4. Consider the control problem from Example 4.2. For the measured output, we assume that $l = 1$, $\omega_1^o = (0, 1/2) \times (0, 1/2)$, and $h_1 = 1$. From Example 4.2, we know that the eigenvalue problem (2.11)-(2.12) has a positive eigenvalue $\lambda_1 = a - 2\mu\pi^2 = \mu\pi^2$ with the eigenfunction $\psi_{11} = 2\sin \pi x \sin \pi y$. Thus the observer is given by

$$\frac{\partial w}{\partial t} = \frac{1}{2} \nabla^2 w + \frac{3}{2} \pi^2 w + \phi + K(x,y) \int_0^{1/2} \int_0^{1/2} (u - w) dx dy, \tag{4.90}$$

$$w(0, y) = w(1, y) = w(x, 0) = w(x, 1) = 0. \tag{4.91}$$

Using (4.74), we calculate

$$h_{111} = \int_0^{1/2} \int_0^{1/2} \psi_{11}(x,y)h_1(x,y)dxdy$$

$$= 2\int_0^{1/2} \int_0^{1/2} \sin\pi x \sin\pi y dxdy$$

$$= \frac{2}{\pi^2}.$$

Thus the conditions of Theorem 4.10 are satisfied. The finite dimensional system (4.73) becomes

$$\frac{dc}{dt} = \mu\pi^2 c - \frac{2}{\pi^2}kc.$$

Then the injection constant $k > \frac{\mu\pi^4}{2}$ makes this system exponentially detectable. It then follows from (4.68) that the injection function

$$K(x,y) = k\psi_{11} = 2k\sin\pi x \sin\pi y$$

makes the observer (4.90) exponentially stable. Combining with Example 4.2, we obtain the observer-based output feedback control system

$$\frac{\partial u}{\partial t} = \frac{1}{2}\nabla^2 u + \frac{3}{2}\pi^2 u - K\int_0^1 \int_0^1 w(x,y,t)\sin\pi x \sin\pi y dxdy,$$

$$\frac{\partial w}{\partial t} = \frac{1}{2}\nabla^2 w + \frac{3}{2}\pi^2 w - K\int_0^1 \int_0^1 w(x,y,t)\sin\pi x \sin\pi y dxdy$$

$$+ 2k\sin\pi x \sin\pi y \int_0^{1/2} \int_0^{1/2} (u-w)dxdy,$$

$$u(0,y) = u(1,y) = u(x,0) = u(x,1) = 0,$$

$$w(0,y) = w(1,y) = w(x,0) = w(x,1) = 0, \ K > \frac{\mu\pi^4}{8}, \ k > \frac{\mu\pi^4}{2},$$

which is exponentially stable.

The structure of the feedback control in (4.20) can be proposed in different ways. For instance, we can consider the interior control problem of the following structure

$$\frac{\partial u}{\partial t} = \mu\nabla^2 u + \nabla\cdot(u\mathbf{v}) + au + \sum_{i=1}^m \phi_i(\mathbf{x})\int_\Gamma b_i(\mathbf{x})u(\mathbf{x},t)dS,$$

$$u = 0 \quad \text{on } \Gamma,$$

$$u(\mathbf{x},0) = u_0(\mathbf{x}),$$

where $b_i \in L^2(\Gamma)$ and $\phi_i \in L^2(\Omega)$ are functions to be designed to stabilize the problem. In this control problem, the measured output is assumed to be the temperature on the boundary. This problem requires advanced mathematics. For details, we refer to [102].

Exercises 4.2

1. Consider the control problem

$$\frac{\partial u}{\partial t} = \frac{\partial^2 u}{\partial x^2} + 2u + \chi_{[0,\pi/2]}(x)\phi_1(t) + \chi_{[\pi/2,\pi]}(x)\phi_2(t),$$
$$u(0,t) = u(\pi,t) = 0,$$
$$u(x,0) = u_0.$$

   a. Design a state feedback controller to stabilize the equilibrium 0.
   b. Assume that the measured output is given by

$$\mathbf{o}_m = \left(\int_0^\varepsilon u(x,t)dx, \int_{\pi-\varepsilon}^\pi u(x,t)dx \right),$$

   where $0 < \varepsilon < \pi/2$. Design an output feedback controller to stabilize the equilibrium 0.

2. Generalize Theorem 4.6 to the Neumann boundary case:

$$\frac{\partial u}{\partial t} = \mu \nabla^2 u + au + \sum_{i=1}^m b_i(\mathbf{x})\phi_i(t) \quad \text{in } \Omega,$$

$$\left. \frac{\partial u}{\partial \mathbf{n}} \right|_\Gamma = 0,$$

$$u(\mathbf{x},0) = u_0(\mathbf{x}),$$

   where $\phi_1, \cdots, \phi_m$ are control inputs, and $b_1, \cdots b_m$ are given functions.
3. Consider the control problem

$$\frac{\partial u}{\partial t} = \frac{\partial^2 u}{\partial x^2} + 2u + \cos(x)\phi_1(t) + \cos(2x)\phi_2(t),$$
$$\frac{\partial u}{\partial x}(0,t) = \frac{\partial u}{\partial x}(\pi,t) = 0,$$
$$u(x,0) = u_0.$$

   Design a state feedback control to stabilize the equilibrium 0. If the control $\cos(x)\phi_1(t) + \cos(2x)\phi_2(t)$ is changed to $\sin(x)\phi_1(t) + \sin(2x)\phi_2(t)$, is it possible to find a state feedback control to stabilize the equilibrium 0?
4. Consider the control problem

$$\frac{\partial u}{\partial t} = \mu \left(\frac{\partial^2 u}{\partial x^2} + \frac{\partial^2 u}{\partial y^2} \right) - v_1 \frac{\partial u}{\partial x} - v_2 \frac{\partial u}{\partial y}$$
$$+ \mu \left(3\pi^2 + \frac{v_1^2 + v_2^2}{4\mu^2} \right) u + \phi(t),$$

$$u(0,y,t) = u(1,y,t) = u(x,0,t) = u(x,1,t) = 0,$$
$$u(x,y,0) = u_0(x,y),$$

where v_1 and v_2 are constants.

a. Show that if $\phi = 0$, the zero equilibrium of the equation is unstable.
b. Find a feedback control ϕ to exponentially stabilize the zero equilibrium.
c. Assume that the measured output is given by

$$\mathbf{o}_m = \left(\int_0^\varepsilon dy \int_0^\varepsilon u(x,y,t)dx, \int_0^\varepsilon dy \int_{1-\varepsilon}^1 u(x,y,t)dx, \right.$$
$$\left. \int_{1-\varepsilon}^1 dy \int_{1-\varepsilon}^1 u(x,y,t)dx, \int_{1-\varepsilon}^1 dy \int_0^\varepsilon u(x,y,t)dx \right),$$

where $0 < \varepsilon < 1/2$. Design an output feedback controller to stabilize the equilibrium 0.
d. Solve the controlled equation numerically to verify the stability.

5. Prove Theorem 4.10.

4.3 Boundary Feedback Stabilization

Consider the boundary control problem

$$\frac{\partial u}{\partial t} = \mu \nabla^2 u + \nabla \cdot (u\mathbf{v}) + au \quad \text{in } \Omega, \tag{4.92}$$

$$u = \sum_{i=1}^m b_i(\mathbf{x})\phi_i(t) \quad \text{on } \Gamma, \tag{4.93}$$

$$u(\mathbf{x},0) = u_0(\mathbf{x}), \tag{4.94}$$

where $b_1, \cdots b_m$ are given functions which prescribe how control actions are distributed over the boundary. If there is a feedback control $\phi_i = F_i(u)$ such that the closed-loop system (4.92)-(4.94) is exponentially stable, we say that the reaction-convection-diffusion equation is *exponentially stabilizable* by boundary feedback. The boundary control problem is more difficult than the interior control.

4.3.1 State Feedback Controls

4.3.1.1 Eigenfunction expansion

We first consider the case where $\mathbf{v} = 0$. In this case, the eigenfunctions of (2.11)-(2.12) are orthogonal. This enables us to use the eigenfunction expansion (4.26) to decompose the control problem into a finite dimensional control system and an infinite dimensional control system. Since u does not satisfy the homogeneous boundary conditions, the term-by-term differentiations with respect to the spatial variables are not justified, that is,

$$\nabla^2 u \neq \sum_{i=1}^{\infty} \sum_{j=1}^{n_i} c_{ij}(t) \nabla^2 \psi_{ij}(\mathbf{x}).$$

Then we cannot substitute (4.26) into the right hand side of (4.92). However, term-by-term time differentiations are valid. Thus we can substitute (4.26) into the left hand side of (4.92) to obtain (noting that we are assuming $\mathbf{v} = 0$)

$$\sum_{i=1}^{\infty} \sum_{j=1}^{n_i} \frac{dc_{ij}(t)}{dt} \psi_{ij}(\mathbf{x}) = \mu \nabla^2 u + au.$$

Multiplying the equation by ψ_{ij} and integrating over Ω by parts, we obtain

$$\begin{aligned}
\frac{dc_{ij}}{dt} &= \int_{\Omega} \left(\mu \nabla^2 u + au \right) \psi_{ij} dV \\
&= \int_{\Omega} \left(-\mu \nabla u \nabla \psi_{ij} + au\psi_{ij} \right) dV \\
&= -\int_{\Gamma} \mu u \frac{\partial \psi_{ij}}{\partial \mathbf{n}} dS + \int_{\Omega} u \left(\mu \nabla^2 \psi_{ij} + a\psi_{ij} \right) dV \\
&= -\sum_{p=1}^{m} \phi_p(t) \int_{\Gamma} \mu b_p \frac{\partial \psi_{ij}}{\partial \mathbf{n}} dS + \lambda_i \int_{\Omega} u\psi_{ij} dV \\
&= \lambda_i c_{ij} + \sum_{p=1}^{m} b_{ijp}\phi_p(t),
\end{aligned}$$

where

$$b_{ijp} = -\int_{\Gamma} \mu b_p \frac{\partial \psi_{ij}}{\partial \mathbf{n}} dS. \tag{4.95}$$

Thus we obtain

$$\frac{dc_{ij}}{dt} = \lambda_i c_{ij} + \sum_{p=1}^{m} b_{ijp}\phi_p(t), \tag{4.96}$$

$$c_{ij}(0) = c_{ij}^0, \quad i = 1, 2, \cdots; j = 1, 2, \cdots, n_i. \tag{4.97}$$

Since this system is the same as (4.30) except that b_{ijp} is defined by (4.95), we have the same result as in the case of the interior control problem.

Theorem 4.11. *Let $\lambda_1 > \lambda_2 > \cdots > \lambda_n > \cdots$ with $\lim_{n \to \infty} \lambda_n = -\infty$ be the eigenvalues of (2.11)-(2.12) and let $\psi_{i1}, \psi_{i2}, \cdots, \psi_{in_i}$ be n_i normalized orthogonal eigenfunctions corresponding to the eigenvalue λ_i. Let $\gamma > 0$ be a given number and N an integer such that $\lambda_N \geq -\gamma > \lambda_{N+1}$. If $\mathrm{Rank}(\mathbf{B}_i) = n_i$ with b_{ijp} being defined by (4.95) for $n = 1, 2, \cdots, N$, then there exists a state feedback control*

$$\phi_p(t) = \sum_{i=1}^{N} \sum_{j=1}^{n_i} K_{p,q(i,j)} \int_{\Omega} \psi_{ij}(\mathbf{x}) u(\mathbf{x},t) dV, \quad p = 1, \cdots, m. \tag{4.98}$$

such that the solution of (4.92)-(4.93) satisfies

$$\int_{\Omega} |u(\mathbf{x},t)|^2 dV \le M e^{-2\gamma t} \int_{\Omega} |u_0(\mathbf{x})|^2 dV, \tag{4.99}$$

where $q(i,j) = \sum_{r=1}^{i-1} n_r + j$, $\mathbf{K} = (K_{pq})$ is an $m \times \sum_{i=1}^{N} n_i$ feedback gain matrix and M is a positive constant.

Example 4.5. Consider the control problem

$$\frac{\partial u}{\partial t} = \frac{\partial^2 u}{\partial x^2} + 2u, \tag{4.100}$$

$$u(0,t) = \phi(t), \ u(\pi,t) = 2\phi(t). \tag{4.101}$$

The eigenvalues and eigenfunctions related to this problem are

$$\lambda_n = 2 - n, \quad \psi_n(x) = \sqrt{\frac{2}{\pi}} \sin(nx), \quad n = 1, 2, \cdots$$

Using (4.95), we calculate b_{ijp} as follows:

$$b_{111} = 3\sqrt{\frac{2}{\pi}}, \ b_{211} = -2\sqrt{\frac{2}{\pi}}.$$

Then the conditions of Theorem 4.11 are satisfied. The finite dimensional control system is

$$\frac{dc_1}{dt} = c_1 + 3\sqrt{\frac{2}{\pi}}\phi,$$

$$\frac{dc_2}{dt} = -2\sqrt{\frac{2}{\pi}}\phi.$$

Let

$$\phi = -k_1 c_1 - k_2 c_2.$$

Then

$$\frac{dc_1}{dt} = \left(1 - 3k_1\sqrt{\frac{2}{\pi}}\right) c_1 - 3k_2\sqrt{\frac{2}{\pi}}c_2,$$

$$\frac{dc_2}{dt} = 2\sqrt{\frac{2}{\pi}}(k_1 c_1 + k_2 c_2).$$

The characteristic polynomial of the coefficient matrix is

$$\lambda^2 + \left(3k_1\sqrt{\frac{2}{\pi}} - 2k_2\sqrt{\frac{2}{\pi}} - 1\right)\lambda + 2k_2\sqrt{\frac{2}{\pi}} = 0.$$

If

$$k_2 > 0, \quad 3k_1\sqrt{\frac{2}{\pi}} > 2k_2\sqrt{\frac{2}{\pi}} + 1,$$

then the real parts of the eigenvalues are negative. Thus the feedback control

$$\phi = -4c_1 - 1c_2$$

stabilizes the finite dimensional control system, and then the feedback control system

$$\frac{\partial u}{\partial t} = \frac{\partial^2 u}{\partial x^2} + 2u,$$

$$u(0,t) = -4\sqrt{\frac{2}{\pi}}\int_0^\pi \sin(x)u(x,t)dx - \sqrt{\frac{2}{\pi}}\int_0^\pi \sin(2x)u(x,t)dx,$$

$$u(\pi,t) = -8\sqrt{\frac{2}{\pi}}\int_0^\pi \sin(x)u(x,t)dx - 2\sqrt{\frac{2}{\pi}}\int_0^\pi \sin(2x)u(x,t)dx.$$

is exponentially stable.

If the velocity $\mathbf{v} \neq 0$, then the decomposition of the equation is difficult and methods from the inertial manifold theory, such as the Lyapunov-Perron method, are needed. For this complex case, we refer to [20, 54, 55, 76, 77], [97, Chapter 8], and [101]. For some simple velocities, we can transform the reaction-convection-diffusion equation to the reaction-diffusion equation. For example, if the ith component of $\mathbf{v}$ is a function of x_i only, that is, $v_i = v_i(x_i)$, then we can use the following change of variable

$$u = w\prod_{i=1}^n \exp\left(-\int_0^{x_i} \frac{v_i(s)}{2\mu}ds\right)$$

to transform the reaction-convection-diffusion equation (4.92) to the following reaction-diffusion equation

$$\frac{\partial w}{\partial t} = \mu\nabla^2 w + bw,$$

where b is a function that depends on $a, \mathbf{v}$, and μ (details are left as an exercise). Thus, in this case, the control problem can be solved by using the above theorem.

If the velocity $\mathbf{v} = \mathbf{v}(\mathbf{x},t)$ is time-dependent, then we no longer have the eigenvalue problem (2.11)-(2.12) and then it is even more difficult to decompose the reaction-convection-diffusion equations into a finite dimensional system and an infinite dimensional system [76]. In some simple cases, we can design feedback controllers to stabilize the equation by using the method of integral transform introduced next. For details, we refer to [78].

4.3.1.2 Integral transform method

We now introduce the *integral transform method* (also called the *backstepping method* in the literature [48]). This method has been mainly used for one-dimensional equations and there have been difficulties in extending it to higher-dimensional equations. Thus we consider the one-dimensional reaction-diffusion equation

$$u_t = \mu u_{xx} + au \quad \text{in } (0,1) \times (0,\infty), \tag{4.102}$$

$$u(0,t) = 0, \quad u(1,t) = \phi \quad \text{in } (0,\infty), \tag{4.103}$$

$$u(x,0) = u_0(x), \tag{4.104}$$

where ϕ is a control to be designed. The subscripts u_t, u_x denote the derivatives of u with respect to t, x, respectively.

Before we design a feedback control for the problem (4.102)-(4.104), we show that the reaction-diffusion equation

$$w_t = \mu w_{xx} - \lambda w \quad \text{in } (0,1) \times (0,\infty), \tag{4.105}$$

$$w(0,t) = w(1,t) = 0 \quad \text{in } (0,\infty), \tag{4.106}$$

$$w(x,0) = w_0(x) \quad \text{in } (0,1) \tag{4.107}$$

is exponentially stable, where λ is a positive constant. We first improve Poincaré's inequality (2.29) as follows

$$\int_0^L |u|^2 dx \leq \frac{L^2}{\pi^2} \int_0^L |u_x|^2 dx \quad \text{for all } u \in H_0^1(0,L) \tag{4.108}$$

and

$$\int_0^L |u_x|^2 dx \leq \frac{L^2}{\pi^2} \int_0^L |u_{xx}|^2 dx \quad \text{for all } u \in H^2(0,L) \cap H_0^1(0,L). \tag{4.109}$$

In fact, u has the eigenfunction expansion

$$u = \sum_{n=1}^{\infty} a_n \sin \frac{n\pi x}{L}.$$

Differentiating u gives

$$u_x = \sum_{n=1}^{\infty} a_n \frac{n\pi}{L} \cos \frac{n\pi x}{L},$$

$$u_{xx} = -\sum_{n=1}^{\infty} a_n \frac{n^2 \pi^2}{L^2} \sin \frac{n\pi x}{L}.$$

It then follows that

$$\int_0^L |u_x|^2 dx = \sum_{n=1}^\infty a_n^2 \frac{n^2 \pi^2}{L^2} \int_0^L \cos^2 \frac{n\pi x}{L} dx$$

$$= \sum_{n=1}^\infty a_n^2 \frac{n^2 \pi^2}{L^2} \frac{L}{2}$$

$$\geq \frac{\pi^2}{L^2} \sum_{n=1}^\infty a_n^2 \frac{L}{2}$$

$$= \frac{\pi^2}{L^2} \int_0^L |u|^2 dx,$$

and

$$\int_0^L |u_{xx}|^2 dx = \sum_{n=1}^\infty a_n^2 \frac{n^4 \pi^4}{L^4} \int_0^L \sin^2 \frac{n\pi x}{L} dx$$

$$= \sum_{n=1}^\infty a_n^2 \frac{n^4 \pi^4}{L^4} \frac{L}{2}$$

$$\geq \frac{\pi^2}{L^2} \sum_{n=1}^\infty a_n^2 \frac{n^2 \pi^2}{L^2} \frac{L}{2}$$

$$= \frac{\pi^2}{L^2} \int_0^L |u_x|^2 dx.$$

Theorem 4.12. *Assume that $\lambda > 0$ is a constant. Then, for arbitrary initial data $w_0(x) \in H_0^1(0,1)$, the solution of the problem (4.105)-(4.107) satisfies*

$$\|w(t)\|_{H^1} \leq \|w_0\|_{H^1} e^{-(\lambda + \mu \pi^2)t}. \tag{4.110}$$

Proof. Multiplying (4.105) by w, integrating over $(0,1)$ by parts, and using the boundary conditions on w, we deduce that

$$\frac{1}{2} \frac{d}{dt} \int_0^1 |w(x,t)|^2 dx = -\mu \int_0^1 w w_{xx} dx - \lambda \int_0^1 w^2 dx \tag{4.111}$$

$$= -\mu \int_0^1 |w_x(t)|^2 dx - \lambda \int_0^1 w^2 dx \quad \text{(use (4.108))}$$

$$\leq -(\lambda + \mu \pi^2) \int_0^1 w^2 dx.$$

Multiplying (4.105) by w_{xx}, integrating over $(0,1)$, and using the boundary conditions on w, we deduce that

$$\frac{1}{2} \frac{d}{dt} \int_0^1 |w_x(x,t)|^2 dx = \int_0^1 w_x(x,t) w_{tx}(x,t) dx \tag{4.112}$$

$$= -\int_0^1 w_t(x,t) w_{xx}(x,t) dx$$

$$= -\mu \int_0^1 |w_{xx}|^2 dx + \lambda \int_0^1 w w_{xx} dx$$

$$= -\mu \int_0^1 |w_{xx}|^2 dx - \lambda \int_0^1 |w_x|^2 dx \quad \text{(use (4.109))}$$

$$\leq -(\lambda + \mu \pi^2) \int_0^1 |w_x|^2 dx.$$

Adding (4.111) and (4.112) together, we obtain

$$\frac{d}{dt}\left(\|w(t)\|_{H^1}^2\right) \leq -2(\lambda + \mu \pi^2)\|w(t)\|_{H^1}^2.$$

It then follows from Lemma 4.1 that

$$\|w(t)\|_{H^1}^2 \leq \|w(0)\|_{H^1}^2 e^{-2(\lambda+\mu\pi^2)t}.$$

The key idea of finding a boundary feedback control to stabilize the problem (4.102)-(4.104) is to construct an integral transformation

$$w(x,t) = u(x,t) + \int_0^x k(x,y)u(y,t)dy \tag{4.113}$$

to convert the problem (4.102)-(4.104) to the exponentially stable problem (4.105)-(4.107), where the kernel k is to be found. To find equations for the kernel k, we need to plug w into (4.105)-(4.107). So we compute the derivatives of w as follows. Differentiating w with respect to t, we obtain

$$w_t(x,t) = u_t(x,t) + \int_0^x k(x,y)u_t(y,t)dy \quad \text{(use (4.102)}$$

$$= u_t(x,t) + \int_0^x k(x,y)[\mu u_{yy}(y,t) + a(y)u(y,t)]dy$$

(Integration by parts twice)

$$= u_t(x,t) + \mu k(x,x)u_x(x,t) - \mu k(x,0)u_x(0,t)$$
$$\quad - \mu k_y(x,x)u(x,t) + \mu k_y(x,0)u(0,t)$$
$$\quad + \int_0^x [\mu k_{yy}(x,y)u(y,t) + k(x,y)a(y)u(y,t)]dy. \tag{4.114}$$

Differentiating w with respect to x, we obtain

$$w_x(x,t) = u_x(x,t) + k(x,x)u(x,t) + \int_0^x k_x(x,y)u(y,t)dy. \tag{4.115}$$

Differentiating w_x with respect to x, we obtain

$$w_{xx}(x,t) = u_{xx}(x,t) + \frac{d}{dx}(k(x,x))u(x,t) + k(x,x)u_x(x,t)$$
$$\quad + k_x(x,x)u(x,t) + \int_0^x k_{xx}(x,y)u(y,t)dy. \tag{4.116}$$

It then follows from (4.102) that

$$
\begin{aligned}
w_t &- \mu w_{xx} + \lambda w \\
= &\ u_t(x,t) + \mu k(x,x)u_x(x,t) - \mu k(x,0)u_x(0,t) \\
&- \mu k_y(x,x)u(x,t) + \mu k_y(x,0)u(0,t) \\
&+ \int_0^x [\mu k_{yy}(x,y)u(y,t) + k(x,y)a(y)u(y,t)]dy \\
&- \mu u_{xx}(x,t) - \mu \frac{d}{dx}(k(x,x))u(x,t) - \mu k(x,x)u_x(x,t) \\
&- \mu k_x(x,x)u(x,t) - \int_0^x \mu k_{xx}(x,y)u(y,t)dy \\
&+ \lambda u(x,t) + \lambda \int_0^x k(x,y)u(y,t)dy \\
= &\ \left(a(x) - \mu k_x(x,x) - \mu k_y(x,x) - \mu \frac{d}{dx}(k(x,x)) + \lambda \right) u(x,t) \\
&+ \mu k_y(x,0)u(0,t) - \mu k(x,0)u_x(0,t) \\
&+ \int_0^x [\mu k_{yy}(x,y) - \mu k_{xx}(x,y,t) + (a(y)+\lambda)k(x,y,t)]u(y,t)dy. \quad (4.117)
\end{aligned}
$$

To make the right hand side equal to zero, the kernel k has to satisfy

$$
\mu k_{xx}(x,y) - \mu k_{yy}(x,y) = (a(y)+\lambda)k(x,y), \quad 0 \le y \le x \le 1, \quad (4.118)
$$
$$
k(x,0) = 0, \quad 0 \le x \le 1, \quad (4.119)
$$
$$
\mu k_x(x,x) + \mu k_y(x,x) + \mu \frac{d}{dx}(k(x,x)) = a(x)+\lambda, \quad 0 \le x \le 1. \quad (4.120)
$$

By the boundary condition (4.103), we deduce that $w(0,t) = 0$. Since

$$
w(1,t) = \phi(t) + \int_0^1 k(1,y)u(y,t)dy,
$$

we derive the feedback control

$$
u(1,t) = \phi(t) = - \int_0^1 k(1,y)u(y,t)dy \quad (4.121)
$$

so that $w(1,t) = 0$. In summary, we have proved the following lemma.

Lemma 4.3. *If $k(x,y)$ is the solution of problem (4.118)-(4.120) and the feedback control is given by (4.121), then the transformation defined by (4.113) converts problem (4.102)-(4.104) into (4.105)-(4.107).*

To study the existence of a solution of problem (4.118)-(4.120), we need results about the uniform convergence of a sequence of functions. To introduce the concept of uniform convergence, we look at the sequence of functions

$$
f_n(x) = x^n, \quad -\frac{1}{2} \le x \le \frac{1}{2}.
$$

For each fixed $x \in [-\frac{1}{2}, \frac{1}{2}]$, it is clear that

$$\lim_{n \to \infty} f_n(x) = 0 = f(x).$$

We now use the $\varepsilon - N$ definition to prove this limit. For any $\varepsilon > 0$, we consider

$$|f_n(x) - f(x)| = |x|^n < \varepsilon.$$

Since $|x| \leq \frac{1}{2}$, we have

$$|x|^n \leq \left(\frac{1}{2}\right)^n < \varepsilon,$$

and then

$$n > -\frac{\ln \varepsilon}{\ln 2}.$$

Taking the integer N such that $N > -\frac{\ln \varepsilon}{\ln 2}$, we then have that

$$|f_n(x) - f(x)| < \varepsilon$$

for all $x \in [-\frac{1}{2}, \frac{1}{2}]$ and all $n > N$. Note that the important fact that N is *independent* of x.

Let us look at another sequence of functions

$$f_n(x) = x^n, \quad 0 \leq x \leq 1.$$

For each fixed $x \in [0, 1]$, it is clear that

$$\lim_{n \to \infty} f_n(x) = \begin{cases} 0 \text{ if } 0 \leq x < 1, \\ 1 \text{ if } x = 1. \end{cases}$$

Define

$$f(x) = \begin{cases} 0 \text{ if } 0 \leq x < 1, \\ 1 \text{ if } x = 1. \end{cases}$$

For a fixed $x \in (0, 1)$ and any $\varepsilon > 0$, we consider

$$|f_n(x) - f(x)| = |x^n| < \varepsilon.$$

Taking the integer N such that $N > -\frac{\ln \varepsilon}{\ln x}$, we then have that

$$|f_n(x) - f(x)| < \varepsilon$$

for all $n > N$. Note that this time N *depends* on x. Also for this sequence, if $\varepsilon < 1$, there is no integer N such that

$$|f_n(x) - f(x)| < \varepsilon$$

for all $x \in [0, 1]$ and all $n > N$ since

$$\max_{0 \le x \le 1} |f_n(x) - f(x)| = 1.$$

The first convergence is called uniform convergence since N does not depend on x. The uniform convergence has many applications in analysis.

Definition 4.5. The sequence $\{f_n(x)\}$ of functions is said to be convergent uniformly on the interval I to the function $f(x)$ if and only if for any $\varepsilon > 0$ there is an integer N independent of x such that

$$|f_n(x) - f(x)| < \varepsilon$$

for all $x \in I$ and all $n > N$.

The importance of uniform convergence with regard to continuous functions is illustrated in the next theorem.

Theorem 4.13. *Suppose that* $\{f_n(x)\}_{n=1}^{\infty}$ *is a sequence of continuous functions on an interval I and that it converges uniformly to $f(x)$ on I. Then $f(x)$ is continuous on I.*

Proof. For any $\varepsilon > 0$, there is an integer N independent of x such that

$$|f_n(x) - f(x)| < \frac{\varepsilon}{3}$$

for all $x \in I$ and all $n > N$. Let $x_0 \in I$. Since f_{N+1} is continuous on I, there is a $\delta > 0$ such that

$$|f_{N+1}(x) - f_{N+1}(x_0)| < \frac{\varepsilon}{3}$$

for all $x \in I$ such that $|x - x_0| < \delta$. It therefore follows that

$$|f(x) - f(x_0)| \le |f(x) - f_{N+1}(x)| + |f_{N+1}(x) - f_{N+1}(x_0)|$$
$$+ |f_{N+1}(x_0) - f(x_0)|$$
$$\le \frac{\varepsilon}{3} + \frac{\varepsilon}{3} + \frac{\varepsilon}{3} = \varepsilon$$

for all $x \in I$ such that $|x - x_0| < \delta$. Thus f is continuous at x_0, an arbitrary point of I.

The next theorem shows that limit and integration can be exchanged for a uniformly convergent sequence of continuous functions.

Theorem 4.14. *Suppose that* $\{f_n(x)\}_{n=1}^{\infty}$ *is a sequence of continuous functions on a bounded interval I and that it converges uniformly to $f(x)$ on I. Let $a \in I$ and define*

$$F_n = \int_a^x f_n(t) dt.$$

Then $f(x)$ is continuous on I and F_n converges uniformly to the function

$$F(x) = \int_a^x f(t)dt.$$

Proof. By Theorem 4.13, we deduce that f is continuous on I. Let L be the length of I. For any $\varepsilon > 0$, it follows from the uniform convergence of $\{f_n(x)\}_{n=1}^\infty$ that there is an integer N independent of x such that

$$|f_n(x) - f(x)| < \frac{\varepsilon}{L}$$

for all $x \in I$ and all $n > N$. It therefore follows that

$$\begin{aligned}
|F_n(x) - F(x)| &= \left| \int_a^x [f_n(t) - f(t)]dt \right| \\
&\leq \int_a^x |f_n(t) - f(t)|\,dt \\
&\leq \frac{\varepsilon}{L}|x - a| \leq \varepsilon
\end{aligned}$$

for all $x \in I$ and all $n > N$. Hence $\{F_n(x)\}_{n=1}^\infty$ converges uniformly on I.

A counterexample for Theorem 4.14 is

$$f_n(x) = nxe^{-nx^2}.$$

The next theorem shows that limit and differentiation can be exchanged for a uniformly convergent sequence of continuous functions.

Theorem 4.15. *Suppose that $\{f_n(x)\}_{n=1}^\infty$ is a sequence of continuous functions each having one continuous derivative on a bounded open interval I. Suppose that $\{f_n(x)\}_{n=1}^\infty$ converges to $f(x)$ for each x on I and $\{f_n'(x)\}_{n=1}^\infty$ converges uniformly to $g(x)$ on I. Then $g(x)$ is continuous on I and $f'(x) = g(x)$ on I.*

Proof. By Theorem 4.13, we deduce that g is continuous on I. Let a be any point of I. We then have

$$\int_a^x f_n'(t)dt = f_n(x) - f_n(a).$$

Since $\{f_n'(x)\}_{n=1}^\infty$ converges uniformly to $g(x)$ on I and $\{f_n(x)\}_{n=1}^\infty$ converges to $f(x)$ for each x on I, it follows from Theorem 4.14 that

$$\int_a^x g(t)dt = f(x) - f(a).$$

Differentiating this equation gives $f'(x) = g(x)$ on I.

The concept of uniform convergence can be extended to series. Let $f_k(x), k = 1, 2, \cdots$, be functions defined on an interval I, and $s_n(x) = \sum_{k=1}^{n} f_k(x)$ be the partial sum.

Definition 4.6. The infinite series $\sum_{k=1}^{\infty} f_k(x)$ is said to converge uniformly on I to a function $s(x)$ if and only if the sequence of partial sums $\{s_n(x)\}$ converges uniformly to $s(x)$ on I.

According to this definition, Theorems on uniform convergence of infinite series can be reduced to corresponding results for the uniform convergence of sequences. The next theorem is the analog to Theorem 4.13.

Theorem 4.16. *Suppose that $f_k(x), k = 1, 2, \cdots$, are continuous functions on an interval I and that $\sum_{k=1}^{\infty} f_k(x)$ converges uniformly to $s(x)$ on I. Then $s(x)$ is continuous on I.*

The next theorem is the analog to Theorem 4.14, which shows that a uniformly convergent series may be integrated term-by-term.

Theorem 4.17. *Suppose that $f_k(x), k = 1, 2, \cdots$, are integrable functions on a bounded interval I and that $\sum_{k=1}^{\infty} f_k(x)$ converges uniformly to $s(x)$ on I. Let $a \in I$. Then*

$$\int_a^x s(t)dt = \int_a^x \sum_{k=1}^{\infty} f_k(t)dt = \sum_{k=1}^{\infty} \int_a^x f_k(t)dt,$$

and $\sum_{k=1}^{\infty} \int_a^x f_k(t)dt$ converges uniformly to $\int_a^x s(t)dt$ on I.

The next theorem is the analog to Theorem 4.15, which shows that a uniformly convergent series may be differentiated term-by-term.

Theorem 4.18. *Suppose that $f_k(x), f_k'(x), k = 1, 2, \cdots$, are continuous functions on a bounded open interval I. Suppose that $\sum_{k=1}^{\infty} f_k(x)$ converges to $s(x)$ for each x on I and $\sum_{k=1}^{\infty} f_k'(x)$ converges uniformly to $u(x)$ on I. Then $s'(x) = u(x)$ on I.*

The following theorem gives a useful indirect test for uniform convergence. It is important to observe that the test can be applied without any knowledge of the sum of the series.

Theorem 4.19. *(Weierstrass M-test)* *Let* $f_k(x), k = 1, 2, \cdots,$ *be functions on a bounded open interval I. Suppose that*

$$|f_k(x)| \le M_k \quad \text{for all } k \text{ and all } x \in I.$$

Suppose that the constant series $\sum_{k=1}^{\infty} M_k$ *converges. Then* $\sum_{k=1}^{\infty} f_k(x)$ *and* $\sum_{k=1}^{\infty} |f_k(x)|$ *converge uniformly on I.*

Proof. Since $\sum_{k=1}^{\infty} M_k$ converges, for any $\varepsilon > 0$, there is an integer N such that

$$\sum_{k=n+1}^{\infty} M_k < \varepsilon$$

for all $n > N$. It then follows that

$$
\begin{aligned}
\left| \sum_{k=1}^{n} f_k(x) - \sum_{k=1}^{\infty} f_k(x) \right| &= \left| \sum_{k=n+1}^{\infty} f_k(x) \right| \\
&\le \sum_{k=n+1}^{\infty} |f_k(x)| \\
&= \left| \sum_{k=1}^{n} |f_k(x)| - \sum_{k=1}^{\infty} |f_k(x)| \right| \\
&\le \sum_{k=n+1}^{\infty} M_k \\
&\le \varepsilon
\end{aligned}
$$

for all $n > N$. Hence $\sum_{k=1}^{\infty} f_k(x)$ and $\sum_{k=1}^{\infty} |f_k(x)|$ converge uniformly on I. $\blacksquare$

All results on the uniform convergence of sequences or series of functions defined an interval can be readily generalized to functions defined on a set of $\mathbb{R}^n$.

The existence of a solution of problem (4.118)-(4.120) can be proved by transforming it to an integral equation using the variable change

$$\xi = x + y, \quad \eta = x - y.$$

Lemma 4.4. *Suppose that* $a \in C^1[0,1]$. *Then problem (4.118)-(4.120) has a unique solution which is twice continuously differentiable in* $0 \le y \le x$.

Proof. We introduce new variables

$$\xi = x + y, \quad \eta = x - y$$

and denote

$$G(\xi,\eta) = k(x,y) = k\left(\frac{\xi+\eta}{2}, \frac{\xi-\eta}{2}\right).$$

To derive equations in terms of these new variables, we compute

$$k_x(x,y) = G_\xi + G_\eta,$$
$$k_y(x,y) = G_\xi - G_\eta,$$
$$k_{xx}(x,y) = G_{\xi\xi} + 2G_{\xi\eta} + G_{\eta\eta},$$
$$k_{yy}(x,y) = G_{\xi\xi} - 2G_{\xi\eta} + G_{\eta\eta},$$
$$k_x(x,x) = G_\xi(\xi,0) + G_\eta(\xi,0), \quad (\text{since } y = x, \eta = 0)$$
$$k_y(x,x) = G_\xi(\xi,0) - G_\eta(\xi,0),$$
$$\frac{d}{dx}(k(x,x)) = \frac{d}{dx}(G(\xi,0))$$
$$= G_\xi(\xi,0)\frac{d\xi}{dx} \quad (\text{since } y = x, \xi = 2x)$$
$$= 2G_\xi(\xi,0).$$

Substituting these derivatives into (4.118)-(4.120), we obtain

$$G_{\xi\eta}(\xi,\eta) = \frac{1}{4\mu}\left(a\left(\frac{\xi-\eta}{2}\right) + \lambda\right)G(\xi,\eta), \quad 0 \le \eta \le \xi \le 2, \quad (4.122)$$

$$G(\xi,\xi) = 0, \quad 0 \le \xi \le 2, \tag{4.123}$$

$$G_\xi(\xi,0) = \frac{1}{4\mu}\left(a\left(\frac{\xi}{2}\right) + \lambda\right), \quad 0 \le \xi \le 2. \tag{4.124}$$

Integrating twice gives the following integral equation

$$G(\xi,\eta) = \frac{1}{4\mu}\int_\eta^\xi \left(a\left(\frac{\tau}{2}\right) + \lambda\right)d\tau$$
$$+ \frac{1}{4\mu}\int_\eta^\xi\int_0^\eta \left(a\left(\frac{\tau-s}{2}\right) + \lambda\right)G(\tau,s)ds d\tau. \tag{4.125}$$

We now use the method of successive approximations to show that this equation has a unique continuous solution. Set

$$G_0(\xi,\eta) = \frac{1}{4\mu}\int_\eta^\xi \left(a\left(\frac{\tau}{2}\right) + \lambda\right)d\tau, \tag{4.126}$$

$$G_n(\xi,\eta) = \frac{1}{4\mu}\int_\eta^\xi\int_0^\eta \left(a\left(\frac{\tau-s}{2}\right) + \lambda\right)G_{n-1}(\tau,s)ds d\tau \tag{4.127}$$

and denote $M = \sup\limits_{0 \le x \le 1} \dfrac{1}{\mu} |a(x) + \lambda|$. Then one can readily show that

$$|G_0(\xi, \eta)| \le \frac{1}{4} M(\xi - \eta) \le M,$$

$$|G_1(\xi, \eta)| \le M^2 \xi \eta,$$

$$|G_2(\xi, \eta)| \le \frac{M^3}{(2!)^2} \xi^2 \eta^2,$$

and by induction,

$$|G_n(\xi, \eta)| \le \frac{M^{n+1}}{(n!)^2} \xi^n \eta^n.$$

It then follows from Weierstrass M-test (Theorem 4.19) that the series

$$G(\xi, \eta) = \sum_{n=0}^{\infty} G_n(\xi, \eta)$$

converges absolutely and uniformly in $0 \le \eta \le \xi \le 2$. Furthermore, by Theorem 4.17, we deduce that

$$
\begin{aligned}
G(\xi, \eta) &= \sum_{n=0}^{\infty} G_n(\xi, \eta) \\
&= \frac{1}{4\mu} \int_{\eta}^{\xi} \left(a\left(\frac{\tau}{2}\right) + \lambda \right) d\tau \\
&\quad + \sum_{n=1}^{\infty} \frac{1}{4\mu} \int_{\eta}^{\xi} \int_{0}^{\eta} \left(a\left(\frac{\tau - s}{2}\right) + \lambda \right) G_{n-1}(\tau, s) ds d\tau \\
&= \frac{1}{4\mu} \int_{\eta}^{\xi} \left(a\left(\frac{\tau}{2}\right) + \lambda \right) d\tau \\
&\quad + \frac{1}{4\mu} \int_{\eta}^{\xi} \int_{0}^{\eta} \left(a\left(\frac{\tau - s}{2}\right) + \lambda \right) \sum_{n=1}^{\infty} G_{n-1}(\tau, s) ds d\tau \\
&= \frac{1}{4\mu} \int_{\eta}^{\xi} \left(a\left(\frac{\tau}{2}\right) + \lambda \right) d\tau \\
&\quad + \frac{1}{4\mu} \int_{\eta}^{\xi} \int_{0}^{\eta} \left(a\left(\frac{\tau - s}{2}\right) + \lambda \right) G(\tau, s) ds d\tau.
\end{aligned}
$$

So $G(\xi, \eta)$ is a continuous solution of equation (4.125).

To prove that the solution is unique, it suffices to prove that the equation

$$G(\xi, \eta) = \frac{1}{4\mu} \int_{\eta}^{\xi} \int_{0}^{\eta} \left(a\left(\frac{\tau - s}{2}\right) + \lambda \right) G(\tau, s) ds d\tau$$

has only zero solution. Thus we estimate $G(\xi, \eta)$ and obtain

$$|G(\xi, \eta)| \leq M\eta|\xi - \eta| \max_{0 \leq \tau \leq 2, 0 \leq s \leq \eta} |G(\tau, s)| \leq 2M\eta \max_{0 \leq \tau \leq 2, 0 \leq s \leq \eta} |G(\tau, s)|.$$

For $\eta \leq \frac{1}{4M}$, we derive that

$$\frac{1}{2} \max_{0 \leq \xi \leq 2, 0 \leq \eta \leq 1/(4M)} |G(\xi, \eta)| \leq 0.$$

Therefore it follows that $G(\xi, \eta) = 0$ for $0 \leq \xi \leq 2$, $0 \leq \eta \leq 1/(4M)$ and then

$$G(\xi, \eta) = \frac{1}{4\mu} \int_\eta^\xi \int_{1/(4M)}^\eta \left(a\left(\frac{\tau - s}{2}\right) + \lambda \right) G(\tau, s) ds d\tau.$$

Repeating the above procedure, we can show that $G(\xi, \eta) = 0$ for $0 \leq \xi \leq 2$, $0 \leq \eta \leq 2$.

Moreover, it follows from (4.125) that G is twice continuously differentiable because $a \in C^1[0,1]$. Indeed, differentiating (4.125) with respect to ξ gives

$$\frac{\partial G(\xi, \eta)}{\partial \xi} = \frac{1}{4\mu} \left(a\left(\frac{\xi}{2}\right) + \lambda \right) + \frac{1}{4\mu} \int_0^\eta \left(a\left(\frac{\xi - s}{2}\right) + \lambda \right) G(\xi, s) ds,$$

which implies that $\frac{\partial G(\xi, \eta)}{\partial \xi}$ is continuous since $G(\xi, \eta)$ is continuous. By analogy, we can show that other derivatives of G are continuous.

The proof of Lemma 4.4 provides a numeric computation scheme of successive approximation to compute the kernel function k in our feedback control (4.121). In fact, if a is a constant, then the kernel k can be explicitly computed using the iteration scheme (4.126)-(4.127). In this case we have

$$G_0(\xi, \eta) = \frac{a + \lambda}{4\mu} (\xi - \eta).$$

By induction, we can show that [98]

$$G_n(\xi, \eta) = \left(\frac{a + \lambda}{4\mu}\right)^{n+1} \frac{(\xi - \eta)\xi^n \eta^n}{(n!)^2 (n+1)},$$

and then

$$G(\xi, \eta) = \sum_{n=0}^\infty \left(\frac{a + \lambda}{4\mu}\right)^{n+1} \frac{(\xi - \eta)\xi^n \eta^n}{(n!)^2 (n+1)}. \tag{4.128}$$

Thus we obtain the kernel

$$k(x, y) = G(x+y, x-y) = \sum_{n=0}^\infty \left(\frac{a + \lambda}{4\mu}\right)^{n+1} \frac{2y(x^2 - y^2)^n}{(n!)^2 (n+1)}. \tag{4.129}$$

Lemma 4.5. *Let $k(x,y)$ be the solution of problem (4.118)-(4.120) and define the linear bounded operator $K : H^i(0,1) \to H^i(0,1)$ $(i = 0,1,2)$ by*

$$w(x) = (Ku)(x) = u(x) + \int_0^x k(x,y)u(y)dy, \quad \text{for } u \in H^i(0,1). \tag{4.130}$$

Then K has a linear bounded inverse $K^{-1} : H^i(0,1) \to H^i(0,1)$ $(i = 0,1,2)$.

Proof. To prove that (4.130) has a bounded inverse, we set

$$v(x) = \int_0^x k(x,y)u(y)dy$$

and then

$$w(x) = u(x) + v(x).$$

Hence we have

$$
\begin{aligned}
v(x) &= \int_0^x k(x,y)u(y)dy \quad (u(y) = w(y) - v(y)) \\
&= \int_0^x k(x,y)[w(y) - v(y)]dy \\
&= \int_0^x k(x,y)w(y)dy - \int_0^x k(x,y)v(y)dy. \tag{4.131}
\end{aligned}
$$

To show that this equation has a unique continuous solution, we set

$$v_0(x) = \int_0^x k(x,y)w(y)dy,$$

$$v_n(x) = -\int_0^x k(x,y)v_{n-1}(y)dy$$

and denote $M = \sup\limits_{0 \le y \le x \le 1} |k(x,y)|$. Then,

$$|v_0(x)| \le M \int_0^x |w(y)|dy \le M\sqrt{\int_0^1 dy}\sqrt{\int_0^1 |w(y)|^2 dy} \le M\|w\|_{L^2},$$

$$|v_1(x)| \le M^2\|w\|_{L^2}x,$$

$$|v_2(x)| \le \frac{M^3\|w\|_{L^2}}{2!}x^2,$$

and by induction,

$$|v_n(x)| \le \frac{M^{n+1}\|w\|_{L^2}}{n!}x^n. \tag{4.132}$$

It then follows from Weierstrass-M-test (Theorem 4.19) that the series

$$v(x) = \sum_{n=0}^\infty v_n(x)$$

converges absolutely and uniformly in $0 \leq x \leq 1$. As in the proof of Lemma 4.4, we can show that the sum is a unique continuous solution of equation (4.131).

Moreover, it follows from (4.132) that

$$\max_{0 \leq x \leq 1} |v(x)| \leq \max_{0 \leq x \leq 1} \sum_{n=0}^{\infty} |v_n(x)| \leq \|w\|_{L^2} \sum_{n=0}^{\infty} \frac{M^{n+1}}{n!}.$$

Hence there exists a constant $C > 0$ such that

$$\|v\|_{L^2} \leq C \|w\|_{L^2}. \tag{4.133}$$

This implies that there exists a bounded linear operator $\Phi : L^2(0,1) \to L^2(0,1)$ such that

$$v(x) = (\Phi w)(x)$$

and then

$$u(x) = w(x) - v(x) = ((I - \Phi)w)(x) = (K^{-1}w)(x). \tag{4.134}$$

It is clear that $K^{-1} : L^2(0,1) \to L^2(0,1)$ is bounded. To show that $K^{-1} : H^1(0,1) \to H^1(0,1)$ is bounded, we take derivative in (4.131) and obtain

$$v_x(x) = k(x,x)w(x) + \int_0^x k_x(x,y)w(y)dy - k(x,x)v(x) - \int_0^x k_x(x,y)v(y)dy,$$

which, combined with (4.133), implies that there exists constant $C > 0$ such that

$$\|v_x\|_{L^2} \leq C \|w\|_{L^2}.$$

Then by (4.134)

$$\|u\|_{H^1} \leq \|w\|_{H^1} + \|v\|_{H^1} \leq C \|w\|_{H^1}.$$

By analogy, we can show that $K^{-1} : H^2(0,1) \to H^2(0,1)$ is bounded.

We now prove that the controlled problem

$$u_t(x,t) = \mu u_{xx}(x,t) + a(x)u(x,t) \quad \text{in } (0,1) \times (0,\infty), \tag{4.135}$$

$$u(0,t) = 0, \quad u(1,t) = -\int_0^1 k(1,y)u(y,t)dy \quad \text{in } (0,\infty), \tag{4.136}$$

$$u(x,0) = u_0(x) \tag{4.137}$$

is exponentially stable.

Theorem 4.20. *Assume that $\lambda > 0$ is any positive constant and $a \in C^1[0,1]$ is any function. For arbitrary initial data $u_0(x) \in H^1(0,1)$ satisfying the compatible condition*

$$u_0(0) = 0, \quad u_0(1) = -\int_0^1 k(1,y)u_0(y)dy,$$

the solutions of problem (4.135)-(4.137) satisfy

$$\|u(t)\|_{H^1} \leq M\|u_0\|_{H^1} e^{-(\lambda+\mu\pi^2)t}, \quad \forall t > 0, \tag{4.138}$$

where M is a positive constant independent of u_0.

Proof. By Lemma 4.5, the problem (4.135)-(4.137) can be transformed to the problem (4.105)-(4.107) via the isomorphism defined by (4.130). Hence there exists a positive constant $C > 0$ such that

$$\|u(t)\|_{H^1} \leq C\|w(t)\|_{H^1},$$
$$\|w_0\|_{H^1} \leq C\|u_0\|_{H^1}.$$

Then (4.138) follows from Theorem 4.12.

The method of integral transform can be applied to the problem of flux control

$$u_t = \mu u_{xx} + au \quad \text{in } (0,1) \times (0,\infty), \tag{4.139}$$

$$u_x(0,t) = 0, \quad \mu u_x(1,t) = \phi \quad \text{in } (0,\infty), \tag{4.140}$$

$$u(x,0) = u_0(x). \tag{4.141}$$

For details, we refer to [75].

4.3.2 Output Feedback Controls

We use the same measured output $\mathbf{o}_m(u)$ defined by (4.24) for feedback. To design an output feedback controller, we consider the following state observer

$$\frac{\partial w}{\partial t} = \mu \nabla^2 w + aw + \mathbf{K} \cdot (\mathbf{o}_m(u) - \mathbf{o}_m(w)) \quad \text{in } \Omega, \tag{4.142}$$

$$w|_\Gamma = \sum_{i=1}^m b_i(\mathbf{x})\phi_i(t), \tag{4.143}$$

$$w(\mathbf{x},0) = 0, \tag{4.144}$$

where $\mathbf{K}$ is an output injection vector. We then use the estimate w for feedback and introduce the observer-based output feedback controller

$$\phi = \mathbf{F}(w) = (F_1(w), \cdots, F_m(w)),$$

which leads to the observer-based output feedback control system

$$\frac{\partial u}{\partial t} = \mu \nabla^2 u + au \quad \text{in } \Omega, \tag{4.145}$$

$$\frac{\partial w}{\partial t} = \mu \nabla^2 w + aw + \mathbf{K} \cdot (\mathbf{o}_m(u) - \mathbf{o}_m(w)) \quad \text{in } \Omega, \tag{4.146}$$

$$u|_\Gamma = \sum_{i=1}^m b_i(\mathbf{x})F_i(w), \tag{4.147}$$

$$w|_\Gamma = \sum_{i=1}^{m} b_i(\mathbf{x}) F_i(w). \tag{4.148}$$

Introducing the error $z = u - w$, we then transform the above system to

$$\frac{\partial w}{\partial t} = \mu \nabla^2 w + aw + \mathbf{K} \cdot \mathbf{o}_m(z) \quad \text{in } \Omega, \tag{4.149}$$

$$\frac{\partial z}{\partial t} = \mu \nabla^2 z + az - \mathbf{K} \cdot \mathbf{o}_m(z) \quad \text{in } \Omega, \tag{4.150}$$

$$w|_\Gamma = \sum_{i=1}^{m} b_i(\mathbf{x}) F_i(w), \tag{4.151}$$

$$z|_\Gamma = 0. \tag{4.152}$$

Theorem 4.21. *Let $\lambda_1 > \lambda_2 > \cdots > \lambda_n > \cdots$ with $\lim_{n \to \infty} \lambda_n = -\infty$ be the eigenvalues of (2.11)-(2.12) and let $\psi_{i1}, \psi_{i2}, \cdots, \psi_{in_i}$ be n_i normalized orthogonal eigenfunctions corresponding to the eigenvalue λ_i. Let $\gamma > 0$ be a given number and N an integer such that $\lambda_N \geq -\gamma > \lambda_{N+1}$. If $\text{Rank}(\mathbf{B}_i) = \text{Rank}(\mathbf{H}_i) = n_i$ for $n = 1, 2, \cdots, N$, where $\mathbf{B}_i$ and $\mathbf{H}_i$ are defined in Theorems 4.11 and 4.9, respectively, then there exists an output injection vector $\mathbf{K}$ in $(H_N)^l$ and an output feedback control $\phi = \mathbf{F}(w)$ such that the solution of (4.145)-(4.148) satisfies*

$$\|u(t)\|_{L^2} \leq M \|u_0\|_{L^2} e^{-\gamma t}, \tag{4.153}$$

where M is a positive constant.

Proof. If the inequality (4.153) holds for w and z, then it also holds for u. Thus it suffices to show that (4.153) holds for w and z. By Theorem 4.9, there exists an output injection vector $\mathbf{K}$ such that

$$\|z(t)\|_{L^2} \leq M \|u_0\|_{L^2} e^{-\gamma t}. \tag{4.154}$$

By Theorem 4.11 and its proof (modifying and using (4.96)), we can show that there exists an output feedback control $\phi = \mathbf{F}(w)$ such that (4.153) holds for w. (The details are left as an exercise.)

Example 4.6. Consider the control problem (4.100) from Example 4.5:

$$\frac{\partial u}{\partial t} = \frac{\partial^2 u}{\partial x^2} + 2u, \tag{4.155}$$

$$u(0,t) = \phi(t), \ u(\pi,t) = 2\phi(t). \tag{4.156}$$

For the measured output, we assume that $l = 1$, $\omega_1^o = [0, \pi/2]$, and $h_1(x) = 1$. The eigenvalues and eigenfunctions related to this problem are

$$\lambda_n = 2 - n, \quad \psi_n(x) = \sqrt{\frac{2}{\pi}} \sin(nx), \quad n = 1, 2, \cdots$$

Then the state observer for this control system is given by

$$\frac{\partial w}{\partial t} = \frac{\partial^2 w}{\partial x^2} + 2w + K(x) \int_0^{\pi/2} (u(x,t) - w(x,t))dx, \qquad (4.157)$$

$$w(0,t) = \phi(t), \ w(\pi,t) = 2\phi(t). \qquad (4.158)$$

Using (4.74), we calculate h_{pij} as follows:

$$h_{111} = \int_0^{\pi/2} \sqrt{\frac{2}{\pi}} \sin x dx = \sqrt{\frac{2}{\pi}}, \ h_{121} = \int_0^{\pi/2} \sqrt{\frac{2}{\pi}} \sin 2x dx = \sqrt{\frac{2}{\pi}}.$$

Then the conditions of Theorem 4.21 are satisfied. The finite dimensional control system (4.73) in this case becomes

$$\frac{dc_1}{dt} = \left(1 - k_1 \sqrt{\frac{2}{\pi}}\right) c_1 - k_1 \sqrt{\frac{2}{\pi}} c_2,$$

$$\frac{dc_2}{dt} = -k_2 \sqrt{\frac{2}{\pi}} (c_1 + c_2).$$

The characteristic polynomial of the coefficient matrix is

$$\lambda^2 + \left(k_1 \sqrt{\frac{2}{\pi}} + k_2 \sqrt{\frac{2}{\pi}} - 1\right) \lambda - k_2 \sqrt{\frac{2}{\pi}} = 0.$$

If

$$k_2 < 0, \quad k_1 + k_2 > \sqrt{\frac{\pi}{2}},$$

then the real parts of the eigenvalues are negative. Hence the injection vector $\mathbf{k} = [4, -1]$ makes the finite dimensional system exponentially detectable. It therefore follows from (4.68) that the injection function

$$K(x) = 4\sqrt{\frac{2}{\pi}} \sin x - \sqrt{\frac{2}{\pi}} \sin 2x$$

makes the observer system (4.157) exponentially stable. Combining with Example 4.5, we have designed the following observer-based output feedback control to stabilize (4.155):

$$\frac{\partial u}{\partial t} = \frac{\partial^2 u}{\partial x^2} + 2u,$$

$$\frac{\partial w}{\partial t} = \frac{\partial^2 w}{\partial x^2} + 2w + \sqrt{\frac{2}{\pi}} [4\sin x - \sin 2x]$$

$$\times \int_0^{\pi/2} (u(x,t) - w(x,t))dx,$$

$$u(0,t) = -4\sqrt{\frac{2}{\pi}}\int_0^\pi \sin(x)w(x,t)dx - \sqrt{\frac{2}{\pi}}\int_0^\pi \sin(2x)w(x,t)dx,$$

$$u(\pi,t) = -8\sqrt{\frac{2}{\pi}}\int_0^\pi \sin(x)w(x,t)dx - 2\sqrt{\frac{2}{\pi}}\int_0^\pi \sin(2x)w(x,t)dx,$$

$$w(0,t) = -4\sqrt{\frac{2}{\pi}}\int_0^\pi \sin(x)w(x,t)dx - \sqrt{\frac{2}{\pi}}\int_0^\pi \sin(2x)w(x,t)dx,$$

$$w(\pi,t) = -8\sqrt{\frac{2}{\pi}}\int_0^\pi \sin(x)w(x,t)dx - 2\sqrt{\frac{2}{\pi}}\int_0^\pi \sin(2x)w(x,t)dx.$$

The output can be injected on the boundary as follows

$$\frac{\partial u}{\partial t} = \mu\nabla^2 u + au \quad \text{in } \Omega, \tag{4.159}$$

$$u|_\Gamma = -\mathbf{K}\cdot\mathbf{o}_m(u). \tag{4.160}$$

This problem is difficult and referred to [55].

We now use the method of integral transformation to design output feedback controllers for the one-dimensional reaction-diffusion equation. These controllers were developed by Krstic et al [48]. Consider the control problem

$$u_t = u_{xx} + au, \tag{4.161}$$

$$u_x(0) = 0, \; u(1) = \phi(t). \tag{4.162}$$

We assume that only $u(0)$ can be measured for feedback. As done for the equation (4.102), we can use the integral transformation (4.113) to transform (4.161)-(4.162) to the following exponentially stable equation

$$w_t = w_{xx},$$

$$w_x(0,t) = w(1,t) = 0,$$

where the state feedback control is given by

$$\phi = -\int_0^1 k(1,y)u(y,t)dy$$

and the kernel satisfies

$$k_{xx}(x,y) - k_{yy}(x,y) = a(y)k(x,y), \tag{4.163}$$

$$k_y(x,0) = 0, \; k(0,0) = 0, \tag{4.164}$$

$$\frac{d}{dx}(k(x,x)) = \frac{a(x)}{2}. \tag{4.165}$$

We design the following observer

$$v_t = v_{xx} + av + p_1(x)[u(0) - v(0)], \tag{4.166}$$

$$v_x(0) = p_{10}[u(0) - v(0)], \; v(1) = \phi(t), \tag{4.167}$$

where $p_1(x)$ is an injection function and p_{10} is an injection constant. In this observer, the output is injected both in the domain and at the end point $x = 0$. Introducing the error $z = u - v$, from (4.161), (4.162), (4.166), and (4.167) we derive the following error equation

$$z_t = z_{xx} + az - p_1(x)z(0), \qquad (4.168)$$
$$z_x(0) = -p_{10}z(0), \ z(1) = 0. \qquad (4.169)$$

Our goal is to find p_1 and p_{10} such that the error equation is exponentially stable. To do so, we use the integral transformation

$$z(x) = h(x) - \int_0^x p(x,y)h(y)dy \qquad (4.170)$$

to transform the error equation into the exponentially stable equation

$$h_t = h_{xx}, \qquad (4.171)$$
$$h_x(0) = h(1) = 0. \qquad (4.172)$$

Differentiating the transformation, we obtain

$$
\begin{aligned}
z_t &= h_t - \int_0^x p(x,y)h_{yy}(y)dy \\
&= h_t - p(x,x)h_x(x) + p(x,0)h_x(0) + p_y(x,x)h(x) \\
&\quad - p_y(x,0)h(0) - \int_0^x p_{yy}(x,y)h(y)dy, \qquad (4.173)
\end{aligned}
$$

$$
\begin{aligned}
z_{xx} &= h_{xx} - h(x)\frac{d}{dx}p(x,x) - p(x,x)h_x(x) \\
&\quad - p_x(x,x)h(x) - \int_0^x p_{xx}(x,y)h(y)dy. \qquad (4.174)
\end{aligned}
$$

Subtracting (4.174) from (4.173), we obtain

$$
a\left(h(x) - \int_0^x p(x,y)h(y)dy\right) - p_1(x)h(0)
$$
$$
= 2h(x)\frac{d}{dx}p(x,x) - p_y(x,0)h(0) + \int_0^x (p_{xx}(x,y) - p_{yy}(x,y))h(y)dy.
$$

For the last equality to hold, three conditions must be satisfied

$$p_{xx}(x,y) - p_{yy}(x,y) = a(x)p(x,y),$$
$$\frac{d}{dx}p(x,x) = \frac{a(x)}{2},$$
$$p_y(x,0) = p_1(x).$$

Moreover, the boundary conditions (4.169) and (4.172) imply that p must also satisfy

$$p(0,0) = p_{10}, \quad p(1,y) = 0.$$

In summary, the kernel p satisfies

$$p_{xx}(x,y) - p_{yy}(x,y) = a(x)p(x,y), \tag{4.175}$$

$$\frac{d}{dx}p(x,x) = \frac{a(x)}{2}, \tag{4.176}$$

$$p(1,y) = 0, \tag{4.177}$$

and the injection function $p_1(x)$ and constant p_{10} are given by

$$p_1(x) = p_y(x,0), \quad p_{10} = p(0,0). \tag{4.178}$$

Since the kernel equations (4.175)-(4.177) are the same as the kernel equations (4.118)-(4.120), they can be solved.

Using the estimate v, we now introduce the output feedback controller

$$\phi = - \int_0^1 k(1,y)v(y,t)dy$$

to obtain the following output feedback control system

$$u_t = u_{xx} + au, \tag{4.179}$$

$$v_t = v_{xx} + av + p_1(x)[u(0) - v(0)], \tag{4.180}$$

$$u_x(0) = 0, \quad u(1) = - \int_0^1 k(1,y)v(y,t)dy, \tag{4.181}$$

$$v_x(0) = p_{10}[u(0) - v(0)], \quad v(1) = - \int_0^1 k(1,y)v(y,t)dy. \tag{4.182}$$

Theorem 4.22. *Let the injection function $p_1(x)$ and the injection constant p_{10} be given by (4.178) and let the state kernel $k(x,y)$ be the solution of (4.163)-(4.165). Then the control system (4.179)-(4.182) is L^2-exponentially stable:*

$$\int_0^1 u^2(x,t)dx \le M \int_0^1 u^2(x,0)dxe^{-t/4}, \tag{4.183}$$

where M is a positive constant.

Proof. If we can prove that $\int_0^1 v^2(x,t)dx$ and $\int_0^1 z^2(x,t)dx = \int_0^1 (u(x,t) - v(x,t))^2dx$ converge to zero exponentially as $t \to \infty$, then $\int_0^1 u^2(x,t)dx$ also converges to zero exponentially. Thus it suffices to show the exponential decay of the following system

$$v_t = v_{xx} + av + p_1(x)z(0), \tag{4.184}$$

$$z_t = z_{xx} + az - p_1(x)z(0), \tag{4.185}$$

$$v_x(0) = p_{10}z(0), \ v(1) = -\int_0^1 k(1,y)v(y,t)dy, \qquad (4.186)$$

$$z_x(0) = -p_{10}z(0), \ z(1) = 0. \qquad (4.187)$$

We have shown that the transformation (4.170) transforms equations (4.185) and (4.187) to the following exponentially stable system

$$h_t = h_{xx},$$
$$h_x(0) = h(1) = 0.$$

As done for the equation (4.102), using (4.163)-(4.165), we can also show that the integral transformation

$$q(x) = v(x) + \int_0^x k(x,y)v(y)dy$$

transforms (4.184)-(4.186) into

$$q_t = q_{xx} + h(0)\left(p_1(x) - p_{10}k(x,0) + \int_0^x k(x,y)p_1(y)dy\right),$$
$$q_x(0,t) = p_{10}h(0), \ q(1,t) = 0.$$

To show that h and w converge to zero exponentially, we construct the Lyapunov function

$$V = \frac{A}{2}\int_0^1 h^2(x,t)dx + \frac{1}{2}\int_0^1 q^2(x,t)dx.$$

Differentiating in t gives

$$\frac{dV}{dt} = -A\int_0^1 h_x^2(x,t)dx - p_{10}q(0,t)h(0,t) - \int_0^1 q^2(x,t)dx$$
$$+ h(0)\int_0^1 q(x,t)\left(p_1(x) - p_{10}k(x,0) + \int_0^x k(x,y)p_1(y)dy\right)dx.$$

Using the Poincaré inequality (4.108) and the Young's inequality (2.4), we deduce that

$$-p_{10}q(0,t)h(0,t) \le \frac{1}{4}q^2(0,t) + p_{10}^2 h^2(0,t) \le \frac{1}{4}\int_0^1 q_x^2(x,t)dx + p_{10}^2\int_0^1 h_x^2(x,0)dx$$

and

$$h(0)\int_0^1 q(x,t)\left(p_1(x) - p_{10}k(x,0) + \int_0^x k(x,y)p_1(y)dy\right)dx$$
$$\le \frac{1}{4}\int_0^1 q_x^2(x,t)dx + B\int_0^1 h_x^2(x,0)dx,$$

where $B = \max_{x \in [0,1]} \left(p_1(x) - p_{10}k(x,0) + \int_0^x k(x,y)p_1(y)dy \right)^2$. It then follows that

$$\frac{dV}{dt} = -(A - B - p_{10}^2) \int_0^1 h_x^2(x,t)dx - \frac{1}{2} \int_0^1 q^2(x,t)dx$$

$$\leq -\frac{1}{4}(A - B - p_{10}^2) \int_0^1 h_x^2(x,t)dx - \frac{1}{8} \int_0^1 q^2(x,t)dx.$$

Taking $A = 2(B + p_{10}^2)$, we obtain

$$\frac{dV}{dt} \leq -\frac{1}{4}V.$$

Hence the (h,p)-system is exponentially stable. Since the (h,p)-system is related to the (z,v)-system by the invertible transformations, the (z,v)-system is also exponentially stable.

Exercises 4.3

1. Consider the control problem

$$\frac{\partial u}{\partial t} = \frac{\partial^2 u}{\partial x^2} + 3u,$$
$$u(0,t) = 0,$$
$$u(\pi,t) = \phi(t),$$
$$u(x,0) = u_0.$$

   a. Design a state feedback controller to stabilize the equilibrium 0.
   b. Assume that the measured output is given by

$$\mathbf{o}_m = \left(\int_0^\varepsilon u(x,t)dx, \int_{\pi-\varepsilon}^\pi u(x,t)dx \right),$$

   where $0 < \varepsilon < \pi/2$. Design an output feedback controller to stabilize the equilibrium 0.

2. Generalize Theorem 4.11 to the Neumann boundary case:

$$\frac{\partial u}{\partial t} = \mu \nabla^2 u + au \quad \text{in } \Omega,$$
$$\left. \frac{\partial u}{\partial \mathbf{n}} \right|_\Gamma = \sum_{i=1}^m b_i(\mathbf{x})\phi_i(t),$$
$$u(\mathbf{x},0) = u_0(\mathbf{x}),$$

where $\phi_1, \cdots, \phi_m$ are control inputs, and $b_1, \cdots b_m$ are given functions.

3. Consider the control problem

$$\frac{\partial u}{\partial t} = \frac{\partial^2 u}{\partial x^2} + u,$$

$$\frac{\partial u}{\partial x}(0,t) = 0,$$

$$\frac{\partial u}{\partial x}(\pi,t) = \phi(t),$$

$$u(x,0) = u_0.$$

Design a state feedback control to stabilize the equilibrium 0.

4. Assume that $\mathbf{v} = (v_1(x_1), v_2(x_2), \cdots, v_n(x_n))$. Use the following change of variable

$$u = w \prod_{i=1}^{n} \exp\left(-\int_0^{x_i} \frac{v_i(s)}{2\mu} ds\right)$$

to transform the reaction-convection-diffusion equation (4.92) to the following reaction-diffusion equation

$$\frac{\partial w}{\partial t} = \mu \nabla^2 w + bw,$$

where b is a function that depends on $a, \mathbf{v}$, and μ.

5. Consider the iteration

$$G_0(\xi,\eta) = \xi - \eta$$

$$G_n(\xi,\eta) = \int_\eta^\xi \int_0^\eta G_{n-1}(\tau,s)dsd\tau, \quad n = 1, 2, \cdots$$

Use the mathematical induction to show that

$$G_n(\xi,\eta) = \frac{(\xi-\eta)\xi^n\eta^n}{(n!)^2(n+1)}, \quad n = 1, 2, \cdots$$

6. Consider the iteration

$$G_0(\xi,\eta) = \xi + \eta,$$

$$G_n(\xi,\eta) = 2\int_0^\eta \int_0^\tau G_{n-1}(\tau,s)dsd\tau + \int_\eta^\xi \int_0^\eta G_{n-1}(\tau,s)dsd\tau.$$

Use the mathematical induction to show that

$$G_n(\xi,\eta) = \frac{(\xi+\eta)\xi^n\eta^n}{(n!)^2(n+1)}, \quad n = 1, 2, \cdots$$

7. Consider the initial boundary value problem of the heat equation

$$u_t = \mu u_{xx}, \tag{4.188}$$

$$u_x(0,t) = -bu(0), \quad u(1,t) = 0, \tag{4.189}$$
$$u(x,0) = u_0(x). \tag{4.190}$$

a. Find the range of the constant b for which this system is unstable, that is, this system has a positive eigenvalue.
b. Prove that the equilibrium 0 of the system

$$w_t = \mu w_{xx}, \tag{4.191}$$
$$w_x(0,t) = 0, \quad w(1,t) = 0 \tag{4.192}$$

is exponentially stable.
c. Derive equations for a kernel k and a boundary feedback control at the right end $x = 1$ such that the transformation

$$w = u + \int_0^x k(x,y)u(y)dy$$

converts the system (4.188)-(4.190) to the target system (4.191)-(4.192)

8. Design a flux control for the reaction convection diffusion equation

$$u_t = \mu u_{xx} + au_x + bu,$$
$$u_x(0,t) = 0, \quad \mu u_x(1,t) = \phi,$$
$$u(x,0) = u_0(x),$$

where a, b are constants.

9. Let the kernel k be the solution of (4.163)-(4.165). Show that the integral transform

$$q(x) = v(x) + \int_0^x k(x,y)v(y)dy$$

transforms the control problem

$$v_t = v_{xx} + av + p_1(x)z(0),$$
$$v_x(0) = p_{10}z(0), \quad v(1) = -\int_0^1 k(1,y)v(y,t)dy$$

into

$$q_t = q_{xx} + z(0)\left(p_1(x) - p_{10}k(x,0) + \int_0^x k(x,y)p_1(y)dy\right),$$
$$q_x(0,t) = p_{10}z(0), \quad q(1,t) = 0.$$

10. ([48, Section 5.3]) Consider the control problem

$$u_t = \mu u_{xx} + au,$$
$$u_x(0,t) = 0, \quad u(1,t) = \phi(t),$$

where μ and a are positive constants. Assume that $u_x(1,t)$ is the measured output. Use the following observer

$$v_t = \mu v_{xx} + av + p_1(x)[u_x(1,t) - v_x(1,t)],$$
$$v_x(0,t) = 0, \quad v(1,t) = \phi(t) + p_{10}[u_x(1,t) - v_x(1,t)],$$

to design an output feedback controller to exponentially stabilize the equilibrium 0, where $p_1(x)$ is an injection function and p_{10} is an injection constant to be designed.

4.4 Optimal Interior Control

Optimal control is an important method to design an optimal feedback controller. We consider the optimal control of

$$\frac{\partial u}{\partial t} = \mu \nabla^2 u + \nabla \cdot (u\mathbf{v}) + au + \phi(\mathbf{x},t) \quad \text{in } \Omega, \qquad (4.193)$$

$$u|_\Gamma = 0, \qquad (4.194)$$

$$u(\mathbf{x},s) = u_0(\mathbf{x}), \qquad (4.195)$$

where ϕ is a control input. In many real problems, the homogenization of concentration of a physical quantity such as reactants in a chemical reaction is preferred. Mathematically this means that $u(\mathbf{x},t) = c$ (constant). Without loss of generality, we may assume that $c = 0$. Hence, for $T > s$, we define the quadratic functional of concentration variance by

$$J(\phi; u_0, s) = \int_s^T \int_\Omega \left(\alpha |u(\mathbf{x},t)|^2 + \beta |\phi(\mathbf{x},t)|^2 \right) dV dt$$

$$+ \gamma \int_\Omega |u(\mathbf{x},T)|^2 dV, \qquad (4.196)$$

where $\alpha, \beta, \gamma \geq 0$ are weight constants. If J is minimized at some control ϕ^*, then the concentration is closest to the target 0. Thus the optimal control problem is to minimize the functional J over $L^2(\Omega \times (s,T))$.

The function ϕ^* such that

$$J(\phi^*; u_0, s) = \min_{\phi \in L^2(\Omega \times (s,T))} J(\phi; u_0, s)$$

is called an *optimal control*. Since the state equation (4.193) is linear and the functional (4.196) is quadratic, the above optimal control problem is called a *linear quadratic optimal control problem* (LQ problem).

4.4.1 Existence and Uniqueness

To get an idea of how to prove the existence and uniqueness of an optimal control, we look at the simple function $f(x) = x^2$. This function has a unique minimum at 0 and has the following properties:

(1) f is strictly convex, i.e., for every $x, y \in \mathbb{R}, x \neq y$, and $\lambda \in (0, 1)$

$$f(\lambda x + (1 - \lambda)y) < \lambda f(x) + (1 - \lambda)f(y).$$

(2) f is coercive, i.e., $\lim\limits_{|x| \to \infty} x^2 = \infty.$

This motivates us to introduce the following definition.

Definition 4.7. Let H be a Hilbert space and $F : H \to [-\infty, +\infty]$ be a functional.

(1) F is said to be strictly convex if for every $x, y \in H, x \neq y$, and $\lambda \in (0, 1)$

$$F(\lambda x + (1 - \lambda)y) < \lambda F(x) + (1 - \lambda)F(y).$$

(2) F is said to be coercive if

$$\lim\limits_{\|x\| \to \infty} F(x) = \infty.$$

(3) F is said to be continuous if $\lim\limits_{x \to y} F(x) = F(y)$ for all $y \in H$.

Theorem 4.23. *If F is convex, continuous, and coercive, then the following minimization problem*

$$\inf_{x \in H} F(x)$$

has a unique solution x_m, that is,

$$F(x_m) = \inf_{x \in H} F(x).$$

The proof of this theorem is referred to Theorem 38.C and Proposition 38.15 of [106]. To apply this theorem, we need to estimate the solution of (4.193)-(4.195). We define

$$a_m = \max_{\mathbf{x} \in \Omega} a(\mathbf{x}), \quad v_m = \frac{1}{2} \max_{\mathbf{x} \in \Omega} \operatorname{div}(\mathbf{v}(\mathbf{x})). \tag{4.197}$$

Lemma 4.6. *The solution u of (4.193)-(4.195) satisfies the following estimate*

$$\int_{\Omega} |u(t)|^2 dV + 2\mu \int_s^t \int_{\Omega} |\nabla u|^2 dV dr$$

$$\leq e^{(2a_m + 2v_m + 1)(t-s)} \left(\int_{\Omega} |u_0|^2 dV + \int_s^t \int_{\Omega} \phi^2 dV dr \right). \tag{4.198}$$

Proof. Multiplying (4.193) by u, integrating over Ω by parts, and using the boundary conditions, we obtain

$$\frac{1}{2}\frac{d}{dt}\int_\Omega |u|^2 dV = \int_\Omega \left(\mu u\nabla^2 u + u(\nabla\cdot(u\mathbf{v})) + au^2 + \phi u\right)dV \quad \text{(use (2.26))}$$

$$= \int_\Gamma \mu u\frac{\partial u}{\partial \mathbf{n}}dS - \mu\int_\Omega |\nabla u|^2 dV + \int_\Gamma u^2\mathbf{n}\cdot\mathbf{v}dS$$

$$- \int_\Gamma u\mathbf{v}\cdot\nabla u dV + \int_\Omega au^2 dV + \int_\Omega \phi u dV$$

(use Young's inequality (2.4) for the last integral)

$$\leq -\mu\int_\Omega |\nabla u|^2 dV - \frac{1}{2}\int_\Omega \mathbf{v}\cdot\nabla(u^2)dV$$

$$+ a_m\int_\Omega u^2 dV + \frac{1}{2}\int_\Omega (\phi^2 + u^2)dV$$

$$= -\mu\int_\Omega |\nabla u|^2 dV - \frac{1}{2}\int_\Gamma u^2\mathbf{v}\cdot\mathbf{n}dV$$

$$+ \frac{1}{2}\int_\Omega u^2\mathrm{div}(\mathbf{v})dV + a_m\int_\Omega u^2 dV + \frac{1}{2}\int_\Omega (\phi^2 + u^2)dV$$

$$\leq \left(a_m + v_m + \frac{1}{2}\right)\int_\Omega u^2 dV + \frac{1}{2}\int_\Omega \phi^2 dV. \tag{4.199}$$

Using Gronwall's inequality (4.14), we derive that

$$\int_\Omega |u(t)|^2 dV \leq e^{(2a_m+2v_m+1)(t-s)}\left(\int_\Omega |u_0|^2 dV + \int_s^t\int_\Omega \phi^2 dV dr\right).$$

Integrating (4.199) over $[s,t]$ gives

$$\int_\Omega |u(t)|^2 dV + 2\mu\int_s^t\int_\Omega |\nabla u|^2 dV dr$$

$$\leq \int_\Omega |u_0|^2 dV + \int_s^t\int_\Omega \phi^2 dV dr + (2a_m + 2v_m + 1)\int_s^t\int_\Omega u^2 dV dr$$

$$\leq \int_\Omega |u_0|^2 dV + \int_s^t\int_\Omega \phi^2 dV dr$$

$$+ (2a_m + 2v_m + 1)\int_s^t e^{(2a_m+2v_m+1)(z-s)}\left(\int_\Omega |u_0|^2 dV + \int_s^z\int_\Omega \phi^2 dV dr\right)dz$$

$$\leq \int_\Omega |u_0|^2 dV + \int_s^t\int_\Omega \phi^2 dV dr$$

$$+ (2a_m + 2v_m + 1)\left(\int_\Omega |u_0|^2 dV + \int_s^t\int_\Omega \phi^2 dV dr\right)\int_s^t e^{(2a_m+2v_m+1)(z-s)}dz$$

$$= e^{(2a_m+2v_m+1)(t-s)}\left(\int_\Omega |u_0|^2 dV + \int_s^t\int_\Omega \phi^2 dV dr\right),$$

which is (4.198).

Theorem 4.24. *The functional J defined by (4.196) is strictly convex, coercive, and continuous.*

Proof. (1) J is strictly convex. We denote the solution of (4.193)-(4.195) corresponding to the control ϕ and the initial condition u_0 by $u(\mathbf{x},t;u_0,\phi)$. Since the equation is linear, we deduce that

$$u(\mathbf{x},t;u_0,\lambda\phi+(1-\lambda)\psi) = \lambda u(\mathbf{x},t;u_0,\phi) + (1-\lambda)u(\mathbf{x},t;u_0,\psi).$$

It therefore follows that

$$
\begin{aligned}
&J(\lambda\phi+(1-\lambda)\psi;u_0,s)\\
&= \int_s^T \int_\Omega \left(\alpha|u(\mathbf{x},t;u_0,\lambda\phi+(1-\lambda)\psi)|^2 + \beta|\lambda\phi+(1-\lambda)\psi|^2\right) dV dt\\
&\quad + \gamma \int_\Omega |u(\mathbf{x},T;u_0,\lambda\phi+(1-\lambda)\psi)|^2 dV\\
&= \int_s^T \int_\Omega \left[\alpha(|\lambda u(\mathbf{x},t;u_0,\phi)+(1-\lambda)u(\mathbf{x},t;u_0,\psi)|^2) + \beta|\lambda\phi+(1-\lambda)\psi|^2\right] dV dt\\
&\quad + \gamma \int_\Omega |\lambda u(\mathbf{x},T;u_0,\phi)+(1-\lambda)u(\mathbf{x},T;u_0,\psi)|^2 dV\\
&< \int_s^T \int_\Omega \alpha \left[\lambda|u(\mathbf{x},t;u_0,\phi)|^2 + (1-\lambda)|u(\mathbf{x},t;u_0,\psi)|^2\right] dV dt\\
&\quad + \int_s^T \int_\Omega \beta \left[\lambda\phi^2 + (1-\lambda)\psi^2\right] dV dt\\
&\quad + \gamma \int_\Omega (\lambda|u(\mathbf{x},T;u_0,\phi)|^2 + (1-\lambda)|u(\mathbf{x},T;u_0,\psi)|^2) dV\\
&= \lambda J(\phi;u_0,s) + (1-\lambda)J(\psi;u_0,s)
\end{aligned}
$$

for $\phi \neq \psi$ and $\lambda \in (0,1)$.

(2) J is coercive because

$$
\begin{aligned}
\lim_{\|\phi\|\to\infty} J(\phi;u_0,s) &= \lim_{\|\phi\|\to\infty} \int_s^T \int_\Omega \left(\alpha|u(\mathbf{x},t;u_0,\phi)|^2 + \beta|\phi|^2\right) dV dt\\
&\quad + \gamma \int_\Omega |u(\mathbf{x},T;u_0,\phi)|^2 dV\\
&\geq \lim_{\|\phi\|\to\infty} \int_s^T \int_\Omega \beta|\phi|^2 dV dt\\
&= \infty.
\end{aligned}
$$

(3) The continuity of J follows from (4.198) and the following equation

$$u(\mathbf{x},t;u_0,\phi) - u(\mathbf{x},t;u_0,\psi) = u(\mathbf{x},t;0,\phi-\psi).$$

From Theorems 4.23 and 4.24 we obtain the following existence and uniqueness theorem.

Theorem 4.25. *Let J be defined by (4.196). Then there exists a unique optimal control ϕ^* such that*

$$J(\phi^*;u_0,s) = \min_{\phi \in L^2(\Omega \times (s,T))} J(\phi;u_0,s).$$

4.4.2 Necessary Conditions

To find the minimum of the function $f(x) = x^3 + 3x^2 - 3x + 1$, we need to find the critical points of the function by setting

$$f'(x) = 3x^2 + 6x - 3 = 0.$$

The method of solving the optimal control problem is similar, but the computation of the derivative of J is more complex. We need to generalize the concept of derivative.

Definition 4.8. Let H be a Hilbert space. A functional $F : H \to [-\infty, +\infty]$ is said to be Gâteaux-differentiable at z if there exists $z^* \in H$ such that

$$(z^*, x) = \lim_{\lambda \to 0^+} \frac{F(z + \lambda x) - F(z)}{\lambda} \tag{4.200}$$

for all $x \in Z$. We call z^* the Gâteaux-differential of F at z and denote it by $F'(z)$.

Consider the function $f(x) = x^2$. Since

$$\lim_{\lambda \to 0^+} \frac{(z + \lambda x)^2 - z^2}{\lambda} = \lim_{\lambda \to 0^+} \frac{2\lambda z x + \lambda^2 x^2}{\lambda} = 2zx,$$

we have $f'(z) = 2z$.

Consider a more general functional $F(x) = \|x\|^2$ in a real Hilbert space. Since

$$\lim_{\lambda \to 0^+} \frac{\|z + \lambda x\|^2 - \|z\|^2}{\lambda} = \lim_{\lambda \to 0^+} \frac{2\lambda (z,x) + \lambda^2 \|x\|^2}{\lambda} = (2z,x),$$

we have $F'(z) = 2z$.

Since the equation (4.193) is linear, we can readily see that

$$u(\mathbf{x},t;u_0,\phi + \lambda \psi) = u(\mathbf{x},t;u_0,\phi) + \lambda u(\mathbf{x},t;0,\psi) \tag{4.201}$$

for any control functions ϕ, ψ and any number λ. We now calculate the Gâteaux-differential of J.

Theorem 4.26. *The Gâteaux-differential of J defined by (4.196) at ϕ is given by*

$$\int_s^T \int_\Omega J'(\phi; u_0, s)\psi \, dV \, dt = 2\alpha \int_s^T \int_\Omega u(\mathbf{x}, t; u_0, \phi)u(\mathbf{x}, t; 0, \psi) \, dV \, dt$$

$$+ 2\beta \int_s^T \int_\Omega \phi \psi \, dV \, dt$$

$$+ 2\gamma \int_\Omega u(\mathbf{x}, T; u_0, \phi)u(\mathbf{x}, T; 0, \psi) \, dV, \quad (4.202)$$

where $u(\mathbf{x}, t; 0, \psi)$ is the solution of

$$\frac{\partial u}{\partial t} = \mu \nabla^2 u + \nabla \cdot (u\mathbf{v}) + au + \psi(\mathbf{x}, t), \qquad (4.203)$$

$$u|_\Gamma = 0, \qquad (4.204)$$

$$u(\mathbf{x}, s) = 0. \qquad (4.205)$$

Proof. Using (4.196) and (4.201), we derive that

$$\lim_{\lambda \to 0^+} \frac{J(\phi + \lambda \psi; u_0, s) - J(\phi; u_0, s)}{\lambda}$$

$$= \lim_{\lambda \to 0^+} \frac{\alpha}{\lambda} \int_s^T \int_\Omega \left(|u(\mathbf{x}, t; u_0, \phi + \lambda \psi)|^2 - |u(\mathbf{x}, t; u_0, \phi)|^2 \right) dV \, dt$$

$$+ \lim_{\lambda \to 0^+} \frac{\beta}{\lambda} \int_s^T \int_\Omega \left(|\phi + \lambda \psi|^2 - |\phi|^2 \right) dV \, dt$$

$$+ \lim_{\lambda \to 0^+} \frac{\gamma}{\lambda} \int_\Omega \left(|u(\mathbf{x}, T; u_0, \phi + \lambda \psi)|^2 - |u(\mathbf{x}, T; u_0, \phi)|^2 \right) dV$$

$$= \lim_{\lambda \to 0^+} \frac{\alpha}{\lambda} \int_s^T \int_\Omega \left(2\lambda u(\mathbf{x}, t; u_0, \phi)u(\mathbf{x}, t; 0, \psi) + \lambda^2 |u(\mathbf{x}, t; 0, \psi)|^2 \right) dV \, dt$$

$$+ \lim_{\lambda \to 0^+} \frac{\beta}{\lambda} \int_s^T \int_\Omega \left(2\lambda \phi \psi + \lambda^2 |\psi|^2 \right) dV \, dt$$

$$+ \lim_{\lambda \to 0^+} \frac{\gamma}{\lambda} \int_\Omega \left(2\lambda u(\mathbf{x}, T; u_0, \phi)u(\mathbf{x}, T; 0, \psi) + \lambda^2 |u(\mathbf{x}, T; 0, \psi)|^2 \right) dV$$

$$= 2\alpha \int_s^T \int_\Omega u(\mathbf{x}, t; u_0, \phi)u(\mathbf{x}, t; 0, \psi) \, dV \, dt$$

$$+ 2\beta \int_s^T \int_\Omega \phi \psi \, dV \, dt + 2\gamma \int_\Omega u(\mathbf{x}, T; u_0, \phi)u(\mathbf{x}, T; 0, \psi) \, dV.$$

If a function $f(x)$ has a minimum at a point x_m, then $f'(x_m) = 0$. This is also true for a functional defined in a Hilbert space.

Theorem 4.27. *Let H be a Hilbert space and F be a functional on H. If F has a minimum at y and is Gâteaux-differentiable at y, then*

$$F'(y) = 0.$$

Proof. For any $x \in H$, we have

$$(F'(y), x) = \lim_{\delta \to 0^+} \frac{F(y + \delta x) - F(y)}{\delta} \geq 0,$$

and

$$(F'(y), -x) = \lim_{\delta \to 0^+} \frac{F(y - \delta x) - F(y)}{\delta} \geq 0.$$

Hence $(F'(y), x) = 0$ for any $x \in H$ and then $F'(y) = 0$.

Corollary 4.1. *If the functional J has a minimum at ϕ^*, then*

$$J'(\phi^*; u_0, s) = 0. \tag{4.206}$$

4.4.3 Optimality Systems

For the function $f(x) = x^3 + 3x^2 - 3x + 1$, we have the equation for its critical points c

$$f'(c) = 3c^2 + 6c - 3 = 0.$$

In analogy, we have the governing system for the optimal control.

Theorem 4.28. *The optimal control ϕ^* satisfies the following system*

$$\frac{\partial u}{\partial t} = \mu \nabla^2 u + \nabla \cdot (u\mathbf{v}) + au + \phi^*, \tag{4.207}$$

$$\frac{\partial v}{\partial t} = -\mu \nabla^2 v + \mathbf{v} \cdot \nabla v - av + \alpha u, \tag{4.208}$$

$$\phi^* = \frac{v}{\beta}, \tag{4.209}$$

$$u|_\Gamma = v|_\Gamma = 0, \tag{4.210}$$

$$u(\mathbf{x}, s) = u_0(\mathbf{x}), \; v(\mathbf{x}, T) = -\gamma u(\mathbf{x}, T). \tag{4.211}$$

Proof. Multiplying (4.203) by the solution v of (4.208) and integrating over $\Omega \times (s, T)$ by parts, we have

$$\int_s^T \int_\Omega v \frac{\partial u}{\partial t}(\mathbf{x}, t; 0, \psi) dV dt \tag{4.212}$$

$$= \mu \int_s^T \int_\Omega v \nabla^2 u(\mathbf{x}, t; 0, \psi) dV dt + \int_s^T \int_\Omega v \nabla \cdot (u(\mathbf{x}, t; 0, \psi)\mathbf{v}) dV dt$$

$$+ \int_s^T \int_\Omega a(\mathbf{x}) u(\mathbf{x}, t; 0, \psi) v dV dt + \int_s^T \int_\Omega v \psi dV dt$$

$$= -\mu \int_s^T \int_\Omega \nabla v \cdot \nabla u(\mathbf{x}, t; 0, \psi) dV dt - \int_s^T \int_\Omega u(\mathbf{x}, t; 0, \psi) \mathbf{v} \cdot \nabla v dV dt$$

$$+ \int_s^T \int_\Omega a(\mathbf{x}) u(\mathbf{x}, t; 0, \psi) v dV dt + \int_s^T \int_\Omega v \psi dV dt.$$

Multiplying (4.208) by the solution $u(\mathbf{x}, t; 0, \psi)$ of (4.203) and integrating over $\Omega \times (s, T)$ by parts, we have

$$\int_s^T \int_\Omega u(\mathbf{x}, t; 0, \psi) \frac{\partial v}{\partial t} dV dt \tag{4.213}$$

$$= -\mu \int_s^T \int_\Omega u(\mathbf{x}, t; 0, \psi) \nabla^2 v dV dt + \int_s^T \int_\Omega u(\mathbf{x}, t; 0, \psi) \mathbf{v} \cdot \nabla v dV dt$$

$$- \int_s^T \int_\Omega a(\mathbf{x}) u(\mathbf{x}, t; 0, \psi) v dV dt + \alpha \int_s^T \int_\Omega u(\mathbf{x}, t; 0, \psi) u(\mathbf{x}, t; u_0, \phi^*) dV dt$$

$$= \mu \int_s^T \int_\Omega \nabla u(\mathbf{x}, t; 0, \psi) \cdot \nabla v dV dt + \int_s^T \int_\Omega u(\mathbf{x}, t; 0, \psi) \mathbf{v} \cdot \nabla v dV dt$$

$$- \int_s^T \int_\Omega a(\mathbf{x}) u(\mathbf{x}, t; 0, \psi) v dV dt + \alpha \int_s^T \int_\Omega u(\mathbf{x}, t; 0, \psi) u(\mathbf{x}, t; u_0, \phi^*) dV dt.$$

Adding (4.212) and (4.213) gives

$$\alpha \int_s^T \int_\Omega u(\mathbf{x}, t; 0, \psi) u(\mathbf{x}, t; u_0, \phi^*) dV dt + \gamma \int_\Omega u(\mathbf{x}, T; 0, \psi) u(\mathbf{x}, T; u_0, \phi^*) dV$$

$$= -\int_s^T \int_\Omega v \psi dV dt. \tag{4.214}$$

On the other hand, it follows from Theorem 4.26 and Corollary 4.1 that

$$\alpha \int_s^T \int_\Omega u(\mathbf{x}, t; u_0, \phi^*) u(\mathbf{x}, t; 0, \psi) dV dt + \beta \int_s^T \int_\Omega \phi^* \psi dV dt$$

$$+ \gamma \int_\Omega u(\mathbf{x}, T; u_0, \phi^*) u(\mathbf{x}, T; 0, \psi) dV = 0$$

for all functions $\psi \in L^2(\Omega \times (s, T))$. Substituting (4.214) into this equation, we obtain

$$\int_s^T \int_\Omega (v - \beta \phi^*) \psi dV dt = 0$$

for all functions $\psi \in L^2(\Omega \times (s, T))$. Thus $v = \beta \phi^*$.

The governing system (4.207)-(4.211) is called an *optimality* (also called *Hamiltonian*) *system*. To decouple it, we define a family of linear operators $P(s)$ by

$$P(s) u_0(\mathbf{x}) = -v(\mathbf{x}, s; u_0) \tag{4.215}$$

for any given function $u_0(\mathbf{x}) \in L^2(\Omega)$, where $v(\mathbf{x}, s; u_0)$ is the solution of (4.207)-(4.211).

Theorem 4.29. *The operator $P(s)$ defined by (4.215) has the following properties:*
(1) $P(s)$ is symmetric, that is,

$$\int_\Omega p_0 P(s) u_0 dV = \int_\Omega u_0 P(s) p_0 dV \quad \text{for any } u_0, p_0 \in L^2(\Omega). \tag{4.216}$$

(2) If ϕ^* is the optimal control of (4.193)-(4.195), then

$$\int_\Omega u_0 P(s) u_0 dV = \int_s^T \int_\Omega \left(\alpha |u|^2 + \beta |\phi^*|^2 \right) dV dt + \gamma \int_\Omega |u(T)|^2) dV. \quad (4.217)$$

Hence $P(s)$ is nonnegative.
(3) $P(T) = \gamma I$.
(4) $P(s)$ is a linear bounded operator in $L^2(\Omega)$.

Proof. (1) Consider the system

$$\frac{\partial p}{\partial t} = \mu \nabla^2 p + \nabla \cdot (p\mathbf{v}) + ap + \psi^*, \quad (4.218)$$

$$\frac{\partial q}{\partial t} = -\mu \nabla^2 q + \mathbf{v} \cdot \nabla q - aq + \alpha p, \quad (4.219)$$

$$\psi^* = \frac{q}{\beta}, \quad (4.220)$$

$$p|_\Gamma = q|_\Gamma = 0, \quad (4.221)$$

$$p(\mathbf{x}, s) = p_0(\mathbf{x}), \quad q(\mathbf{x}, T) = -\gamma p(\mathbf{x}, T). \quad (4.222)$$

Multiplying (4.207) by the solution q of (4.219) and integrating over $\Omega \times (s, T)$ by parts, we have

$$\begin{aligned}
\int_s^T \int_\Omega q \frac{\partial u}{\partial t} dV dt &= \mu \int_s^T \int_\Omega q \nabla^2 u dV dt + \int_s^T \int_\Omega q \nabla \cdot (u\mathbf{v}) dV dt \\
&\quad + \int_s^T \int_\Omega auq dV dt + \int_s^T \int_\Omega q\phi^* dV dt \\
&= -\mu \int_s^T \int_\Omega \nabla q \cdot \nabla u dV dt - \int_s^T \int_\Omega u\mathbf{v} \cdot \nabla q dV dt \\
&\quad + \int_s^T \int_\Omega auq dV dt + \int_s^T \int_\Omega q\phi^* dV dt.
\end{aligned} \quad (4.223)$$

Multiplying (4.219) by the solution u of (4.207) and integrating over $\Omega \times (s, T)$ by parts, we have

$$\begin{aligned}
\int_s^T \int_\Omega u \frac{\partial q}{\partial t} dV dt &= -\mu \int_s^T \int_\Omega u \nabla^2 q dV dt + \int_s^T \int_\Omega u\mathbf{v} \cdot \nabla q dV dt \\
&\quad - \int_s^T \int_\Omega auq dV dt + \alpha \int_s^T \int_\Omega up dV dt \\
&= \mu \int_s^T \int_\Omega \nabla q \cdot \nabla u dV dt + \int_s^T \int_\Omega u\mathbf{v} \cdot \nabla q dV d \\
&\quad - \int_s^T \int_\Omega auq dV dt + \alpha \int_s^T \int_\Omega up dV dt.
\end{aligned} \quad (4.224)$$

Adding (4.223) and (4.224) gives

$$\int_\Omega u_0 P(s) p_0 dV = \gamma \int_\Omega u(T) p(T) dV + \int_s^T \int_\Omega (q\phi^* + \alpha up) dV dt$$

$$= \gamma \int_\Omega u(T) p(T) dV + \int_s^T \int_\Omega (\beta \phi^* \psi^* + \alpha up) dV dt. \quad (4.225)$$

Repeating the above procedure by multiplying (4.218) by the solution v of (4.208) and multiplying (4.208) by the solution p of (4.218), we can obtain

$$\int_\Omega p_0 P(s) u_0 dV = \gamma \int_\Omega u(T) p(T) dV + \int_s^T \int_\Omega (v\psi^* + \alpha up) dV dt$$

$$= \gamma \int_\Omega u(T) p(T) dV + \int_s^T \int_\Omega (\beta \phi^* \psi^* + \alpha up) dV dt. \quad (4.226)$$

Then (4.216) follows from (4.225) and (4.226).

(2) If $p_0 = u_0$, then $u = p, v = q$, and $\phi^* = \psi^*$. It then follows from (4.226) that

$$\int_\Omega u_0 P(s) u_0 dV = \gamma \int_\Omega |u(T)|^2) dV + \int_s^T \int_\Omega (\beta |\phi^*|^2 + \alpha |u|^2) dV dt.$$

(3) Letting $s \to T$ in (4.226), we derive that

$$\int_\Omega p_0 P(T) u_0 dV = \gamma \int_\Omega u_0 p_0 dV \quad \text{for all } p_0 \in L^2(\Omega).$$

Thus $P(T) u_0 = \gamma u_0$ for all $u_0 \in L^2(\Omega)$.

(4) By (4.226), we deduce from Hölder's inequality (2.5) and Young's inequality (2.4) that

$$\left| \int_\Omega p_0 P(s) u_0 dV \right|$$

$$\leq \gamma \left(\int_\Omega |u(T)|^2 dV \right)^{1/2} \left(\int_\Omega |p(T)|^2 dV \right)^{1/2}$$

$$+ \beta \left(\int_s^T \int_\Omega |\phi^*|^2 dV dt \right)^{1/2} \left(\int_s^T \int_\Omega |\psi^*|^2 dV dt \right)^{1/2}$$

$$+ \alpha \left(\int_s^T \int_\Omega |u|^2 dV dt \right)^{1/2} \left(\int_s^T \int_\Omega |p|^2 dV dt \right)^{1/2}$$

$$\leq \left(\gamma \int_\Omega |u(T)|^2 dV + \beta \int_s^T \int_\Omega |\phi^*|^2 dV dt + \alpha \int_s^T \int_\Omega |u|^2 dV dt \right)^{1/2}$$

$$\times \left(\gamma \int_\Omega |p(T)|^2 dV + \beta \int_s^T \int_\Omega |\psi^*|^2 dV dt + \alpha \int_s^T \int_\Omega |p|^2 dV dt \right)^{1/2}$$

$$= (J(\phi^*; u_0, s))^{1/2} (J(\psi^*; p_0, s))^{1/2}. \quad (4.227)$$

Moreover, by (4.15), we deduce that there exists a constant $C > 0$ such that

$$J(\phi^*;u_0,s) \le J(0;u_0,s)$$
$$= \gamma \int_\Omega |u(T)|^2 dV + \alpha \int_s^T \int_\Omega |u|^2 dV dt$$
$$\le C \int_\Omega |u_0|^2 dV$$

and

$$J(\psi^*;p_0,s) \le J(0;p_0,s)$$
$$= \gamma \int_\Omega |p(T)|^2 dV + \alpha \int_s^T \int_\Omega |p|^2 dV dt$$
$$\le C \int_\Omega |p_0|^2 dV.$$

It therefore follows from (4.227) that

$$\left| \int_\Omega p_0 P(s) u_0 dV \right| \le C \left(\int_\Omega |u_0|^2 dV \int_\Omega |p_0|^2 dV \right)^{1/2}$$

for all $p_0, u_0 \in L^2(\Omega)$. Taking $p_0 = P(s)u_0$, we obtain

$$\|P(s)u_0\|_{L^2} \le C\|u_0\|_{L^2}.$$

Thus $P(s)$ is bounded. Since the optimality system (4.207)-(4.211) is linear, $P(s)$ is linear.

4.4.4 Optimal Interior State Feedback Controls

To use the state u to express the optimal control ϕ^*, we consider the optimal control problem

$$\frac{\partial u}{\partial t} = \mu \nabla^2 u + \nabla \cdot (u\mathbf{v}) + au + \phi(\mathbf{x},t), \qquad (4.228)$$

$$u|_\Gamma = 0, \qquad (4.229)$$

$$u(\mathbf{x},0) = w_0(\mathbf{x}). \qquad (4.230)$$

Note that the initial time $t = 0$. By Theorem 4.28, we derive that the optimal control ϕ^* satisfies the following system:

$$\frac{\partial u}{\partial t} = \mu \nabla^2 u + \nabla \cdot (u\mathbf{v}) + au + \phi^*, \qquad (4.231)$$

$$\frac{\partial v}{\partial t} = -\mu \nabla^2 v + \mathbf{v} \cdot \nabla v - av + \alpha u, \qquad (4.232)$$

$$\phi^* = \frac{v}{\beta}, \qquad (4.233)$$

$$u|_\Gamma = v|_\Gamma = 0, \tag{4.234}$$

$$u(\mathbf{x},0) = w_0(\mathbf{x}), \; v(\mathbf{x},T) = -\mu u(\mathbf{x},T). \tag{4.235}$$

If we restrict the system (4.231)-(4.235) on $[s,T]$, then the initial condition (4.235) becomes

$$u(\mathbf{x},s) = u(\mathbf{x},s), \; v(\mathbf{x},T) = -\gamma u(\mathbf{x},T). \tag{4.236}$$

If we compare the system (4.207)-(4.211) with the system (4.231)-(4.235) on $[s,T]$, it follows from the uniqueness of the optimal control that

$$u_0(\mathbf{x}) \text{ (in (4.211))} = u(\mathbf{x},s) \text{ (solution of (4.231))},$$

$$v(\mathbf{x},s;u_0) \text{ (solution of (4.208))} = v(\mathbf{x},s;w_0) \text{ (solution of (4.232))}.$$

Substituting these equations into (4.215) gives

$$-v(\mathbf{x},s;w_0) = P(s)u(\mathbf{x},s). \tag{4.237}$$

We then obtain the optimal state feedback control

$$\phi^* = \frac{v(\mathbf{x},s;w_0)}{\beta} = -\frac{1}{\beta}P(s)u(\mathbf{x},s). \tag{4.238}$$

We note that P also depends on T: $P = P(s,T)$. Next we show that $P(s,T)$ converges to an operator Π independent of s as $T \to \infty$ and the above feedback controller with P replaced by Π exponentially stabilizes the reaction-convection-diffusion equation.

We first derive an equation for the operator P by performing a formal computation. Let $u_1,u_2 \in H^2(\Omega) \cap H_0^1(\Omega)$. Multiplying (4.231) and (4.232) by u_1,u_2, respectively, and integrating over Ω, we obtain

$$\int_\Omega \frac{\partial u}{\partial t} u_1 dV = \int_\Omega \left(\mu \nabla^2 u u_1 + u_1 \nabla \cdot (u\mathbf{v}) + auu_1 + \frac{1}{\beta}vu_1 \right) dV, \tag{4.239}$$

$$\int_\Omega \frac{\partial v}{\partial t} u_2 dV = \int_\Omega \left(-\mu \nabla^2 v u_2 + u_2 \mathbf{v} \cdot \nabla v - avu_2 + \alpha uu_2 \right) dV$$

$$= \int_\Omega \left(-\mu v \nabla^2 u_2 - v \nabla \cdot (u_2 \mathbf{v}) - avu_2 + \alpha uu_2 \right) dV. \tag{4.240}$$

Substituting (4.237) into (4.240) gives

$$-\int_\Omega \frac{\partial(Pu)}{\partial t} u_2 dV = \int_\Omega \left(\mu P(u) \nabla^2 u_2 + P(u) \nabla \cdot (u_2 \mathbf{v}) + aP(u)u_2 + \alpha uu_2 \right) dV,$$

and then

$$-\int_\Omega \left(\frac{dP}{dt}(u)u_2 + P\left(\frac{\partial u}{\partial t} \right) u_2 \right) dV$$

$$= \int_\Omega \left(\mu P(u) \nabla^2 u_2 + P(u) \nabla \cdot (u_2 \mathbf{v}) + aP(u)u_2 + \alpha uu_2 \right) dV.$$

It then follows from the symmetry of P that

$$-\int_{\Omega} \left(\frac{dP}{dt}(u)u_2 + \frac{\partial u}{\partial t} P u_2 \right) dV$$

$$= \int_{\Omega} \left(\mu u P \nabla^2 u_2 + u P \nabla \cdot (u_2 \mathbf{v}) + aP(u)u_2 + \alpha u u_2 \right) dV.$$

Substituting (4.239) into this equation for $\int_{\Omega} \frac{\partial u}{\partial t} P(u_2) dV$ gives

$$-\int_{\Omega} \frac{dP}{dt}(u)u_2 dV$$

$$= \int_{\Omega} \left(\mu \nabla^2 u P(u_2) + P(u_2) \nabla \cdot (u\mathbf{v}) + auP(u_2) - \frac{1}{\beta} P(u)P(u_2) \right) dV$$

$$+ \int_{\Omega} \left(\mu u P \nabla^2 u_2 + u P \nabla \cdot (u_2 \mathbf{v}) + aP(u)u_2 + \alpha u u_2 \right) dV.$$

Since the solution u can be arbitrary, using the property (3) of Theorem 4.29 and replacing u with u_1, we then obtain

$$\int_{\Omega} \frac{dP}{dt}(u_1)u_2 dV = -\int_{\Omega} \left(\mu \nabla^2 u_1 P(u_2) + \mu u_1 P \nabla^2 u_2 \right) dV$$

$$- \int_{\Omega} \left(P(u_2) \nabla \cdot (u_1 \mathbf{v}) + u_1 P \nabla \cdot (u_2 \mathbf{v}) \right) dV$$

$$- \int_{\Omega} \left(au_1 P(u_2) + aP(u_1)u_2 + \alpha u_1 u_2 \right) dV,$$

$$+ \int_{\Omega} \frac{1}{\beta} P(u_1)P(u_2) dV, \tag{4.241}$$

$$\int_{\Omega} u_2 P(T)(u_1) dV = \gamma \int_{\Omega} u_2 u_1 dV \tag{4.242}$$

for all $u_1, u_2 \in H^2(\Omega) \cap H_0^1(\Omega)$. This equation is called a *differential Riccati equation*. The rigorous mathematical justification of this equation is beyond the scope of this text and is referred to one of the advanced control books [6, 24, 62, 64, 67, 69, 99].

The Riccati equation is difficult to solve and we look at a simple example adapted from [24, p.284, Example 6.1.10].

Example 4.7. Consider the control problem

$$\frac{\partial u}{\partial t} = \frac{\partial^2 u}{\partial x^2} + \phi(x,t),$$

$$u(0,t) = u(\pi,t) = 0,$$

$$u(x,0) = u_0(x),$$

with the performance functional

$$J(\phi; u_0, 0) = \int_0^T \int_0^\pi (u^2 + \phi^2) \, dx \, dt + \int_0^\pi u(T)^2 \, dx.$$

The eigenvalue problem

$$\frac{d^2 \psi}{dx^2} = \lambda \psi, \quad \psi(0) = \psi(\pi)$$

has eigenfunctions $\psi_n(x) = \sin(nx)$ $(n = 1, 2, \cdots)$. Let

$$P(t) \sin(nx) = \sum_{m=1}^\infty p_{mn}(t) \sin(mx).$$

Taking $u_1 = \sin(nx), u_2 = \sin(mx)$ in the Riccati equation (4.241), we obtain the system of infinite differential equations

$$\frac{dp_{mn}}{dt} = (m^2 + n^2) p_{mn} + \sum_{l=1}^\infty p_{ml} p_{ln} - \delta_{mn}, \qquad (4.243)$$

$$p_{mn}(T) = \delta_{mn}, \qquad (4.244)$$

where

$$\delta_{mn} = \begin{cases} 1 \text{ if } m = n, \\ 0 \text{ if } m \neq n. \end{cases}$$

If $m \neq n$, $p_{mn} \equiv 0$ is a solution. Then the above system is reduced to

$$\frac{dp_{nn}}{dt} = 2n^2 p_{nn} + p_{nn}^2 - 1, \qquad (4.245)$$

$$p_{nn}(T) = 1. \qquad (4.246)$$

Solving the system, we obtain the solution

$$p_{nn}(t) = \frac{a_n(1 + b_n) e^{c_n(t-T)} - b_n(1 + a_n)}{1 + a_n - (1 + b_n) e^{c_n(t-T)}},$$

where

$$a_n = n^2 + \sqrt{n^4 + 1}, \quad b_n = n^2 - \sqrt{n^4 + 1}, \quad c_n = 2\sqrt{n^4 + 1}.$$

Let

$$u(x, t) = \sum_{n=1}^\infty c_n(t) \sin(nx),$$

where

$$c_n(t) = \frac{2}{\pi} \int_0^\pi u(x, t) \sin(nx) \, dx.$$

We then obtain the optimal state feedback controller

$$
\begin{aligned}
\phi &= -P(t)u(x,t) \\
&= -\sum_{n=1}^{\infty} c_n(t)P(t)\sin(nx) \\
&= -\frac{2}{\pi}\sum_{n=1}^{\infty} p_{nn}(t)\sin(nx)\int_0^{\pi} u(x,t)\sin(nx)dx.
\end{aligned}
$$

Note that the optimal feedback controller contains all eigenmodes while the controller designed in Section 4.2 contains finite ones.

We introduce Lyapunov's method to find the optimal feedback operator Π for the reaction-convection-diffusion equation

$$
\frac{\partial u}{\partial t} = \mu\nabla^2 u + \nabla\cdot(u\mathbf{v}) + au + \phi(\mathbf{x},t), \tag{4.247}
$$

$$
u|_{\Gamma} = 0, \tag{4.248}
$$

$$
u(\mathbf{x},0) = u_0(\mathbf{x}), \tag{4.249}
$$

Definition 4.9. Let u be the solution of (4.247)-(4.249). If a continuously differentiable functional $V(u)$ satisfies

(1) $V(0) = 0$,
(2) $V(u) > 0$ for $u \neq 0$,
(3) $\frac{dV}{dt} \leq 0$,

it is called a Lyapunov functional.

The Lyapunov's procedure of designing a feedback controller to stabilize a given system is to construct a Lyapunov functional for the system. In achieving this, we construct a linear symmetric operator $\Pi : L^2(\Omega) \to L^2(\Omega)$ such that

$$
V(u) = \int_{\Omega} u(t)\Pi u(t)dV
$$

is a Lyapunov functional of (4.247)-(4.249). Using the equation (4.247), we compute

$$
\begin{aligned}
\frac{dV}{dt} &= \int_{\Omega} u_t \Pi u dV + \int_{\Omega} u\Pi u_t dV \\
&= \int_{\Omega} \left(\mu\nabla^2 u + \nabla\cdot(u\mathbf{v}) + au + \phi\right)\Pi u dV \\
&\quad + \int_{\Omega} u\Pi\left(\mu\nabla^2 u + \nabla\cdot(u\mathbf{v}) + au + \phi\right)dV \\
&= \int_{\Omega} \left(\mu\nabla^2 u\Pi u + \mu u\Pi\nabla^2 u + \nabla\cdot(u\mathbf{v})\Pi u + u\Pi\nabla\cdot(u\mathbf{v})\right)dV \\
&\quad + \int_{\Omega} \left(au\Pi u + u\Pi(au) + \phi\Pi u + u\Pi\phi\right)dV.
\end{aligned}
$$

Our goal is to construct Π such that $\frac{dV}{dt}$ is negative. Thus we set

$$\int_{\Omega} \left(\mu \nabla^2 u \Pi u + \mu u \Pi \nabla^2 u + \nabla \cdot (u\mathbf{v}) \Pi u + u \Pi \nabla \cdot (u\mathbf{v}) + a u \Pi u + u \Pi (au) \right) dV$$

$$= \int_{\Omega} \left(\frac{1}{\beta} \Pi u \Pi u - \alpha u^2 \right) dV,$$

where α and β are positive constants. It then follows that

$$
\begin{aligned}
\frac{dV}{dt} &= -\alpha \int_{\Omega} u^2 dV + \int_{\Omega} \left(\frac{1}{\beta} \Pi u \Pi u + \phi \Pi u + u \Pi \phi \right) dV \\
&= -\int_{\Omega} \left(\alpha u^2 + \beta \phi^2 \right) dV + \int_{\Omega} \left(\frac{1}{\beta} \Pi u \Pi u + \phi \Pi u + \phi \Pi u + \beta \phi^2 \right) dV \\
&= -\int_{\Omega} \left(\alpha u^2 + \beta \phi^2 \right) dV + \frac{1}{\beta} \int_{\Omega} (\Pi u + \beta \phi)^2 \, dV.
\end{aligned}
\tag{4.250}
$$

If we take

$$\phi = -\frac{1}{\beta} \Pi u, \tag{4.251}$$

then $\frac{dV}{dt}$ is negative. Hence the stabilization problem is changed to the study of the operator equation

$$\int_{\Omega} \left(\mu u_1 \Pi \nabla^2 u_2 + \mu u_2 \Pi \nabla^2 u_1 + \nabla \cdot (u_1 \mathbf{v}) \Pi u_2 + u_1 \Pi \nabla \cdot (u_2 \mathbf{v}) + a u_1 \Pi (u_2) \right) dV$$

$$+ \int_{\Omega} \left(a u_2 \Pi (u_1) - \frac{1}{\beta} \Pi (u_1) \Pi (u_2) + \alpha u_1 u_2 \right) dV = 0 \tag{4.252}$$

for all $u_1, u_2 \in H^2(\Omega) \cap H_0^1(\Omega)$. This equation is called an *algebraic Riccati equation*.

The study of existence and uniqueness of solutions of the Riccati equation is beyond the scope of this text and we state just a theorem from [24, p.290, Theorem 6.1.13] without proof. Consider the differential Riccati equation

$$\frac{d}{dt} \int_{\Omega} u_1 P u_2 dV = \int_{\Omega} \left(\mu P u_1 \nabla^2 u_2 + \mu P u_2 \nabla^2 u_1 \right) dV$$

$$+ \int_{\Omega} \left(\nabla \cdot (u_1 \mathbf{v}) P u_2 + u_1 P \nabla \cdot (u_2 \mathbf{v}) + a u_1 P u_2 + a u_2 P(u_1) \right) dV$$

$$+ \int_{\Omega} \left(-\frac{1}{\beta} P u_2 P u_1 + \alpha u_1 u_2 \right) dV, \tag{4.253}$$

$$\int_{\Omega} u_1 P(0) u_2 dV = \gamma \int_{\Omega} u_1 u_2 dV \tag{4.254}$$

for $u_1, u_2 \in H^2(\Omega) \cap H_0^1(\Omega)$. The algebraic Riccati equation (4.252) is the steady state equation of the differential equation.

Theorem 4.30. *The differential Riccati equation (4.253)-(4.254) has a unique continuously differentiable solution that has the following properties:*

(1) $\int_\Omega u_0 P(t) u_0 dV = \min\limits_{\phi \in L^2(\Omega \times (0,t))} J(\phi, u_0, 0)$, *where J is defined by (4.196).*

(2) If $\gamma = 0$, then $\int_\Omega u_0 P(t_1) u_0 dV \leq \int_\Omega u_0 P(t_2) u_0 dV$ for $t_1 \leq t_2$.

(3) If $\gamma = 0$, then the limit

$$\lim_{t \to \infty} P(t) u_0 = \Pi u_0 \quad \text{for all } u_0 \in L^2(\Omega) \tag{4.255}$$

exists and Π is the minimal nonnegative symmetric solution of (4.252). That is, any other solution Π_1 of (4.252) satisfies

$$\int_\Omega u_0 \Pi u_0 dV \leq \int_\Omega u_0 \Pi_1 u_0 dV.$$

We now show that the feedback control system

$$\frac{\partial u}{\partial t} = \mu \nabla^2 u + \nabla \cdot (u\mathbf{v}) + au - \frac{1}{\beta} \Pi u, \tag{4.256}$$

$$u|_\Gamma = 0, \tag{4.257}$$

$$u(\mathbf{x}, 0) = u_0(\mathbf{x}) \tag{4.258}$$

is exponentially stable.

Theorem 4.31. *Let $\beta > 0$ and Π be the nonnegative symmetric solution of (4.252). Then the feedback control system (4.256)-(4.258) is exponentially stable.*

Proof. Since Π is a bounded operator, it satisfies the condition of Theorem 2.18 with $a = 0$. Thus the operator $\mu \nabla^2 u + \nabla \cdot (u\mathbf{v}) + au - \frac{1}{\beta} \Pi u$ generates an analytic semigroup $T(t)$ on $L^2(\Omega)$, which is given by

$$T(t) u_0 = u(\mathbf{x}, t; u_0),$$

where $u(\mathbf{x}, t; u_0)$ is the solution of (4.256)-(4.258). By (4.250), we deduce that

$$\int_0^\infty \int_\Omega \left(\alpha u^2 + \beta \phi^2 \right) dV dt \leq C \int_\Omega |u_0|^2 dV$$

for some $C > 0$, and then

$$\int_0^\infty \int_\Omega \alpha \|T(t)\|^2 dt \leq C.$$

Thus the theorem follows from Theorem 4.5.

We now show that the feedback controller (4.251) is optimal. For this we define the quadratic performance functional over an infinite time interval

$$J(\phi; u_0) = \int_0^\infty \int_\Omega \left(\alpha |u(\mathbf{x}, t)|^2 + \beta |\phi(\mathbf{x}, t)|^2 \right) dV dt, \tag{4.259}$$

where u is the solution of (4.247)-(4.249) and $\alpha, \beta \geq 0$ are weight constants. We prove that the controller (4.251) minimizes the functional J over $L^2(\Omega \times (0, \infty))$. Since the control problem (4.247)-(4.249) is exponentially stabilizable, the optimal control problem is well posed, that is, for every $u_0 \in L^2(\Omega)$, there exists a control input $\phi \in L^2(\Omega \times (0, \infty))$ such that $J(\phi; u_0) < \infty$.

Using the solutions of the algebraic Riccati equation (4.252), we can prove the existence of an optimal control. From (4.250), we first have the following technical lemma.

Lemma 4.7. *Let Π be the nonnegative symmetric solution of (4.252) and define*

$$J(\phi; u_0, t) = \int_0^t \int_\Omega \left(\alpha |u(\mathbf{x}, t)|^2 + \beta |\phi(\mathbf{x}, t)|^2 \right) dV \, dt. \tag{4.260}$$

Then

$$J(\phi; u_0, t) = \int_\Omega u(0) \Pi u(0) dV - \int_\Omega u(t) \Pi u(t) dV$$
$$+ \frac{1}{\beta} \int_0^t \int_\Omega (\Pi u + \beta \phi)^2 \, dV. \tag{4.261}$$

Theorem 4.32. *Let J be defined by (4.259). Then there exists a unique optimal control ϕ^* such that*

$$\min_{\phi \in L^2(\Omega \times (0, \infty))} J(\phi; u_0) = J(\phi^*; u_0) = \int_\Omega u_0 \Pi u_0 dV, \tag{4.262}$$

where Π is the nonnegative symmetric solution of (4.252) and

$$\phi^* = -\frac{1}{\beta} \Pi u.$$

Proof. We note that the optimal control problem is well posed and then the minimum below is finite. It follows from Theorem 4.30 that

$$\min_{\phi \in L^2(\Omega \times (0, \infty))} J(\phi; u_0) \geq \min_{\phi \in L^2(\Omega \times (0, \infty))} J(\phi; u_0, t)$$
$$= \min_{\phi \in L^2(\Omega \times (0, t))} J(\phi; u_0, t)$$
$$= \int_\Omega u_0 P(t) u_0 dV,$$

which, combined with (4.255), implies that

$$\min_{\phi \in L^2(\Omega \times (0, \infty))} J(\phi, u_0) \geq \int_\Omega u_0 \Pi u_0 dV. \tag{4.263}$$

On the other hand, by (4.261), we deduce that

$$
\begin{aligned}
J(\phi;u_0,t) &= \int_\Omega u_0 \Pi u_0 dV - \int_\Omega u(t) \Pi u(t) dV \\
&\quad + \frac{1}{\beta} \int_0^t \int_\Omega (\Pi u + \beta \phi)^2 \, dV \\
&\leq \int_\Omega u_0 \Pi u_0 dV + \frac{1}{\beta} \int_0^t \int_\Omega (\Pi u + \beta \phi)^2 \, dV.
\end{aligned}
$$

Taking $\phi = \phi^* = -\frac{1}{\beta}\Pi u$, we obtain

$$
J(\phi^*;u_0,t) \leq \int_\Omega u_0 \Pi u_0 dV
$$

and then

$$
\min_{\phi \in L^2(\Omega \times (0,\infty))} J(\phi,u_0) \leq J(\phi^*;u_0) \leq \int_\Omega u_0 \Pi u_0 dV. \tag{4.264}
$$

Putting (4.263) and (4.264) together yields (4.262). In addition, we can readily show that J is strictly convex and then the optimal control is unique.

Example 4.8. Consider the control problem

$$
\begin{aligned}
\frac{\partial u}{\partial t} &= \frac{\partial^2 u}{\partial x^2} + \phi(x,t), \\
u(0,t) &= u(\pi,t) = 0, \\
u(x,0) &= u_0(x)
\end{aligned}
$$

with the performance functional

$$
J(\phi;u_0) = \int_0^\infty \int_0^\pi (u^2 + \phi^2) dx dt.
$$

Let

$$
\Pi \sin(nx) = \sum_{m=1}^\infty p_{mn} \sin(mx).
$$

As in Example 4.7, taking $u_1 = \sin(mx), u_2 = \sin(nx)$ in the algebraic Riccati equation (4.252), we obtain the infinite system

$$
(m^2 + n^2) p_{mn} + \sum_{l=1}^\infty p_{ml} p_{ln} - \delta_{mn} = 0. \tag{4.265}
$$

If $m \neq n$, $p_{mn} \equiv 0$ is a solution. Then the above system is reduced to

$$
2n^2 p_{nn} + p_{nn}^2 - 1 = 0. \tag{4.266}
$$

Solving the system, we obtain the solution

$$p_{nn} = -n^2 \pm \sqrt{n^4 + 1}.$$

Since P is nonnegative, we have

$$p_{nn} = -n^2 + \sqrt{n^4 + 1}.$$

Let

$$u(x,t) = \sum_{n=1}^{\infty} c_n(t) \sin(nx),$$

where

$$c_n(t) = \frac{2}{\pi} \int_0^{\pi} u(x,t) \sin(nx) dx.$$

We then obtain the optimal state feedback control

$$\begin{aligned}
\phi &= -\Pi u(x,t) \\
&= -\sum_{n=1}^{\infty} c_n(t) \Pi \sin(nx) \\
&= -\frac{2}{\pi} \sum_{n=1}^{\infty} (-n^2 + \sqrt{n^4 + 1}) \sin(nx) \int_0^{\pi} u(x,t) \sin(nx) dx.
\end{aligned}$$

Exercises 4.4

1. Verify that the operator $P(s)$ defined by (4.215) is linear.
2. Show that the functional $J(\phi; u_0, s)$ defined by (4.196) satisfies

$$J(0; u_0, s) \leq C \int_{\Omega} |u_0|^2 dV,$$

where C is a positive constant.
3. Consider the control problem

$$\frac{\partial u}{\partial t} = \frac{\partial^2 u}{\partial x^2} + \phi(x,t),$$

$$\frac{\partial u}{\partial x}(0,t) = \frac{\partial u}{\partial x}(\pi,t) = 0,$$

$$u(x,0) = u_0(x)$$

with the performance functional

$$J(\phi; u_0, 0) = \int_0^T \int_0^{\pi} (u^2 + \phi^2) dx dt + \int_0^{\pi} u(T)^2 dx.$$

a. Calculate the Gâteaux-differential of J.
b. Derive the optimality system for the optimal control.
c. Derive a Riccati equation.
d. Design an optimal state feedback control.

4. Consider the optimal control problem

$$\frac{\partial u}{\partial t} = \mu \nabla^2 u + \nabla \cdot (u\mathbf{v}) + au + \sum_{i=1}^{m} b_i(\mathbf{x}) \phi_i(t),$$

$$u|_\Gamma = 0,$$

$$u(\mathbf{x}, s) = u_0(\mathbf{x})$$

with the quadratic performance functional

$$J(\phi_1, \cdots, \phi_m; u_0, s) = \alpha \int_s^T \int_\Omega |u|^2 dV dt + \beta \sum_{i=1}^{m} \int_s^T |\phi_i|^2 dt + \gamma \int_\Omega |u(T)|^2 dV,$$

where $\alpha, \beta, \gamma \geq 0$ are weight constants and b_i ($i = 1, 2, \cdots, m$) are given functions.

a. Calculate the Gâteaux-differential of J.
b. Derive the optimality system for the optimal control.
c. Derive a Riccati equation.

5. Let H be a Hilbert space and $F(x) = \|x\|$ for any $x \in H$. Prove that $F'(x) = \frac{x}{\|x\|}$ for $x \neq 0$.

6. Consider the control problem

$$\frac{\partial u}{\partial t} = \frac{\partial^2 u}{\partial x^2} + \phi(x, t),$$

$$\frac{\partial u}{\partial x}(0, t) = \frac{\partial u}{\partial x}(\pi, t) = 0,$$

$$u(x, 0) = u_0(x).$$

Use Lyapunov's procedure to derive an algebraic Riccati equation and then design an optimal state feedback control.

4.5 Optimal Boundary Control

Consider the boundary control problem

$$\frac{\partial u}{\partial t} = \mu \nabla^2 u + \nabla \cdot (u\mathbf{v}) + au, \tag{4.267}$$

$$u|_\Gamma = \phi, \tag{4.268}$$

$$u(\mathbf{x}, s) = u_0(\mathbf{x}), \tag{4.269}$$

where ϕ is a control input. For $T > s$, we define the quadratic functional of concentration variance by

$$J(\phi;u_0,s,T) = \alpha \int_s^T \int_\Omega |u|^2 dvdt + \beta \int_s^T \int_\Gamma |\phi|^2 dSdt$$
$$+ \gamma \int_\Omega |u(T)|^2 dV, \qquad (4.270)$$

where $\alpha, \beta, \gamma \geq 0$ are weight constants. The optimal control problem is to minimize the functional J over $L^2(\Gamma \times (s,T))$ with respect to ϕ, where u_0 and s are parameters. This problem can be solved in the same way as for the interior control problem, but the existence is handled in a different way at the end of the section.

4.5.1 Necessary Conditions

We denote the solution of (4.267)-(4.269) corresponding to the control ϕ and the initial condition u_0 by $u(\mathbf{x},t;u_0,\phi)$. Since the equation is linear, we can readily see that

$$u(\mathbf{x},t;u_0,\phi + \lambda \psi) = u(\mathbf{x},t;u_0,\phi) + \lambda u(\mathbf{x},t;0,\psi) \qquad (4.271)$$

for any control functions ϕ, ψ and any number λ. We now calculate the Gâteaux-differential of J.

Theorem 4.33. *The Gâteaux-differential of J defined by (4.270) at ϕ is given by*

$$\int_s^T \int_\Gamma J'(\phi;u_0,s,T)\psi dSdt = 2\alpha \int_s^T \int_\Omega u(\mathbf{x},t;u_0,\phi)u(\mathbf{x},t;0,\psi)dVdt$$
$$+ 2\gamma \int_\Omega u(\mathbf{x},T;u_0,\phi)u(\mathbf{x},T;0,\psi)dV$$
$$+ 2\beta \int_s^T \int_\Gamma \phi \psi dSdt, \qquad (4.272)$$

where $u(\mathbf{x},t;0,\psi)$ is the solution of

$$\frac{\partial u}{\partial t} = \mu \nabla^2 u + \nabla \cdot (u\mathbf{v}) + au, \qquad (4.273)$$
$$u|_\Gamma = \psi, \qquad (4.274)$$
$$u(\mathbf{x},s) = 0. \qquad (4.275)$$

Proof. Using (4.270) and (4.271), we derive that

$$\int_s^T \int_\Gamma J'(\phi;u_0,s,T)\psi dSdt$$
$$= \lim_{\lambda \to 0^+} \frac{J(\phi + \lambda \psi;u_0,s) - J(\phi;u_0,s)}{\lambda}$$

$$= \lim_{\lambda \to 0^+} \frac{\alpha}{\lambda} \int_s^T \int_\Omega \left(|u(\mathbf{x},t;u_0,\phi+\lambda\psi)|^2 - |u(\mathbf{x},t;u_0,\phi)|^2 \right) dVdt$$

$$+ \lim_{\lambda \to 0^+} \frac{\beta}{\lambda} \int_s^T \int_\Gamma \left(|\phi+\lambda\psi|^2 - |\phi|^2 \right) dSdt$$

$$+ \lim_{\lambda \to 0^+} \frac{\gamma}{\lambda} \int_\Omega \left(|u(\mathbf{x},T;u_0,\phi+\lambda\psi)|^2 - |u(\mathbf{x},T;u_0,\phi)|^2 \right) dV$$

$$= \lim_{\lambda \to 0^+} \frac{\alpha}{\lambda} \int_s^T \int_\Omega \left(2\lambda u(\mathbf{x},t;u_0,\phi)u(\mathbf{x},t;0,\psi) + \lambda^2 |u(\mathbf{x},t;0,\psi)|^2 \right) dVdt$$

$$+ \lim_{\lambda \to 0^+} \frac{\beta}{\lambda} \int_s^T \int_\Gamma \left(2\lambda\phi\psi + \lambda^2 |\psi|^2 \right) dSdt$$

$$+ \lim_{\lambda \to 0^+} \frac{\gamma}{\lambda} \int_\Omega \left(2\lambda u(\mathbf{x},T;u_0,\phi)u(\mathbf{x},T;0,\psi) + \lambda^2 |u(\mathbf{x},T;0,\psi)|^2 \right) dV$$

$$= 2\alpha \int_s^T \int_\Omega u(\mathbf{x},t;u_0,\phi)u(\mathbf{x},t;0,\psi)dVdt$$

$$+ 2\beta \int_s^T \int_\Gamma \phi\psi dSdt + 2\gamma \int_\Omega u(\mathbf{x},T;u_0,\phi)u(\mathbf{x},T;0,\psi)dV.$$

By Theorem 4.27, we deduce the following necessary condition for the optimal control.

Theorem 4.34. *If the functional J has a minimum at ϕ^*, then*

$$J'(\phi^*;u_0,s,T) = 0. \tag{4.276}$$

4.5.2 Optimality Systems

The following theorem is the boundary control version of Theorem 4.28.

Theorem 4.35. *The optimal control ϕ^* satisfies the following system*

$$\frac{\partial u}{\partial t} = \mu \nabla^2 u + \nabla \cdot (u\mathbf{v}) + au, \tag{4.277}$$

$$\frac{\partial v}{\partial t} = -\mu \nabla^2 v + \mathbf{v} \cdot \nabla v - av + \alpha u, \tag{4.278}$$

$$u|_\Gamma = \phi^*, \quad v|_\Gamma = 0, \tag{4.279}$$

$$\phi^* = -\frac{\mu}{\beta} \frac{\partial v}{\partial \mathbf{n}}\bigg|_\Gamma, \tag{4.280}$$

$$u(\mathbf{x},s) = u_0(\mathbf{x}), \, v(\mathbf{x},T) = -\gamma u(\mathbf{x},T). \tag{4.281}$$

Proof. Multiplying (4.273) by the solution v of (4.278) and integrating over $\Omega \times (s,T)$ by parts, we have

$$\int_s^T \int_\Omega v \frac{\partial u}{\partial t}(\mathbf{x},t;0,\psi)dVdt \tag{4.282}$$

$$= \mu \int_s^T \int_\Omega v\nabla^2 u(\mathbf{x},t;0,\psi)dVdt + \int_s^T \int_\Omega v\nabla \cdot (u(\mathbf{x},t;0,\psi)\mathbf{v})dVdt$$

$$+ \int_s^T \int_\Omega au(\mathbf{x},t;0,\psi)vdVdt$$

$$= -\mu \int_s^T \int_\Omega \nabla v \cdot \nabla u(\mathbf{x},t;0,\psi)dVdt - \int_s^T \int_\Omega u(\mathbf{x},t;0,\psi)\mathbf{v} \cdot \nabla v dVdt$$

$$+ \int_s^T \int_\Omega au(\mathbf{x},t;0,\psi)vdVdt.$$

Multiplying (4.278) by the solution $u(\mathbf{x},t;0,\psi)$ of (4.273) and integrating over $\Omega \times (s,T)$ by parts, we have

$$\int_s^T \int_\Omega u(\mathbf{x},t;0,\psi)\frac{\partial v}{\partial t}dVdt$$

$$= -\mu \int_s^T \int_\Omega u(\mathbf{x},t;0,\psi)\nabla^2 v dVdt + \int_s^T \int_\Omega u(\mathbf{x},t;0,\psi)\mathbf{v} \cdot \nabla v dVdt$$

$$- \int_s^T \int_\Omega au(\mathbf{x},t;0,\psi)vdVdt + \alpha \int_s^T \int_\Omega u(\mathbf{x},t;0,\psi)u(\mathbf{x},t;u_0,\phi^*)dVdt$$

$$= -\mu \int_s^T \int_\Gamma \psi \frac{\partial v}{\partial \mathbf{n}}dSdt + \mu \int_s^T \int_\Omega \nabla u(\mathbf{x},t;0,\psi) \cdot \nabla v dVdt$$

$$+ \int_s^T \int_\Omega u(\mathbf{x},t;0,\psi)\mathbf{v} \cdot \nabla v dVdt - \int_s^T \int_\Omega au(\mathbf{x},t;0,\psi)vdVdt$$

$$+\alpha \int_s^T \int_\Omega u(\mathbf{x},t;0,\psi)u(\mathbf{x},t;u_0,\phi^*)dVdt. \tag{4.283}$$

Adding (4.282) and (4.283) gives

$$\alpha \int_s^T \int_\Omega u(\mathbf{x},t;0,\psi)u(\mathbf{x},t;u_0,\phi^*)dVdt + \gamma \int_\Omega u(\mathbf{x},T;0,\psi)u(\mathbf{x},T;u_0,\phi^*)dV$$

$$= \mu \int_s^T \int_\Gamma \psi \frac{\partial v}{\partial \mathbf{n}}dSdt. \tag{4.284}$$

On the other hand, it follows from Theorems 4.33 and 4.34 that

$$\alpha \int_s^T \int_\Omega u(\mathbf{x},t;u_0,\phi^*)u(\mathbf{x},t;0,\psi)dVdt + \beta \int_s^T \int_\Gamma \phi^* \psi dSdt$$

$$+\gamma \int_\Omega u(\mathbf{x},T;u_0,\phi^*)u(\mathbf{x},T;0,\psi)dV = 0$$

for all functions $\psi \in L^2(\Gamma \times (s,T))$. Substituting (4.284) into this equation, we obtain

$$\int_s^T \int_\Gamma \left(\mu \frac{\partial v}{\partial \mathbf{n}} + \beta \phi^* \right) \psi dV dt = 0$$

for all functions $\psi \in L^2(\Gamma \times (s,T))$. Thus $\beta \phi^* = -\mu \left. \frac{\partial v}{\partial \mathbf{n}} \right|_\Gamma$.

To decouple the optimality system (4.277)-(4.281), we define a family of linear operators $P(s)$ by

$$P(s)u_0(\mathbf{x}) = -v(\mathbf{x},s;u_0) \tag{4.285}$$

for any given function $u_0(\mathbf{x}) \in L^2(\Omega)$, where $v(\mathbf{x},s;u_0)$ is the solution of (4.277)-(4.281).

Theorem 4.36. *The operator $P(s)$ defined by (4.285) has the following properties:*

(1) $P(s)$ is symmetric, that is,

$$\int_\Omega p_0 P(s)u_0 dV = \int_\Omega u_0 P(s)p_0 dV \quad \text{for any } u_0, p_0 \in L^2(\Omega). \tag{4.286}$$

(2) If ϕ^* is the optimal control of (4.267)-(4.269), then

$$\int_\Omega u_0 P(s)u_0 dV = \alpha \int_s^T \int_\Omega |u|^2 dV dt + \beta \int_s^T \int_\Gamma |\phi^*|^2 dS dt$$
$$+ \gamma \int_\Omega |u(T)|^2) dV. \tag{4.287}$$

Hence $P(s)$ is nonnegative.
(3) $P(T) = \gamma I$.
(4) $P(s)$ is a linear bounded operator in $L^2(\Omega)$.

Proof. (1) Consider the system

$$\frac{\partial p}{\partial t} = \mu \nabla^2 p + \nabla \cdot (p\mathbf{v}) + ap, \tag{4.288}$$

$$\frac{\partial q}{\partial t} = -\mu \nabla^2 q + \mathbf{v} \cdot \nabla q - aq + \alpha p, \tag{4.289}$$

$$p|_\Gamma = \psi^*, \quad q|_\Gamma = 0, \tag{4.290}$$

$$\psi^* = -\frac{\mu}{\beta} \left. \frac{\partial q}{\partial \mathbf{n}} \right|_\Gamma, \tag{4.291}$$

$$p(\mathbf{x},s) = p_0(\mathbf{x}), \quad q(\mathbf{x},T) = -\gamma p(\mathbf{x},T). \tag{4.292}$$

Multiplying (4.277) by the solution q of (4.289) and integrating over $\Omega \times (s,T)$ by parts, we have

$$\int_s^T \int_\Omega q \frac{\partial u}{\partial t} dV dt = \mu \int_s^T \int_\Omega q \nabla^2 u dV dt + \int_s^T \int_\Omega q \nabla \cdot (u\mathbf{v}) dV dt$$

$$+ \int_s^T \int_\Omega auq dV dt$$

$$= -\mu \int_s^T \int_\Omega \nabla q \cdot \nabla u dV dt - \int_s^T \int_\Omega u\mathbf{v} \cdot \nabla q dV dt$$

$$+ \int_s^T \int_\Omega auq dV dt. \tag{4.293}$$

Multiplying (4.289) by the solution u of (4.277) and integrating over $\Omega \times (s,T)$ by parts, we have

$$\int_s^T \int_\Omega u \frac{\partial q}{\partial t} dV dt = -\mu \int_s^T \int_\Omega u \nabla^2 q dV dt + \int_s^T \int_\Omega u\mathbf{v} \cdot \nabla q dV dt$$

$$+ \int_s^T \int_\Omega (\alpha u p - auq) dV dt$$

$$= -\mu \int_s^T \int_\Gamma \phi^* \frac{\partial q}{\partial \mathbf{n}} dS dt + \mu \int_s^T \int_\Omega \nabla q \cdot \nabla u dV dt$$

$$+ \int_s^T \int_\Omega (u\mathbf{v} \cdot \nabla q - auq + \alpha u p) dV dt. \tag{4.294}$$

Adding (4.293) and (4.294) gives

$$\int_\Omega u_0 P(s) p_0 dV$$

$$= \alpha \int_s^T \int_\Omega u p dV dt - \mu \int_s^T \int_\Gamma \phi^* \frac{\partial q}{\partial \mathbf{n}} dS dt + \gamma \int_\Omega u(T) p(T) dV$$

$$= \alpha \int_s^T \int_\Omega u p dV dt + \beta \int_s^T \int_\Gamma \phi^* \psi^* dS dt + \gamma \int_\Omega u(T) p(T) dV. \tag{4.295}$$

Repeating the above procedure by multiplying (4.288) by the solution v of (4.278) and multiplying (4.278) by the solution p of (4.288), we can obtain

$$\int_\Omega p_0 P(s) u_0 dV$$

$$= \alpha \int_s^T \int_\Omega u p dV dt + \beta \int_s^T \int_\Gamma \phi^* \psi^* dS dt + \gamma \int_\Omega u(T) p(T) dV. \tag{4.296}$$

Then (4.286) follows from (4.295) and (4.296).

(2) If $p_0 = u_0$, then $u = p, v = q$, and $\phi^* = \psi^*$. It then follows from (4.296) that

$$\int_\Omega u_0 P(s) u_0 dV = \alpha \int_s^T \int_\Omega u^2 dV dt + \beta \int_s^T \int_\Gamma |\phi^*|^2 dS dt + \gamma \int_\Omega u(T)^2 dV.$$

(3) Letting $s \to T$ in (4.296), we derive that

$$\int_\Omega p_0 P(T) u_0 dV = \gamma \int_\Omega u_0 p_0 dV \quad \text{for all } p_0 \in L^2(\Omega).$$

Thus $P(T) u_0 = \gamma u_0$ for all $u_0 \in L^2(\Omega)$.

(4) By (4.296), we deduce from Hölder's inequality (2.5) and Young's inequality (2.4) that

$$\left| \int_\Omega p_0 P(s) u_0 dV \right|$$

$$\leq \gamma \left(\int_\Omega |u(T)|^2 dV \right)^{1/2} \left(\int_\Omega |p(T)|^2 dV \right)^{1/2}$$

$$+ \beta \left(\int_s^T \int_\Gamma |\phi^*|^2 dS dt \right)^{1/2} \left(\int_s^T \int_\Gamma |\psi^*|^2 dS dt \right)^{1/2}$$

$$+ \alpha \left(\int_s^T \int_\Omega |u|^2 dV dt \right)^{1/2} \left(\int_s^T \int_\Omega |p|^2 dV dt \right)^{1/2}$$

$$\leq \left(\gamma \int_\Omega |u(T)|^2 dV + \beta \int_s^T \int_\Gamma |\phi^*|^2 dS dt + \alpha \int_s^T \int_\Omega |u|^2 dV dt \right)^{1/2}$$

$$\times \left(\gamma \int_\Omega |p(T)|^2 dV + \beta \int_s^T \int_\Gamma |\psi^*|^2 dS dt + \alpha \int_s^T \int_\Omega |p|^2 dV dt \right)^{1/2}$$

$$= (J(\phi^*; u_0, s, T))^{1/2} (J(\psi^*; p_0, s, T))^{1/2}. \tag{4.297}$$

Moreover, by (4.15), we deduce that there exists a constant $C > 0$ such that

$$J(\phi^*; u_0, s, T) \leq J(0; u_0, s, T)$$
$$= \gamma \int_\Omega |u(T)|^2 dV + \alpha \int_s^T \int_\Omega |u|^2 dV dt$$
$$\leq C \int_\Omega |u_0|^2 dV$$

and

$$J(\psi^*; p_0, s, T) \leq J(0; p_0, s, T)$$
$$= \gamma \int_\Omega |p(T)|^2 dV + \alpha \int_s^T \int_\Omega |p|^2 dV dt$$
$$\leq C \int_\Omega |p_0|^2 dV.$$

It therefore follows from (4.297) that

$$\left| \int_\Omega p_0 P(s) u_0 dV \right| \leq C \left(\int_\Omega |u_0|^2 dV \int_\Omega |p_0|^2 dV \right)^{1/2}$$

for all $p_0, u_0 \in L^2(\Omega)$. Taking $p_0 = P(s)u_0$, we obtain

$$\|P(s)u_0\|_{L^2} \leq C\|u_0\|_{L^2}.$$

Thus $P(s)$ is bounded. Since the optimality system (4.277)-(4.281) is linear, $P(s)$ is linear.

Consider the optimal control problem

$$\frac{\partial u}{\partial t} = \mu \nabla^2 u + \nabla \cdot (u\mathbf{v}) + au, \tag{4.298}$$

$$u|_\Gamma = \phi, \tag{4.299}$$

$$u(\mathbf{x}, 0) = w_0(\mathbf{x}). \tag{4.300}$$

Note that the initial time $t = 0$. By Theorem 4.35, we derive that the optimal control ϕ^* satisfies the following system:

$$\frac{\partial u}{\partial t} = \mu \nabla^2 u + \nabla \cdot (u\mathbf{v}) + au, \tag{4.301}$$

$$\frac{\partial v}{\partial t} = -\mu \nabla^2 v + \mathbf{v} \cdot \nabla v - av + \alpha u, \tag{4.302}$$

$$u|_\Gamma = \phi^*, \quad v|_\Gamma = 0, \tag{4.303}$$

$$\phi^* = -\frac{\mu}{\beta} \frac{\partial v}{\partial \mathbf{n}}\bigg|_\Gamma, \tag{4.304}$$

$$u(\mathbf{x}, 0) = w_0(\mathbf{x}), \ v(\mathbf{x}, T) = -\gamma u(\mathbf{x}, T). \tag{4.305}$$

If we restrict the system (4.301)-(4.305) on $[s, T]$, then the initial condition (4.305) becomes

$$u(\mathbf{x}, s) = u(\mathbf{x}, s), \ v(\mathbf{x}, T) = -\gamma u(\mathbf{x}, T). \tag{4.306}$$

If we compare the system (4.277)-(4.281) with the system (4.301)-(4.305) on $[s, T]$, we find that

$$u_0(\mathbf{x}) \text{ (in (4.281))} = u(\mathbf{x}, s) \text{ (solution of (4.301))},$$
$$v(\mathbf{x}, s; u_0) \text{ (solution of (4.278))} = v(\mathbf{x}, s; w_0) \text{ (solution of (4.302))}.$$

Substituting these equations into (4.285) gives

$$-v(\mathbf{x}, s; w_0) = P(s)u(\mathbf{x}, s). \tag{4.307}$$

We then obtain the optimal state feedback control

$$\phi^* = -\frac{\mu}{\beta} \frac{\partial v}{\partial \mathbf{n}}\bigg|_\Gamma = \frac{\mu}{\beta} \frac{\partial P(s)u(\mathbf{x}, s)}{\partial \mathbf{n}}\bigg|_\Gamma. \tag{4.308}$$

We perform a formal computation to derive an equation for the operator P. Let $u_1, u_2 \in H^2(\Omega)$. Multiplying (4.301) and (4.302) by u_1, u_2, respectively, and integrating over Ω, we obtain

$$\int_\Omega \frac{\partial u}{\partial t} u_1 dV = \int_\Omega \left(\mu u_1 \nabla^2 u + u_1 \nabla \cdot (u\mathbf{v}) + auu_1 \right) dV$$

$$= \mu \int_\Gamma \frac{\partial u}{\partial \mathbf{n}} u_1 dS - \mu \int_\Gamma u \frac{\partial u_1}{\partial \mathbf{n}} dS + \mu \int_\Omega u \nabla^2 u_1 dV$$

$$+ \int_\Omega (u_1 \nabla \cdot (u\mathbf{v}) + auu_1) dV, \qquad (4.309)$$

$$\int_\Omega \frac{\partial v}{\partial t} u_2 dV = \int_\Omega \left(-\mu u_2 \nabla^2 v + u_2 \mathbf{v} \cdot \nabla v - avu_2 + \alpha uu_2 \right) dV. \qquad (4.310)$$

Substituting (4.307) into (4.310) gives

$$-\int_\Omega \frac{\partial (Pu)}{\partial t} u_2 dV = \int_\Omega \left(\mu u_2 \nabla^2 P(u) - u_2 \mathbf{v} \cdot \nabla P(u) + aP(u)u_2 + \alpha uu_2 \right) dV,$$

and then

$$-\int_\Omega \frac{dP}{dt}(u)u_2 dV - \int_\Omega P\left(\frac{\partial u}{\partial t} \right) u_2 dV$$

$$= \int_\Omega \left(\mu u_2 \nabla^2 P(u) - u_2 \mathbf{v} \cdot \nabla P(u) + aP(u)u_2 + \alpha uu_2 \right) dV.$$

It then follows from the symmetry of P that

$$-\int_\Omega \frac{dP}{dt}(u)u_2 dV - \int_\Omega \frac{\partial u}{\partial t} P(u_2) dV$$

$$= \int_\Omega \left(\mu u_2 \nabla^2 P(u) - u_2 \mathbf{v} \cdot \nabla P(u) + aP(u)u_2 + \alpha uu_2 \right) dV.$$

Substituting (4.309) into this equation for $\int_\Omega \frac{\partial u}{\partial t} P(u_2) dV$ gives

$$-\int_\Omega \frac{dP}{dt}(u)u_2 dV - \mu \int_\Gamma \frac{\partial u}{\partial \mathbf{n}} P(u_2) dS + \mu \int_\Gamma u \frac{\partial P(u_2)}{\partial \mathbf{n}} dS$$

$$-\int_\Omega \left(\mu u \nabla^2 P(u_2) + P(u_2) \nabla \cdot (u\mathbf{v}) + auP(u_2) \right) dV$$

$$= \int_\Omega \left(\mu u_2 \nabla^2 P(u) - u_2 \mathbf{v} \cdot \nabla P(u) + aP(u)u_2 + \alpha uu_2 \right) dV.$$

Since $P(u_2)|_\Gamma = 0, u|_\Gamma = \frac{\mu}{\beta} \frac{\partial P(u)}{\partial \mathbf{n}}$, and u is arbitrary, replacing u by u_1, we deduce from the property (3) of Theorem 4.36 that

$$\int_\Omega u_2 \frac{dP}{dt}(u_1)dV = \int_\Gamma \frac{\mu^2}{\beta} \frac{\partial P(u_1)}{\partial \mathbf{n}} \frac{\partial P(u_2)}{\partial \mathbf{n}} dS$$

$$- \int_\Omega \left(\mu u_1 \nabla^2 P(u_2) + \mu u_2 \nabla^2 P(u_1) \right) dV$$

$$+ \int_\Omega \left(u_2 \mathbf{v} \cdot \nabla P(u_1) - P(u_2)\nabla \cdot (u_1 \mathbf{v}) \right) dV$$

$$- \int_\Omega \left(au_1 P(u_2) + au_2 P(u_1) + \alpha u_1 u_2 \right) dV, \quad (4.311)$$

$$\int_\Omega u_1 P(T)(u_2)dV = \gamma \int_\Omega u_1 u_2 dV \tag{4.312}$$

for all $u_1, u_2 \in H^2(\Omega)$. This equation is called a *differential Riccati equation*.

The study of the Riccati equation requires advanced mathematics and is referred to one of the advanced control books [6, 24, 64, 67, 69, 99]. We state a result from [6, p.449, Theorem 2.2] without proof.

Theorem 4.37. *The differential Riccati equation (4.311)-(4.312) has a unique, non-negative, symmetric, strongly continuous solution $P(t)$ (that is, $P(t)u_0$ is continuous for any $u_0 \in L^2(\Omega)$).*

4.5.3 Existence and Uniqueness

Using the solutions of the Riccati equation, we can prove the existence of an optimal control. For this we define the Riccati variance

$$V(u) = \int_\Omega u P u \, dV. \tag{4.313}$$

Lemma 4.8. *Let P be the nonnegative symmetric solution of (4.311)-(4.312). Then*

$$\frac{dV}{dt} = -\alpha \int_\Omega u^2 dV - \beta \int_\Gamma \phi^2 dS + \frac{1}{\beta} \int_\Gamma \left[\mu \frac{\partial Pu}{\partial \mathbf{n}} - \beta \phi \right]^2 dS. \tag{4.314}$$

Proof. Using the equation (4.267), we compute

$$\frac{dV}{dt} = \int_\Omega u P' u \, dV + \int_\Omega u_t P u \, dV + \int_\Omega u P u_t \, dV$$

$$= \int_\Omega u P' u \, dV + \int_\Omega \left(\mu \nabla^2 u + \nabla \cdot (u\mathbf{v}) + au \right) P u \, dV$$

$$+ \int_\Omega u P \left(\mu \nabla^2 u + \nabla \cdot (u\mathbf{v}) + au \right) dV$$

(integrate twice by parts and use the symmetry of P and $Pu|_\Gamma = 0$)

$$= \int_\Omega u P' u \, dV - 2\mu \int_\Gamma \phi \frac{\partial Pu}{\partial \mathbf{n}} dS + \int_\Omega \left(2\mu u \nabla^2 Pu + 2au Pu \right) dV$$

$$+ \int_\Omega (u P \nabla \cdot (u\mathbf{v}) - u\mathbf{v} \cdot \nabla Pu) dV$$

$$= -\alpha \int_\Omega u^2 dV - 2\mu \int_\Gamma \phi \frac{\partial Pu}{\partial \mathbf{n}} dS + \frac{\mu^2}{\beta} \int_\Gamma \left(\frac{\partial Pu}{\partial \mathbf{n}} \right)^2 dS \quad \text{(by (4.311))}$$

$$= -\alpha \int_\Omega u^2 dV - \beta \int_\Gamma \phi^2 dS$$

$$+ \int_\Gamma \left[\frac{\mu^2}{\beta} \left(\frac{\partial Pu}{\partial \mathbf{n}} \right)^2 - 2\mu \phi \frac{\partial Pu}{\partial \mathbf{n}} + \beta \phi^2 \right] dS$$

$$= -\alpha \int_\Omega u^2 dV - \beta \int_\Gamma \phi^2 dS$$

$$+ \frac{1}{\beta} \int_\Gamma \left[\mu \frac{\partial Pu}{\partial \mathbf{n}} - \beta \phi \right]^2 dS.$$

Lemma 4.9. *Let P be the nonnegative symmetric solution of (4.311)-(4.312). Then*

$$J(\phi; u_0, s, T) = \int_\Omega u_0 P(s) u_0 dV + \frac{1}{\beta} \int_s^T \int_\Gamma \left[\mu \frac{\partial Pu}{\partial \mathbf{n}} - \beta \phi \right]^2 dS dt. \quad (4.315)$$

Proof. The identity (4.315) can be obtained by integrating (4.314) from s to T and using $V(T) = \int_\Omega u(T) P(T) u(T) dV = \gamma \int_\Omega u(T)^2 dV$.

Theorem 4.38. *Let J be defined by (4.270). Then there exists a unique optimal control ϕ^* such that*

$$\min_{\phi \in L^2(\Gamma \times (s,T))} J(\phi; u_0, s, T) = J(\phi^*; u_0, s, T) = \int_\Omega u_0 P(s) u_0 dV, \quad (4.316)$$

where P is the nonnegative symmetric solution of (4.311)-(4.312) and

$$\phi^* = \frac{\mu}{\beta} \frac{\partial Pu}{\partial \mathbf{n}} \bigg|_\Gamma. \quad (4.317)$$

Proof. It follows from (4.315) that

$$J(\phi^*; u_0, s, T) \geq \min_{\phi \in L^2(\Gamma \times (s,T))} J(\phi; u_0, s, T)$$

$$\geq \int_\Omega u_0 P(s) u_0 dV$$

$$= J(\phi^*; u_0, s, T)$$

if we take

$$\phi = \phi^* = \frac{\mu}{\beta} \frac{\partial Pu}{\partial \mathbf{n}}.$$

In addition, we can readily show that J is strictly convex and then the optimal control is unique.

Example 4.9. Consider the control problem

$$\frac{\partial u}{\partial t} = \frac{\partial^2 u}{\partial x^2},$$
$$u(0,t) = \phi_1(t), \quad u(\pi,t) = \phi_2(t),$$
$$u(x,0) = u_0(x),$$

with the performance functional

$$J(\phi;u_0,0) = \int_0^T \int_0^\pi u^2 dxdt + \int_0^T (\phi_1^2 + \phi_2^2)dt + \int_0^\pi u(T)^2 dx.$$

Let

$$P(t)\sin(nx) = \sum_{m=1}^\infty p_{mn}(t)\sin(mx).$$

Then

$$p_{mn}(t) = \frac{2}{\pi}\int_0^\pi \sin(mx)P(t)\sin(nx)dx.$$

Taking $u_1 = \sin(mx)$ and $u_2 = \sin(nx)$ in the Riccati equation (4.311), we derive that

$$\frac{dp_{mn}}{dt} = (m^2+n^2)p_{mn} + \frac{2}{\pi}\sum_{i,j=1}^\infty p_{im}p_{jn}ij((-1)^{i+j}-1) - \delta_{mn}, \qquad (4.318)$$

$$p_{mn}(T) = \delta_{mn}. \qquad (4.319)$$

Unlike Example 4.7, this infinite system is coupled and cannot be solved explicitly. Let

$$u(x,t) = \sum_{n=1}^\infty c_n(t)\sin(nx),$$

where

$$c_n(t) = \frac{2}{\pi}\int_0^\pi u(x,t)\sin(nx)dx.$$

We then obtain the optimal boundary state feedback control

$$\phi_1(t) = -\left.\frac{\partial P(t)u(x,t)}{\partial x}\right|_{x=0}$$
$$= -\sum_{n=1}^\infty c_n(t)\left.\frac{\partial P(t)\sin(nx)}{\partial x}\right|_{x=0}$$
$$= -\frac{2}{\pi}\sum_{n=1}^\infty\sum_{m=1}^\infty mp_{mn}(t)\int_0^\pi u(x,t)\sin(nx)dx,$$

$$\phi_2(t) = \left.\frac{\partial P(t)u(x,t)}{\partial x}\right|_{x=\pi}$$

$$= \sum_{n=1}^{\infty} c_n(t) \left.\frac{\partial P(t)\sin(nx)}{\partial x}\right|_{x=\pi}$$

$$= \frac{2}{\pi} \sum_{n=1}^{\infty} \sum_{m=1}^{\infty} m\cos(m\pi)p_{mn}(t) \int_0^{\pi} u(x,t)\sin(nx)dx.$$

4.5.4 Optimal Boundary State Feedback Controls

We now show that the optimal feedback controller (4.317) actually exponentially stabilizes the equation when $T \to \infty$. Consider the boundary control problem

$$\frac{\partial u}{\partial t} = \mu\nabla^2 u + \nabla \cdot (u\mathbf{v}) + au, \tag{4.320}$$

$$u|_{\Gamma} = \phi, \tag{4.321}$$

$$u(\mathbf{x},0) = u_0(\mathbf{x}) \tag{4.322}$$

with the quadratic performance functional over an infinite time interval

$$J(\phi;u_0) = \alpha \int_0^{\infty} \int_{\Omega} |u(\mathbf{x},t)|^2 dV dt + \beta \int_0^{\infty} \int_{\Gamma} |\phi(\mathbf{x},t)|^2 dS dt, \tag{4.323}$$

where $\alpha, \beta \geq 0$ are weight constants. The optimal control problem is to minimize the functional J over $L^2(\Gamma \times (0,\infty))$. Since the control problem (4.320)-(4.322) is exponentially stabilizable, it is well posed; that is, for each $u_0 \in L^2(\Omega)$, there exists a control ϕ such that $J(\phi,u_0) < \infty$.

The following optimality theorem is the infinite time version of Theorem 4.35. Its proof is beyond the scope of the text and is referred to [69].

Theorem 4.39. *The optimal control ϕ^* satisfies the following system:*

$$\frac{\partial u}{\partial t} = \mu\nabla^2 u + \nabla \cdot (u\mathbf{v}) + au, \tag{4.324}$$

$$\frac{\partial v}{\partial t} = -\mu\nabla^2 v + \mathbf{v} \cdot \nabla v - av + \alpha u, \tag{4.325}$$

$$u|_{\Gamma} = \phi^*, \quad v|_{\Gamma} = 0, \tag{4.326}$$

$$\phi^* = -\frac{\mu}{\beta} \left.\frac{\partial v}{\partial \mathbf{n}}\right|_{\Gamma}, \tag{4.327}$$

$$u(\mathbf{x},0) = u_0(\mathbf{x}), \; v(\mathbf{x},\infty) = 0. \tag{4.328}$$

We define a linear operator Π by

$$\Pi u_0(\mathbf{x}) = -v(\mathbf{x},0;u_0) \tag{4.329}$$

for any given function $u_0(\mathbf{x}) \in L^2(\Omega)$, where $v(\mathbf{x}, 0; u_0)$ is the solution of (4.324)-(4.328). It can be seen that

$$\Pi u(\mathbf{x}, t; u_0) = -v(\mathbf{x}, t; u_0). \tag{4.330}$$

In the same way as the derivation of the differential Riccati equation (4.311), we can derive that

$$\int_\Gamma \frac{\mu^2}{\beta} \frac{\partial \Pi(u_1)}{\partial \mathbf{n}} \frac{\partial \Pi(u_2)}{\partial \mathbf{n}} dS - \int_\Omega \left(\mu u_1 \nabla^2 \Pi(u_2) + \mu u_2 \nabla^2 \Pi(u_1) \right) dV$$

$$+ \int_\Omega \left(u_2 \mathbf{v} \cdot \nabla \Pi(u_1) - \Pi(u_2) \nabla \cdot (u_1 \mathbf{v}) \right) dV$$

$$- \int_\Omega \left(a u_1 \Pi(u_2) + a u_2 \Pi(u_1) + \alpha u_1 u_2 \right) dV = 0 \tag{4.331}$$

for all $u_1, u_2 \in H^2(\Omega)$. This equation is called an *algebraic Riccati equation*. The study of the Riccati equation is beyond the scope of this text and we state a result from [6, p.522, Proposition 2.2] without proof. Consider the differential Riccati equation

$$\int_\Omega u_2 \frac{dP}{dt}(u_1) dV = - \int_\Gamma \frac{\mu^2}{\beta} \frac{\partial P(u_1)}{\partial \mathbf{n}} \frac{\partial P(u_2)}{\partial \mathbf{n}} dS$$

$$+ \int_\Omega \left(\mu u_1 \nabla^2 P(u_2) + \mu u_2 \nabla^2 P(u_1) \right) dV$$

$$- \int_\Omega \left(u_2 \mathbf{v} \cdot \nabla P(u_1) - P(u_2) \nabla \cdot (u_1 \mathbf{v}) \right) dV$$

$$+ \int_\Omega \left(a u_1 P(u_2) + a u_2 P(u_1) + \alpha u_1 u_2 \right) dV, \tag{4.332}$$

$$\int_\Omega u_1 P(0)(u_2) dV = \gamma \int_\Omega u_1 u_2 dV \tag{4.333}$$

for all $u_1, u_2 \in H^2(\Omega)$. The algebraic Riccati equation (4.331) is the steady state equation of the differential equation.

Theorem 4.40. *The differential Riccati equation (4.332)-(4.333) has a unique strongly continuous solution that has the following properties:*

1. $\int_\Omega u_0 P(t) u_0 dV = \min\limits_{\phi \in L^2(\Gamma \times (0,t))} J(\phi, u_0, 0, T)$, where J is defined by (4.270).
2. *If $\gamma = 0$, then the limit*

$$\lim_{t \to \infty} P(t) u_0 = \Pi u_0 \quad \text{for all } u_0 \in L^2(\Omega) \tag{4.334}$$

exists and Π is the minimal nonnegative symmetric solution of (4.331). That is, any other solution Π_1 of (4.331) satisfies

$$\int_\Omega u_0 \Pi u_0 dV \le \int_\Omega u_0 \Pi_1 u_0 dV.$$

Using the solutions of the algebraic Riccati equation (4.331), we can prove the existence of an optimal control.

Lemma 4.10. *Let Π be the nonnegative symmetric solution of (4.331) and define*

$$J(\phi; u_0, 0, t) = \alpha \int_0^t \int_\Omega |u(\mathbf{x}, t)|^2 dV dt + \beta \int_0^t \int_\Gamma |\phi(\mathbf{x}, t)|^2 dS dt. \qquad (4.335)$$

Then

$$J(\phi; u_0, 0, t) = \int_\Omega u(0) \Pi u(0) dV - \int_\Omega u(t) \Pi u(t) dV$$
$$+ \frac{1}{\beta} \int_0^t \int_\Gamma \left[\mu \frac{\partial \Pi u}{\partial \mathbf{n}} - \beta \phi \right]^2 dS dt. \qquad (4.336)$$

Proof. Define

$$V(u) = \int_\Omega u(t) \Pi u(t) dV.$$

Using the equation (4.320), we compute

$$\frac{dV}{dt} = \int_\Omega u_t \Pi u dV + \int_\Omega u \Pi u_t dV$$
$$= \int_\Omega \left(\mu \nabla^2 u + \nabla \cdot (u\mathbf{v}) + au \right) \Pi u dV + \int_\Omega u \Pi \left(\mu \nabla^2 u + \nabla \cdot (u\mathbf{v}) + au \right) dV$$

(integrate twice by parts and use the symmetry of Π and $\Pi u|_\Gamma = 0$)

$$= -2\mu \int_\Gamma \phi \frac{\partial \Pi u}{\partial \mathbf{n}} dS + \int_\Omega \left(2\mu u \nabla^2 \Pi u + u\Pi \nabla \cdot (u\mathbf{v}) - u\mathbf{v} \cdot \nabla \Pi u + 2au\Pi u \right) dV$$
$$= -\alpha \int_\Omega u^2 dV - 2\mu \int_\Gamma \phi \frac{\partial \Pi u}{\partial \mathbf{n}} dS + \frac{\mu^2}{\beta} \int_\Gamma \left(\frac{\partial \Pi u}{\partial \mathbf{n}} \right)^2 dS. \quad \text{(by (4.331)}$$

Adding $\beta \int_\Gamma |\phi|^2 dS$ to this equations and integrating from 0 to t, we obtain

$$J(\phi; u_0, 0, t) = \int_\Omega u(0) \Pi u(0) dV - \int_\Omega u(t) \Pi u(t) dV$$
$$+ \int_\Gamma \left[\frac{\mu^2}{\beta} \left(\frac{\partial \Pi u}{\partial \mathbf{n}} \right)^2 - 2\mu \phi \frac{\partial \Pi u}{\partial \mathbf{n}} + \beta \phi^2 \right] dS$$
$$= \int_\Omega u(0) \Pi u(0) dV - \int_\Omega u(t) \Pi u(t) dV$$
$$+ \frac{1}{\beta} \int_0^t \int_\Gamma \left[\mu \frac{\partial \Pi u}{\partial \mathbf{n}} - \beta \phi \right]^2 dS dt.$$

Theorem 4.41. *Let J be defined by (4.323). Then there exists a unique optimal control ϕ^* such that*

$$\min_{\phi \in L^2(\Gamma \times (0, \infty))} J(\phi; u_0) = J(\phi^*; u_0) = \int_\Omega u_0 \Pi u_0 dV, \qquad (4.337)$$

where Π is the nonnegative symmetric solution of (4.331) and

$$\phi^* = \frac{\mu}{\beta} \frac{\partial \Pi u}{\partial \mathbf{n}} \bigg|_{\Gamma}. \tag{4.338}$$

Proof. We note that the optimal control problem is well posed and then the minimum below is finite. It follows from Theorem 4.40 that

$$\min_{\phi \in L^2(\Gamma \times (0,\infty))} J(\phi; u_0) \geq \min_{\phi \in L^2(\Gamma \times (0,\infty))} J(\phi; u_0, 0, t)$$

$$= \min_{\phi \in L^2(\Gamma \times (0,t))} J(\phi; u_0, 0, t)$$

$$= \int_{\Omega} u_0 P(t) u_0 dV,$$

which, combined with (4.334), implies that

$$\min_{\phi \in L^2(\Gamma \times (0,\infty))} J(\phi, u_0) \geq \int_{\Omega} u_0 \Pi u_0 dV. \tag{4.339}$$

On the other hand, by (4.336), we deduce that

$$J(\phi; u_0, 0, t) = \int_{\Omega} u_0 \Pi u_0 dV - \int_{\Omega} u(t) \Pi u(t) dV$$

$$+ \frac{1}{\beta} \int_{\Gamma} \left[\mu \frac{\partial \Pi u}{\partial \mathbf{n}} - \beta \phi \right]^2 dS$$

$$\leq \int_{\Omega} u_0 \Pi u_0 dV + \frac{1}{\beta} \int_{\Gamma} \left[\mu \frac{\partial \Pi u}{\partial \mathbf{n}} - \beta \phi \right]^2 dS.$$

Taking $\phi = \phi^* = \frac{\mu}{\beta} \frac{\partial \Pi u}{\partial \mathbf{n}}$, we obtain

$$J(\phi^*; u_0, 0, t) \leq \int_{\Omega} u_0 \Pi u_0 dV$$

and then

$$\min_{\phi \in L^2(\Gamma \times (0,\infty))} J(\phi, u_0) \leq J(\phi^*; u_0) \leq \int_{\Omega} u_0 \Pi u_0 dV. \tag{4.340}$$

Putting (4.339) and (4.340) together yields (4.337). In addition, we can readily show that J is strictly convex and then the optimal control is unique.

With the optimal feedback controller (4.338), we obtain the feedback control system

$$\frac{\partial u}{\partial t} = \mu \nabla^2 u + \nabla \cdot (u\mathbf{v}) + au, \tag{4.341}$$

$$u|_{\Gamma} = \frac{\mu}{\beta} \frac{\partial \Pi u}{\partial \mathbf{n}} \bigg|_{\Gamma}, \tag{4.342}$$

$$u(\mathbf{x},0) = u_0(\mathbf{x}). \tag{4.343}$$

Define

$$T(t)u_0 = u(t;u_0).$$

We can show that $T(t)$ is a strongly continuous semigroup on $L^2(\Omega)$, but the proof requires the advanced semigroup theory and is referred to [6, p.525, Proposition 3.2]. It follows from Theorem 4.41 that the solution of the system satisfies

$$\int_0^\infty \int_\Omega \alpha u^2 dV dt \le J(\phi^*;u_0) < \infty,$$

and then

$$\int_0^\infty \|T(t)\|^2 dt < \infty.$$

Thus, from Theorem 4.5, we obtain the following theorem.

Theorem 4.42. *Let $\beta > 0$ and Π be the nonnegative symmetric solution of (4.331). Then the feedback control system (4.341)-(4.343) is exponentially stable.*

Example 4.10. Consider the control problem

$$\frac{\partial u}{\partial t} = \frac{\partial^2 u}{\partial x^2},$$
$$u(0,t) = \phi_1(t), \quad u(\pi,t) = \phi_2(t),$$
$$u(x,0) = u_0(x)$$

with the performance functional

$$J(\phi;u_0,0) = \int_0^\infty \int_0^\pi u^2 dx dt + \int_0^\infty (\phi_1^2 + \phi_2^2) dt.$$

Let

$$\Pi \sin(nx) = \sum_{m=1}^\infty p_{mn} \sin(mx).$$

Then

$$p_{mn} = \frac{2}{\pi} \int_0^\pi \sin(mx) \Pi \sin(nx) dx.$$

Taking $u_1 = \sin(mx)$ and $u_2 = \sin(nx)$ in the Riccati equation (4.331), we derive that

$$(m^2 + n^2)\, p_{mn} + \frac{2}{\pi} \sum_{i,j=1}^\infty p_{im} p_{jn} i j ((-1)^{i+j} - 1) - \delta_{mn} = 0. \tag{4.344}$$

This system cannot be solved explicitly. Let

$$u(x,t) = \sum_{n=1}^\infty c_n(t) \sin(nx),$$

where

$$c_n(t) = \frac{2}{\pi} \int_0^\pi u(x,t) \sin(nx) dx.$$

We then obtain the optimal boundary state feedback control

$$\phi_1(t) = -\left.\frac{\partial \Pi u(x,t)}{\partial x}\right|_{x=0}$$

$$= -\sum_{n=1}^\infty c_n(t) \left.\frac{\partial \Pi \sin(nx)}{\partial x}\right|_{x=0}$$

$$= -\frac{2}{\pi} \sum_{n=1}^\infty \sum_{m=1}^\infty m p_{mn} \int_0^\pi u(x,t) \sin(nx) dx,$$

$$\phi_2(t) = \left.\frac{\partial \Pi u(x,t)}{\partial x}\right|_{x=\pi}$$

$$= \sum_{n=1}^\infty c_n(t) \left.\frac{\partial \Pi \sin(nx)}{\partial x}\right|_{x=\pi}$$

$$= \frac{2}{\pi} \sum_{n=1}^\infty \sum_{m=1}^\infty m \cos(m\pi) p_{mn} \int_0^\pi u(x,t) \sin(nx) dx.$$

Exercises 4.5

1. Prove that the operator $P(s)$ defined by (4.285) is linear.
2. Consider the control problem

$$\frac{\partial u}{\partial t} = \frac{\partial^2 u}{\partial x^2},$$

$$\frac{\partial u}{\partial x}(0,t) = \phi_1(t), \quad \frac{\partial u}{\partial x}(\pi,t) = \phi_2(t),$$

$$u(x,0) = u_0(x)$$

with the performance functional

$$J(\phi; u_0, 0, T) = \int_0^T \int_0^\pi u^2 dx dt + \int_0^T (\phi_1(t)^2 + \phi_2(t)^2) dx dt + \int_0^\pi u(T)^2 dx.$$

   a. Calculate the Gâteaux-differential of J.
   b. Derive the optimality system for the optimal control.
   c. Derive a Riccati equation.
   d. Design an optimal state feedback control.

3. Consider the optimal control problem

$$\frac{\partial u}{\partial t} = \mu \nabla^2 u + \nabla \cdot (u\mathbf{v}) + au,$$

$$u|_\Gamma = \sum_{i=1}^{m} b_i(\mathbf{x})\phi_i(t),$$

$$u(\mathbf{x},s) = u_0(\mathbf{x})$$

with the quadratic performance functional

$$J(\phi_1,\cdots,\phi_m;u_0,s,T) = \alpha \int_s^T \int_\Omega |u|^2 dV dt + \beta \sum_{i=1}^{m} \int_s^T |\phi_i|^2 dt + \gamma \int_\Omega |u(T)|^2 dV,$$

where $\alpha,\beta,\gamma \geq 0$ are weight constants and b_i ($i=1,2,\cdots,m$) are given functions.

a. Calculate the Gâteaux-differential of J.
b. Derive the optimality system for the optimal control.
c. Derive a Riccati equation.

4. Use the system (4.324)-(4.328) to show that the operator Π defined by (4.329) has the following properties:

(1) Π is symmetric, that is,

$$\int_\Omega p_0 \Pi u_0 dV = \int_\Omega u_0 \Pi p_0 dV \quad \text{for any } u_0, p_0 \in L^2(\Omega). \qquad (4.345)$$

(2) If ϕ^* is the optimal control of (4.320)-(4.322), then

$$\int_\Omega u_0 \Pi u_0 dV = \alpha \int_0^\infty \int_\Omega |u|^2 dV dt + \beta \int_0^\infty \int_\Gamma |\phi^*|^2 dS dt. \qquad (4.346)$$

5. Derive the algebraic Riccati equation (4.331) by following the procedure of deriving the differential Riccati equation (4.311).

6. Consider the control problem

$$\frac{\partial u}{\partial t} = \frac{\partial^2 u}{\partial x^2},$$

$$\frac{\partial u}{\partial x}(0,t) = \phi_1(t), \quad \frac{\partial u}{\partial x}(\pi,t) = \phi_2(t),$$

$$u(x,0) = u_0(x)$$

with the performance functional

$$J(\phi_1,\phi_2;u_0) = \int_0^\infty \int_0^\pi u^2 dx dt + \int_0^\infty (\phi_1(t)^2 + \phi_2(t)^2) dx dt.$$

Define

$$V(t) = \int_0^\pi u \Pi u dx,$$

where Π is a linear operator to be designed.

a. Derive a Riccati equation for Π such that

$$\frac{dV}{dt} = -\int_0^\pi u^2 dx - \phi_1(t)^2 - \phi_2(t)^2 + ([\Pi u](0,t) - \phi_1(t))^2$$
$$+ ([\Pi u](\pi,t) + \phi_2(t))^2.$$

(The optimal control is given by

$$\phi_1(t) = [\Pi u](0,t), \quad \phi_2(t) = [\Pi u](\pi,t).$$

The proof is not required and beyond the text).
b. Follow Example 4.10 to reduce the Riccati equation to an infinite system.

4.6 Generalization to Abstract Dynamical Systems

We briefly mention that the control theory for the reaction-convection-diffusion equations has been extended to the abstract dynamical systems

$$\frac{du}{dt} = Au + B\phi, \quad u(0) = u_0, \tag{4.347}$$
$$v = Cu + D\phi, \tag{4.348}$$

where A is the infinitesimal generator of an analytic semigroup $T(t)$ on a Hilbert space H, B is a linear operator from a Hilbert space X to H, C is a linear operator from a Hilbert space H to Y, D is a linear operator from X to Y, ϕ is a *control input*, and v is an *output*. To see that the control problem of the reaction-convection-diffusion equations is an example of (4.347)-(4.348), we define the operator A on the Hilbert space $L^2(\Omega)$ by

$$Au = \mu\nabla^2 u + \nabla \cdot (u\mathbf{v}) + au \tag{4.349}$$

with the domain $D(A) = H^2(\Omega) \cap H_0^1(\Omega)$, where I is the identity operator. The control operator $B : \mathbb{R}^m \to L^2(\Omega)$ is defined by

$$B = [b_1(x), \cdots, b_m(x)]. \tag{4.350}$$

The output operator $C : L^2(\Omega) \to \mathbb{R}^l$ is defined by

$$Cu = \left(\int_{\omega_1^o} h_1(\mathbf{x})u(\mathbf{x})dV, \cdots, \int_{\omega_l^o} h_l(\mathbf{x})u(\mathbf{x})dV \right).$$

Then the problem (4.20)-(4.22) is formulated into (4.347)-(4.348) with $D = 0$.

Definition 4.10. If there is a linear operator F from H to X such that $A + BF$ generates an exponentially stable C_0 semigroup, then we say that the system (4.347) is exponentially stabilizable and F is called a feedback operator. If there is a linear operator L from Y to H such that $A + LC$ generates an exponentially stable C_0 semigroup, then we say that the system (4.347)-(4.348) is exponentially detectable and L is called an output injection operator.

As in the case of the reaction-convection-diffusion equations, the stabilization of the problem (4.347)-(4.348) can be solved by decomposing it into a finite dimensional control system and an infinite dimensional system. For details, we refer to [24, 99].

The optimal control theory of the problem (4.347)-(4.348) has been also well established and is referred to [6, Part IV], [64], and [67, Chapter 9] for details. Given $T > 0$, we want to minimize the cost functional

$$J(u) = \int_0^T \left(\|Cu(t)\|^2 + \|\phi(t)\|^2 \right) dt + (P_0 u(T), u(T)) \qquad (4.351)$$

over all controls $\phi \in L^2(0, T; X)$ subject to the differential equation (4.347), where P_0 is a hermitian and nonnegative operator on H. The optimal control problem can be solved by the dynamic programming approach in the following two steps:

1. Solve the Riccati equation

$$\frac{dP}{dt} = A^* P + PA - PBB^* P + C^* C, \quad P(0) = P_0.$$

2. Prove that the optimal control ϕ^* is given by the optimal state feedback controller

$$\phi^* = -B^* P(T - t) u(t)$$

and that u is the solution of the closed loop equation

$$\frac{du}{dt} = (A - BB^* P(T - t)) u, \quad u(0) = u_0.$$

The operator B can be either bounded or unbounded. The bounded case corresponds to the interior control of the reaction-convection-diffusion equations while the unbounded case corresponds to the boundary control. In formulating the boundary control problem into (4.347)-(4.348), we need the following Dirichlet "lifting" operator $L : L^2(\Gamma) \to L^2(\Omega)$ defined as follows. Let λ_0 be a real number such that the equation

$$\mu \nabla^2 u + \nabla \cdot (u\mathbf{v}) + au + \lambda_0 u = 0, \quad u|_\Gamma = f$$

has a unique solution. We then define $L(f) = u$. Using this operator, (4.267)-(4.268) can be written as

$$\frac{du}{dt} = A_{\lambda_0} u - \lambda_0 u - A_{\lambda_0} L(\phi(t)),$$

where

$$A_{\lambda_0} u = \mu \nabla^2 u + \nabla \cdot (u\mathbf{v}) + au + \lambda_0 u.$$

Then the control operator is given by $B = -A_{\lambda_0} L$, which is unbounded.

4.7 References and Notes

The material presented in this chapter is just a simplification of the material from the advanced control books and papers [6, 24, 62, 64, 67, 69, 99]. Here are the resources: Theorem 4.6 is adopted from [24]; the material about uniform convergence from [92]; Exercises 4.3 from [48]; the material about optimal control from [6, 24, 62, 64, 67, 69, 99].

There have been numerous references on feedback control of the parabolic equations and we mention some of them for further studies: Amamm [2], Boskovic, Balogh, and Krstic [8], Boskovic, Krstic, and Liu [7], Burns, Rubio, and King [9], Christofieds [20], Krstic [48], and Lasiecka and Triggiani [54, 55, 57, 58, 64]. For more references, we refer to the above mentioned control books.

Chapter 5
One-dimensional Wave Equation

In this chapter, we study the control problem of the one-dimensional wave equation

$$\frac{\partial^2 u}{\partial t^2} = c^2 \frac{\partial^2 u}{\partial x^2}.$$

A typical physical problem modeled by the wave equation is the vibration of a string. In this problem, $u = u(x,t)$ represents the vertical displacement of the string from its equilibrium and the positive constant c (m/s) is a wave speed.

In what follows, for convenience, we will use the subscripts u_t, u_x or $\frac{\partial u}{\partial t}, \frac{\partial u}{\partial x}$ interchangeably to denote the derivatives of u with respect to t, x, respectively.

5.1 Stability

Consider the wave equation

$$\frac{\partial^2 u}{\partial t^2} = c^2 \frac{\partial^2 u}{\partial x^2} \quad \text{in } (0,L) \times (0,\infty), \tag{5.1}$$

$$u(0,t) = 0, \quad u(L,t) = 0, \quad t \geq 0, \tag{5.2}$$

$$u(x,0) = u_0(x), \quad \frac{\partial u}{\partial t}(x,0) = u_1(x), \quad x \in (0,L), \tag{5.3}$$

where u_0, u_1 are initial conditions. We define the energy of the system (5.1)-(5.3) by

$$E(t) = \frac{1}{2} \int_0^L \left(\left| \frac{\partial u}{\partial t}(x,t) \right|^2 + c^2 \left| \frac{\partial u}{\partial x}(x,t) \right|^2 \right) dx. \tag{5.4}$$

The following theorem shows that the equilibrium 0 is stable, but not exponentially stable.

W. Liu, *Elementary Feedback Stabilization of the Linear Reaction-Convection-Diffusion Equation and the Wave Equation*, Mathématiques et Applications 66, DOI 10.1007/978-3-642-04613-1_5, © Springer-Verlag Berlin Heidelberg 2010

Theorem 5.1. *The energy of the system (5.1)-(5.3) satisfies*

$$E(t) = E(0) \quad \text{for all } t \geq 0. \tag{5.5}$$

Proof. Using the equation (5.1) and integrating by parts, we derive that

$$
\begin{aligned}
\frac{dE}{dt} &= \int_0^L \left(\frac{\partial u}{\partial t}(x,t) \frac{\partial^2 u}{\partial t^2}(x,t) + c^2 \frac{\partial u}{\partial x}(x,t) \frac{\partial^2 u}{\partial x \partial t}(x,t) \right) dx \\
&= \int_0^L \left(c^2 \frac{\partial u}{\partial t}(x,t) \frac{\partial^2 u}{\partial x^2}(x,t) + c^2 \frac{\partial u}{\partial x}(x,t) \frac{\partial^2 u}{\partial x \partial t}(x,t) \right) dx \\
&= c^2 \frac{\partial u}{\partial t}(x,t) \frac{\partial u}{\partial x}(x,t) \Big|_0^L \\
&\quad + \int_0^L \left(-c^2 \frac{\partial^2 u}{\partial t \partial x}(x,t) \frac{\partial u}{\partial x}(x,t) + c^2 \frac{\partial u}{\partial x}(x,t) \frac{\partial^2 u}{\partial x \partial t}(x,t) \right) dx \\
&= 0.
\end{aligned}
$$

So $E(t) = E(0)$ for all $t \geq 0$.

Analogous to finite dimensional control systems, the stability of the system (5.1)-(5.3) is determined by its eigenvalues. To see this, we define the operator A on the Hilbert space $\mathcal{H} = H_0^1(0,L) \times L^2(0,L)$ by

$$A = \begin{bmatrix} 0 & I \\ c^2 \frac{d^2}{dx^2} & 0 \end{bmatrix} \tag{5.6}$$

with the domain $D(A) = (H^2(0,L) \cap H_0^1(0,L)) \times H_0^1(0,L)$, where I denotes the identity operator. Set $v = u_t$ and

$$\mathbf{u} = \begin{bmatrix} u \\ v \end{bmatrix}, \quad \mathbf{u}_0 = \begin{bmatrix} u_0 \\ u_1 \end{bmatrix}.$$

Then the problem (5.1)-(5.3) can be formulated as an abstract system

$$\frac{d\mathbf{u}}{dt} = A\mathbf{u}, \tag{5.7}$$

$$\mathbf{u}(0) = \mathbf{u}_0, \tag{5.8}$$

in the state space $\mathcal{H}$.

Theorem 5.2. *The eigenvalues of A defined by (5.6) and their corresponding eigenfunctions are given by*

$$\lambda_{\pm n} = \pm \frac{nc\pi i}{L}, \quad \mathbf{u}_{\pm n} = \begin{bmatrix} \sin\left(\frac{n\pi x}{L}\right) \\ \pm \frac{nc\pi i}{L} \sin\left(\frac{n\pi x}{L}\right) \end{bmatrix}, \quad n = 1, 2, \cdots, \tag{5.9}$$

where i denotes the imaginary unit. Furthermore, $\{\mathbf{u}_{\pm n}\}$ is an orthogonal basis in $H_0^1(0,L) \times L^2(0,L)$ and

$$\omega_0 = 0 = \sup\{\mathrm{Re}(\lambda), \ \lambda \in \sigma(A)\}, \qquad (5.10)$$

where ω_0 is the growth bound of the solution of the wave equation defined by (2.56) and $\sigma(A)$ denotes the spectrum of A.

Proof. The eigenvalue problem

$$A\mathbf{u} = \lambda\mathbf{u}$$

is equivalent to

$$v = \lambda u, \quad c^2\frac{d^2u}{dx^2} = \lambda v, \quad u(0) = u(L) = 0.$$

Then (5.9) follows from Theorem 2.4. To prove that $\{\mathbf{u}_{\pm n}\}$ is an orthogonal basis in $H_0^1(0,L) \times L^2(0,L)$, it suffices to show that any real function $(u,v) \in H_0^1(0,L) \times L^2(0,L)$ can be expanded in terms of $\{\mathbf{u}_{\pm n}\}$. Let

$$\begin{bmatrix} u \\ v \end{bmatrix} = \sum_{n=1}^{\infty} c_n \begin{bmatrix} \sin\left(\frac{n\pi x}{L}\right) \\ \frac{nc\pi i}{L}\sin\left(\frac{n\pi x}{L}\right) \end{bmatrix} + \sum_{n=1}^{\infty} d_n \begin{bmatrix} \sin\left(\frac{n\pi x}{L}\right) \\ -\frac{nc\pi i}{L}\sin\left(\frac{n\pi x}{L}\right) \end{bmatrix}.$$

Since $\{\sin\left(\frac{n\pi x}{L}\right)\}$ is an orthogonal basis in $L^2(0.L)$, we have the eigenfunction expansions

$$u = \sum_{n=1}^{\infty} a_n \sin\left(\frac{n\pi x}{L}\right), \quad v = \sum_{n=1}^{\infty} b_n \sin\left(\frac{n\pi x}{L}\right).$$

Then c_n and d_n must satisfy the system

$$c_n + d_n = a_n, \quad \frac{nc\pi i}{L}c_n - \frac{nc\pi i}{L}d_n = b_n.$$

Since this system has a unique solution, (u,v) can be expanded in terms of $\{\mathbf{u}_{\pm n}\}$. In addition, (5.10) follows from either (5.5) or Theorem 2.22.

Exercises 5.1

1. Consider the wave equation

$$\frac{\partial^2 u}{\partial t^2} = c^2\frac{\partial^2 u}{\partial x^2}.$$

What happens to the energy $E(t)$ if

a. $u(0,t) = 0$ and $\frac{\partial u}{\partial x}(L,t) = 0$;

b. $u(0,t) = 0$ and $\frac{\partial u}{\partial x}(L,t) = -au(L,t) - m\frac{\partial^2 u}{\partial t^2}(L,t)$ with $a,m > 0$.

2. Consider the wave equation

$$\frac{\partial^2 u}{\partial t^2} = \frac{\partial^2 u}{\partial x^2} + a\frac{\partial^2 u}{\partial x \partial t},$$
$$u(0,t) = 0, \quad u(1,t) = 0,$$
$$u(x,0) = u_0(x), \quad \frac{\partial u}{\partial t}(x,0) = u_1(x),$$

where a is a constant. Show that

$$E(t) \equiv E(0).$$

3. Consider the wave equation

$$\frac{\partial^2 u}{\partial t^2} = \frac{\partial^2 u}{\partial x^2} - a\frac{\partial u}{\partial x},$$
$$u(0,t) = 0, \quad u(1,t) = 0,$$
$$u(x,0) = u_0(x), \quad \frac{\partial u}{\partial t}(x,0) = u_1(x),$$

where a is a constant. Define the weighted energy

$$E_w(t) = \int_0^1 e^{-ax}\left[\left|\frac{\partial u}{\partial t}\right|^2 + \left|\frac{\partial u}{\partial x}\right|^2\right]dx.$$

Show that

$$E_w(t) \equiv E_w(0).$$

4. Use the method of separation of variable to solve the wave equation

$$\frac{\partial^2 u}{\partial t^2} = c^2\frac{\partial^2 u}{\partial x^2} + au \quad \text{in } (0,L) \times (0,\infty),$$
$$u(0,t) = 0, \quad u(L,t) = 0, \quad t \geq 0,$$
$$u(x,0) = u_0(x), \quad \frac{\partial u}{\partial t}(x,0) = u_1(x), \quad x \in (0,L),$$

where a is a constant. Determine the range of a for which the equilibrium 0 is not stable.

5.2 Linear Interior Feedback Stabilization

Consider the wave equation with an interior control

$$\frac{\partial^2 u}{\partial t^2} = c^2\frac{\partial^2 u}{\partial x^2} + \phi \quad \text{in } (0,L) \times (0,\infty), \tag{5.11}$$

$$u(0,t) = 0, \quad u(L,t) = 0, \quad t \geq 0, \tag{5.12}$$

$$u(x,0) = u_0(x), \quad \frac{\partial u}{\partial t}(x,0) = u_1(x), \quad x \in (0,L), \tag{5.13}$$

where $\phi = \phi(x,t)$ is a control to be found, and u_0, u_1 are initial states. Physically the control ϕ represents an external force exerted on a string. In this control system, the state variables are u and $\frac{\partial u}{\partial t}$. If there is a feedback control $\phi = F(u, u_t)$ such that the above closed-loop system is exponentially stable, we say that the wave equation is *exponentially stabilizable* by interior feedback.

The output of the system can be proposed in many ways in accordance with specific physical problems. Since our interest is in the vibration of a string, the controlled output is the displacement

$$o_c = u(x,t). \tag{5.14}$$

If the velocity $u_t(x,t)$ can be measured on $[0,L]$, then the measured output is

$$o_m = u_t(x,t). \tag{5.15}$$

Let the reference output $o_r(x) = r(x)$. Since $u(0,t) = u(L,t) = 0$, $r(x)$ should satisfy $r(0) = r(L) = 0$. If the control ϕ regulates u to r, then $\phi(x,t), u(x,t)$ will converge to $\bar{\phi}(x)$ and $r(x)$, respectively, and the measured output $u_t(x,t)$ will converge to zero. Thus the control steady state $\bar{\phi}$ satisfies

$$c^2 \frac{d^2 r}{dx^2} + \bar{\phi} = 0.$$

Introducing new variables

$$w = u - r, \quad \psi = \phi - \bar{\phi}, \quad \eta_c = o_c - o_r, \quad \eta_m = o_m, \tag{5.16}$$

we then transform the control problem into

$$\frac{\partial^2 w}{\partial t^2} = c^2 \frac{\partial^2 w}{\partial x^2} + \psi,$$
$$\eta_m = w_t(x,t),$$
$$\eta_c = w(x,t)$$

with the zero reference output. Therefore, in what follows, we assume that the reference output is zero.

The control problem is to design a feedback control ϕ to regulate the controlled output o_c to zero, that is, the solution u of the problem (5.11)-(5.13) converges to zero in some norm. Note that zero is the equilibrium of the problem (5.11)-(5.13) with zero steady-state control. Hence, as in the case of finite dimensional systems, the control problem is transformed into the stabilization of the zero equilibrium.

Unlike the reaction-convection-diffusion equation, the control problem (5.11)-(5.13) of the wave equation cannot be decomposed into a finite dimensional control system and a stable infinite dimensional system because all eigenvalues have zero real parts. By Theorem 2.5, the set of eigenfunctions $\{\sin\left(\frac{i\pi x}{L}\right)\}$ is a basis in $L^2(0,L)$. Then we can assume that

$$u = \sum_{i=1}^{\infty} c_i(t) \sin\left(\frac{i\pi x}{L}\right), \tag{5.17}$$

$$u_0 = \sum_{i=1}^{\infty} c_i^0 \sin\left(\frac{i\pi x}{L}\right), \tag{5.18}$$

$$u_1 = \sum_{i=1}^{\infty} c_i^1 \sin\left(\frac{i\pi x}{L}\right), \tag{5.19}$$

where

$$c_i(t) = \frac{2}{L} \int_0^L u(x,t) \sin\left(\frac{i\pi x}{L}\right) dx, \tag{5.20}$$

$$c_i^0 = \frac{2}{L} \int_0^L u_0(x) \sin\left(\frac{i\pi x}{L}\right) dx, \tag{5.21}$$

$$c_i^1 = \frac{2}{L} \int_0^L u_1(x) \sin\left(\frac{i\pi x}{L}\right) dx. \tag{5.22}$$

Substituting (5.17) into (5.11), we obtain

$$\sum_{i=1}^{\infty} \ddot{c}_i(t) \sin\left(\frac{i\pi x}{L}\right) = -c^2 \sum_{i=1}^{\infty} c_i(t) \left(\frac{i\pi}{L}\right)^2 \sin\left(\frac{i\pi x}{L}\right) + \phi(x,t).$$

Multiplying the equation by $\sin\left(\frac{i\pi x}{L}\right)$ and integrating from 0 to L, we obtain

$$\ddot{c}_i = -\left(\frac{ic\pi}{L}\right)^2 c_i + \phi_i(t), \tag{5.23}$$

$$c_i(0) = c_i^0, \quad \dot{c}_i(0) = c_i^1, \quad i = 1, 2, \cdots, \tag{5.24}$$

where

$$\phi_i = \frac{2}{L} \int_0^L \sin\left(\frac{i\pi x}{L}\right) \phi(x) dx.$$

Evidently, the real parts of eigenvalues for each equation are zero. Thus the original control problem (5.11)-(5.13) cannot be decomposed into a finite dimensional control system and a stable infinite dimensional system. So we need a different method to design a feedback controller.

If a control ϕ drives the energy to zero, the energy should be decreasing, and then its derivative should be negative. This motivates us to examine the derivative of the energy

$$\frac{dE}{dt} = \int_0^L \left(\frac{\partial u}{\partial t}(x,t) \frac{\partial^2 u}{\partial t^2}(x,t) + c^2 \frac{\partial u}{\partial x}(x,t) \frac{\partial^2 u}{\partial x \partial t}(x,t) \right) dx$$

$$= \int_0^L \left(c^2 \frac{\partial u}{\partial t}(x,t) \frac{\partial^2 u}{\partial x^2}(x,t) + c^2 \frac{\partial u}{\partial x}(x,t) \frac{\partial^2 u}{\partial x \partial t}(x,t) \right) dx$$

$$+ \int_0^L \frac{\partial u}{\partial t}(x,t) \phi(x,t) dx$$

$$= \int_0^L \frac{\partial u}{\partial t}(x,t) \phi(x,t) dx.$$

To make it negative, we take

$$\phi(x,t) = -k \frac{\partial u}{\partial t}(x,t) \tag{5.25}$$

so that

$$\frac{dE}{dt} = -k \int_0^L \left| \frac{\partial u}{\partial t}(x,t) \right|^2 dx \le 0, \tag{5.26}$$

and then the energy is decreasing, where k is a positive constant called *control gain*. In fact, we will show that this damping force will make the vibration die out exponentially. The feedback control (5.25) is called a *velocity feedback controller*. Since the state u is not used for feedback, this feedback can be regarded as an output feedback if the velocity can be physically measured on the whole domain. The above method of designing a feedback controller is referred as the *method of energy*.

We now use the Fourier method to prove that the equilibrium 0 of the system (5.11)-(5.13) is exponentially stable.

Theorem 5.3. *Assume that $\frac{2Nc\pi}{L} < k \le \frac{2(N+1)c\pi}{L}$ for some integer $N \ge 0$. Then the solution of the problem (5.11)-(5.13) with the feedback control (5.25) is given by*

$$u(x,t) = \sum_{n=1}^{\infty} c_n(t) \sin \left(\frac{n\pi x}{L} \right), \tag{5.27}$$

where

$$a_n = \frac{2}{L} \int_0^L u_0(x) \sin \left(\frac{n\pi x}{L} \right) dx, \tag{5.28}$$

$$b_n = \frac{2}{L} \int_0^L u_1(x) \sin \left(\frac{n\pi x}{L} \right) dx, \tag{5.29}$$

$$c_n = \frac{ka_n + b_n + a_n \sqrt{\omega_n}}{2\sqrt{\omega_n}} \exp((-k + \sqrt{\omega_n})t/2) \quad 1 \le n < N+1,$$

$$+ \frac{a_n \sqrt{\omega_n} - ka_n - b_n}{2\sqrt{\omega_n}} \exp((-k - \sqrt{\omega_n})t/2), \tag{5.30}$$

$$c_{N+1} = \begin{cases} e^{-kt/2} \left(a_{N+1} + (ka_{N+1} + b_{N+1})t \right), & k = \frac{2(N+1)c\pi}{L} \\ a_{N+1} e^{-kt/2} \cos \left(\sqrt{-\omega_{N+1}} t/2 \right) \\ + \frac{ka_{N+1} + b_{N+1}}{\sqrt{-\omega_{N+1}}} e^{-kt/2} \sin \left(\sqrt{-\omega_{N+1}} t/2 \right), & k \ne \frac{2(N+1)c\pi}{L} \end{cases} \tag{5.31}$$

$$c_n = e^{-kt/2}\left(a_n \cos\left(\sqrt{-\omega_n}t/2\right) + \frac{ka_n + b_n}{\sqrt{-\omega_n}} \sin\left(\sqrt{-\omega_n}t/2\right)\right),$$
$$n > N + 1, \tag{5.32}$$

$$\omega_n = k^2 - \frac{4n^2c^2\pi^2}{L^2}. \tag{5.33}$$

Proof. Since

$$u_0(x) = \sum_{n=1}^{\infty} a_n \sin\left(\frac{n\pi x}{L}\right), \tag{5.34}$$

$$u_1(x) = \sum_{n=1}^{\infty} b_n \sin\left(\frac{n\pi x}{L}\right), \tag{5.35}$$

we have

$$c_n(0) = a_n, \quad \frac{dc_n}{dt}(0) = b_n. \tag{5.36}$$

Substituting (5.27) into the equation (5.11), we obtain

$$\sum_{n=1}^{\infty} \left(\frac{d^2c_n}{dt^2}(t) + k\frac{dc_n}{dt}(t) + \frac{n^2c^2\pi^2c_n}{L^2}\right) \sin\left(\frac{n\pi x}{L}\right) = 0,$$

which implies that

$$\frac{d^2c_n}{dt^2}(t) + k\frac{dc_n}{dt}(t) + \frac{n^2c^2\pi^2c_n}{L^2} = 0. \tag{5.37}$$

Solving the equations (5.36)-(5.37) gives (5.30)-(5.32).

Corollary 5.1. *Assume that $\frac{2Nc\pi}{L} \le k < \frac{2(N+1)c\pi}{L}$ for some integer $N \ge 0$. Then there exists $M > 0$ such that the energy of the problem (5.11)-(5.13) with the feedback control (5.25) satisfies*

$$E(t) \le ME(0)e^{-\sigma t} \quad \text{for } t \ge 0, \tag{5.38}$$

where

$$\sigma = \begin{cases} \frac{k}{2}, & 0 < k \le \frac{2c\pi}{L}, \\ \frac{k}{2} - \frac{1}{2}\sqrt{k^2 - \frac{4c^2\pi^2}{L^2}}, & k > \frac{2c\pi}{L}. \end{cases} \tag{5.39}$$

Proof. By (5.34) and (5.35), we deduce that

$$\left\|\frac{\partial u_0}{\partial x}\right\|_{L^2}^2 = \sum_{n=1}^{\infty} \frac{n^2\pi^2 a_n^2}{2L},$$

$$\|u_1\|_{L^2}^2 = \sum_{n=1}^{\infty} \frac{Lb_n^2}{2}.$$

From (5.30)-(5.32), we derive that there exists a positive constant C such that

$$|c'_n(t)|^2 \leq C(n^2 a_n^2 + b_n^2)e^{-\sigma t},$$

$$|c_n(t)|^2 \leq C\left(a_n^2 + \frac{b_n^2}{n^2}\right)e^{-\sigma t}.$$

It therefore follows that

$$
\begin{aligned}
E(t) &= \frac{1}{2}\int_0^L \left|\sum_{n=1}^\infty c'_n(t)\sin\left(\frac{n\pi x}{L}\right)\right|^2 dx + \frac{1}{2}\int_0^L c^2 \left|\sum_{n=1}^\infty c_n(t)\frac{n\pi}{L}\cos\left(\frac{n\pi x}{L}\right)\right|^2 dx \\
&= \frac{L}{4}\sum_{n=1}^\infty |c'_n(t)|^2 + \frac{L}{4}c^2 \sum_{n=1}^\infty |c_n(t)|^2 \frac{n^2\pi^2}{L^2} \\
&\leq \frac{L}{4}\sum_{n=1}^\infty C(n^2 a_n^2 + b_n^2)e^{-\sigma t} + \frac{L}{4}c^2 \sum_{n=1}^\infty \frac{n^2\pi^2}{L^2}C\left(a_n^2 + \frac{b_n^2}{n^2}\right)e^{-\sigma t} \\
&\leq ME(0)e^{-\sigma t},
\end{aligned}
$$

where M is a new positive constant.

Since the function $f(k) = \frac{k}{2} - \frac{1}{2}\sqrt{k^2 - \frac{4c^2\pi^2}{L^2}} = \dfrac{2c^2\pi^2}{L^2\left(k+\sqrt{k^2 - \frac{4c^2\pi^2}{L^2}}\right)}$ is decreasing

function on $[0,\infty)$, the maximum decay rate σ_m that can be achieved by a velocity feedback is $\frac{2c\pi}{L} - \varepsilon$ for any small $\varepsilon > 0$ when the control gain $k = \frac{2c\pi}{L}$. Hence, a larger control gain does not give a larger decay rate.

Define the operator A on the Hilbert space $\mathscr{H} = H_0^1(0,L) \times L^2(0,L)$ by

$$
A = \begin{bmatrix} 0 & I \\ c^2\frac{d^2}{dx^2} & -kI \end{bmatrix} \tag{5.40}
$$

with the domain $D(A) = (H^2(0,L)\cap H_0^1(0,L)) \times H_0^1(0,L)$, where I denotes the identity operator.

Theorem 5.4. *The eigenvalues of A defined by (5.40) and their corresponding eigenfunctions are given by*

$$
\lambda_{\pm n} = \frac{-k \pm \sqrt{k^2 - 4n^2 c^2\pi^2/L^2}}{2}, \tag{5.41}
$$

$$
\mathbf{u}_{\pm n} = \begin{bmatrix} \sin\left(\frac{n\pi x}{L}\right) \\ \lambda_{\pm n}\sin\left(\frac{n\pi x}{L}\right) \end{bmatrix}, \quad n = 1,2,\cdots. \tag{5.42}
$$

Moreover

$$
\omega_0 = \sup\{\mathrm{Re}(\lambda), \ \lambda \in \sigma(A)\}, \tag{5.43}
$$

where ω_0 is the growth bound of the solution of the wave equation defined by (2.56) and $\sigma(A)$ denotes the spectrum of A.

Proof. The eigenvalue problem

$$A\mathbf{u} = \lambda \mathbf{u}, \quad \mathbf{u} = (u,v)^*$$

is equivalent to

$$v = \lambda u, \quad c^2 \frac{d^2 u}{dx^2} - kv = \lambda v, \quad u(0) = u(L) = 0.$$

It then follows from Theorem 2.4 that

$$\lambda_n^2 + k\lambda_n = -\frac{c^2 n^2 \pi^2}{L^2}, \quad u_n = \sin\left(\frac{n\pi x}{L}\right),$$

which implies (5.41). Then (5.43) follows from Corollary 5.1.

If $k = \frac{c\pi n}{L}$, then the algebraic multiplicity of λ_n is 2. We define $\mathbf{u}_{+n}$ as in (5.42) and the generalized eigenfunction $\mathbf{u}_{-n}$ by

$$(A - \lambda_n I)\mathbf{u}_{-n} = \left(A + \frac{k}{2}I\right)\mathbf{u}_{-n} = \mathbf{u}_{+n}, \quad (\mathbf{u}_{+n}, \mathbf{u}_{-n}) = 0.$$

Solving this equation, we obtain

$$\mathbf{u}_{-n} = \begin{bmatrix} \frac{1}{k}\sin\left(\frac{n\pi x}{L}\right) \\ \frac{1}{2}\sin\left(\frac{n\pi x}{L}\right) \end{bmatrix}.$$

Evidently, $\{\mathbf{u}_{\pm n}\}$ is not orthogonal. Define the linear operator L by

$$L\begin{bmatrix} \sin\left(\frac{n\pi x}{L}\right) \\ \pm\frac{n\pi i}{L}\sin\left(\frac{n\pi x}{L}\right) \end{bmatrix} = \begin{bmatrix} \sin\left(\frac{n\pi x}{L}\right) \\ \lambda_{\pm n}\sin\left(\frac{n\pi x}{L}\right) \end{bmatrix}.$$

We can show that L is an invertible linear bounded operator. So $\{\mathbf{u}_{\pm n}\}$ is a Riesz basis. (By definition, a Riesz basis is the image of an orthogonal basis under an invertible linear bounded operator.)

When the feedback gain k is a nonnegative function of x and strictly positive on some subinterval, Corollary 5.1 still holds, but its proof is sophisticated and referred to [22].

Exercises 5.2

1. Consider the wave equation with a velocity feedback control

$$\frac{\partial^2 u}{\partial t^2} = c^2 \frac{\partial^2 u}{\partial x^2} + au - k\frac{\partial u}{\partial t}(x,t) \quad \text{in } (0,L) \times (0,\infty),$$

$$u(0,t) = 0, \quad u(L,t) = 0 \quad t \geq 0,$$

$$u(x,0) = u_0(x), \quad \frac{\partial u}{\partial t}(x,0) = u_1(x) \quad x \in (0,L),$$

where a, k are constants with $k > 0$. Use the method of separation of variables to solve the problem and then determine the range of a such that the equation is exponentially stable.

2. Consider the wave equation with a velocity feedback control

$$\frac{\partial^2 u}{\partial t^2} = c^2 \frac{\partial^2 u}{\partial x^2} - k \frac{\partial u}{\partial t} \quad \text{in } (0,L) \times (0,\infty),$$

$$u(0,t) = 0, \quad \frac{\partial u}{\partial x}(L,t) = 0, \quad t \geq 0,$$

$$u(x,0) = u_0(x), \quad \frac{\partial u}{\partial t}(x,0) = u_1(x), \quad x \in (0,L),$$

where k is a positive constant. Define the energy functional $E(t)$ by

$$E(t) = \frac{1}{2} \int_0^L \left(\left| \frac{\partial u}{\partial t}(x,t) \right|^2 + c^2 \left| \frac{\partial u}{\partial x}(x,t) \right|^2 \right) dx,$$

and perturbed energy functional $E_\varepsilon(t)$ by

$$E_\varepsilon(t) = E(t) + \varepsilon \int_0^L u(x,t) \frac{\partial u}{\partial t}(x,t) dx.$$

a. Show that if ε is small enough, then there exist positive constants c_1 and c_2 such that
$$c_1 E(t) \leq E_\varepsilon(t) \leq c_2 E(t).$$

b. Show that if ε is small enough, then there exists a positive constant δ such that
$$\frac{dE_\varepsilon(t)}{dt} \leq -\delta E_\varepsilon(t).$$

c. Solve the above differential equation and then show that
$$E(t) \leq M E(0) e^{-\delta t},$$

where M is a positive constant.

5.3 Linear Boundary Feedback Stabilization

In the interior control problem (5.11), a damping force is exerted on a string. Such a damping force can be also applied at ends of the string. In fact, this boundary control mechanism is easier to implement in real problems.

Consider the wave equation with a boundary control

$$\frac{\partial^2 u}{\partial t^2} = c^2 \frac{\partial^2 u}{\partial x^2} \quad \text{in } (0,L) \times (0,\infty), \tag{5.44}$$

$$u(0,t) = 0, \quad c^2 \frac{\partial u}{\partial x}(L,t) = \phi(t), \quad t \geq 0, \tag{5.45}$$

$$u(x,0) = u_0(x), \quad \frac{\partial u}{\partial t}(x,0) = u_1(x), \quad x \in (0,L), \tag{5.46}$$

where $\phi = \phi(t)$ represents a control force exerted at the right end of the string. The controlled output is the displacement

$$o_c = u(x,t),$$

and the measured output is assumed to be the velocity at the right end $x = L$

$$o_m = u_t(L,t).$$

If there is a feedback control $\phi = F(u,u_t)$ such that the above closed-loop system is exponentially stable, we say that the wave equation is *exponentially stabilizable* by boundary feedback.

To find a feedback control to regulate u to zero, we examine the derivative of the energy

$$\begin{aligned}
\frac{dE}{dt} &= \int_0^L \left(\frac{\partial u}{\partial t}(x,t) \frac{\partial^2 u}{\partial t^2}(x,t) + c^2 \frac{\partial u}{\partial x}(x,t) \frac{\partial^2 u}{\partial x \partial t}(x,t) \right) dx \\
&= \int_0^L \left(c^2 \frac{\partial u}{\partial t}(x,t) \frac{\partial^2 u}{\partial x^2}(x,t) + c^2 \frac{\partial u}{\partial x}(x,t) \frac{\partial^2 u}{\partial x \partial t}(x,t) \right) dx \\
&= \frac{\partial u}{\partial t}(L,t) \phi(t).
\end{aligned}$$

This leads us to take

$$\phi = -k \frac{\partial u}{\partial t}(L,t) \tag{5.47}$$

so that

$$\frac{dE}{dt} = -k \left| \frac{\partial u}{\partial t}(L,t) \right|^2 \leq 0, \tag{5.48}$$

and then the energy is decreasing, where k is a positive constant, called control gain. Thus we have designed the feedback control $\phi = F(u,u_t) = -k \frac{\partial u}{\partial t}(L,t)$. Since only the velocity at the right-end point is used for feedback, the feedback can be regarded as an output feedback. Using the perturbed energy method developed in [14, 15, 16, 17, 18, 74, 80, 81], we prove that the feedback controller exponentially stabilizes the equilibrium 0 of the system (5.44)-(5.46).

We construct the following perturbed energy functional E_δ

$$F(t) = \int_0^L 2u_t x \frac{\partial u}{\partial x} dx, \tag{5.49}$$

$$E_\delta(t) = E(t) + \delta F(t), \tag{5.50}$$

where δ is a positive constant. We first show that, by choosing δ sufficiently small, E_δ and E are equivalent.

Lemma 5.1. *The perturbed energy satisfies*

$$\left(1 - \frac{2L\delta}{c}\right) E(t) \le E_\delta(t) \le \left(1 + \frac{2L\delta}{c}\right) E(t). \tag{5.51}$$

Proof. Using Young's inequality (2.4), we derive that

$$\begin{aligned}
|F(t)| &= \left| \int_0^L 2u_t x \frac{\partial u}{\partial x} dx \right| \\
&\le \frac{1}{c} \int_0^L 2|x||u_t| \left| c \frac{\partial u}{\partial x} \right| dx \\
&\le \frac{L}{c} \int_0^L 2|u_t| \left| c \frac{\partial u}{\partial x} \right| dx \\
&\le \frac{L}{c} \int_0^L \left(|u_t|^2 + c^2 \left| \frac{\partial u}{\partial x} \right|^2 \right) dx \\
&= \frac{2L}{c} E(t).
\end{aligned}$$

It therefore follows that

$$E_\delta(t) \le E(t) + \delta |F(t)| \le \left(1 + \frac{2L\delta}{c}\right) E(t)$$

and

$$E_\delta(t) \ge E(t) - \delta |F(t)| \ge \left(1 - \frac{2L\delta}{c}\right) E(t).$$

Theorem 5.5. *The solution of (5.44)-(5.46) with the feedback (5.47) satisfies*

$$E(t) \le ME(0)e^{-\sigma t} \quad \text{for } t \ge 0, \tag{5.52}$$

where

$$\delta = \frac{1}{2} \min \left\{ \frac{c}{2L}, \frac{kc^2}{L(c^2 + k^2)} \right\}, \tag{5.53}$$

$$\sigma = 2\delta \left(1 - \frac{2L\delta}{c}\right), \tag{5.54}$$

$$M = \frac{c + 2L\delta}{c - 2L\delta}. \qquad (5.55)$$

Proof. By (5.49), we have

$$F'(t) = \int_0^L 2u_{tt}x\frac{\partial u}{\partial x}dx + \int_0^L 2u_t x\frac{\partial^2 u}{\partial x\partial t}dx. \qquad (5.56)$$

Moreover

$$\begin{aligned}
\int_0^L 2u_{tt}x\frac{\partial u}{\partial x}dx &= \int_0^L 2c^2\frac{\partial^2 u}{\partial x^2}x\frac{\partial u}{\partial x}dx \\
&= Lc^2 u_x^2(L,t) - c^2\int_0^L u_x^2(x,t)dx \\
&= \frac{Lk^2}{c^2}u_t^2(L,t) - c^2\int_0^L u_x^2(x,t)dx \qquad (5.57)
\end{aligned}$$

and

$$\int_0^L 2u_t x\frac{\partial^2 u}{\partial x\partial t}dx = Lu_t^2(L,t) - \int_0^L u_t^2(x,t)dx. \qquad (5.58)$$

It therefore follows from (5.56), (5.57), and (5.58) that

$$F'(t) = L\left(1 + \frac{k^2}{c^2}\right)u_t^2(L,t) - 2E(t). \qquad (5.59)$$

We then derive from (5.48) that

$$\begin{aligned}
E_\delta'(t) &= E'(t) + \delta F'(t) \\
&= -2\delta E(t) + \left(\delta L\left(1 + \frac{k^2}{c^2}\right) - k\right)u_t^2(L,t) \\
&\leq -2\delta E(t).
\end{aligned}$$

It then follows from (5.51) that

$$E_\delta'(t) \leq -2\delta\left(1 - \frac{2L\delta}{c}\right)E_\delta(t) = -\sigma E_\delta(t). \qquad (5.60)$$

Using Gronwall's inequality (4.14), we derive that

$$E_\delta(t) \leq E_\delta(0)\exp(-\sigma t). \qquad (5.61)$$

It therefore follows from (5.51) that

$$\begin{aligned}
E(t) &\leq \frac{1}{1 - \frac{2L\delta}{c}}E_\delta(t) \\
&\leq \frac{1}{1 - \frac{2L\delta}{c}}E_\delta(0)\exp(-\sigma t)
\end{aligned}$$

$$\leq \frac{1 + \frac{2L\delta}{c}}{1 - \frac{2L\delta}{c}} E(0)\exp(-\sigma t).$$

Since the function $f(k) = \frac{kc^2}{L(c^2+k^2)}$ attains the maximum $\frac{c}{2L}$ at $k = c$, δ attains the maximum $\frac{c}{4L}$ at $k = c$. In addition, the decay rate σ attains the maximum $\frac{c}{4L}$ at $\delta = \frac{c}{4L}$. Thus the maximum decay rate is achieved when the control gain $k = c$.

Exercises 5.3

1. Consider the boundary control problem

$$\frac{\partial^2 u}{\partial t^2} = c^2 \frac{\partial^2 u}{\partial x^2} \quad \text{in } (0,L) \times (0,\infty),$$

$$u(0,t) = 0, \quad c^2 \frac{\partial u}{\partial x}(L,t) = -k\frac{\partial u}{\partial t}(L,t), \quad t \geq 0,$$

$$u(x,0) = u_0(x), \quad \frac{\partial u}{\partial t}(x,0) = u_1(x), \quad x \in (0,L).$$

Derive the eigenvalue problem and then solve it by assuming that $u = e^{\lambda t}\varphi(x)$.

2. Let $E : [0,\infty) \to [0,\infty)$ be a non-increasing function and assume that there exists a constant $K > 0$ such that

$$\int_t^\infty E(s)ds \leq KE(t), \quad t \geq 0.$$

Show that

$$E(t) \leq eE(0)e^{-t/K}, \quad t \geq K.$$

(Hint: Define $f(t) = e^{t/K} \int_t^\infty E(s)ds$ and then show that $f(t)$ is non-increasing).

3. Consider the wave equation with a boundary velocity feedback control

$$\frac{\partial^2 u}{\partial t^2} = c^2 \frac{\partial^2 u}{\partial x^2} \quad \text{in } (0,L) \times (0,\infty), \tag{5.62}$$

$$u(0,t) = 0, \quad c^2 \frac{\partial u}{\partial x}(L,t) = -k\frac{\partial u}{\partial t}(L,t), \quad t \geq 0, \tag{5.63}$$

$$u(x,0) = u_0(x), \quad \frac{\partial u}{\partial t}(x,0) = u_1(x), \quad x \in (0,L), \tag{5.64}$$

where k is a positive constant. Define the energy functional $E(t)$ by

$$E(t) = \frac{1}{2}\int_0^L \left(\left|\frac{\partial u}{\partial t}(x,t)\right|^2 + c^2\left|\frac{\partial u}{\partial x}(x,t)\right|^2 \right) dx.$$

a. Show that

$$E(T) + k \int_s^T \left| \frac{\partial u}{\partial t}(L,t) \right|^2 dt = E(s) \quad \text{for any } 0 < s < T.$$

b. Use this energy identity and the multiplier technique (multiply the equation (5.62) by $x\frac{\partial u}{\partial x}$ and then integrate over $(0,L) \times (s,T)$) to show that there exists a constant $K > 0$ such that

$$\int_s^T E(t) \le KE(s) \quad \text{for any } 0 \le s < T.$$

c. Use the result of Problem 2 to show that there exist positive constants M and δ such that

$$E(t) \le ME(0)e^{-\delta t}.$$

4. Prove the Poincaré's inequality of the transmission form: There exists a positive constant C such that

$$\int_0^{x_0} |u_1(x)|^2 dx + \int_{x_0}^L |u_2(x)|^2 dx \tag{5.65}$$

$$\le C \left(\int_0^{x_0} \left| \frac{du_1(x)}{dx} \right|^2 dx + \int_{x_0}^L \left| \frac{du_2(x)}{dx} \right|^2 dx \right)$$

for all $u_1 \in H^1(0,x_0)$ and $u_2 \in H^1(x_0,L)$ with $u_1(x_0) = u_2(x_0)$, where $x_0 \in (0,L)$.
5. Consider the transmission problem of the wave equation with a boundary velocity feedback control

$$\frac{\partial^2 u_1}{\partial t^2} = c_1^2 \frac{\partial^2 u_1}{\partial x^2} \quad \text{in } (0,L/2) \times (0,\infty), \tag{5.66}$$

$$\frac{\partial^2 u_2}{\partial t^2} = c_2^2 \frac{\partial^2 u_2}{\partial x^2} \quad \text{in } (L/2,L) \times (0,\infty), \tag{5.67}$$

$$u_1(0,t) = 0, \quad c^2 \frac{\partial u_2}{\partial x}(L,t) = -k\frac{\partial u_2}{\partial t}(L,t), \tag{5.68}$$

$$u_1(L/2,t) = u_2(L/2,t), \quad c_1^2 \frac{\partial u_1}{\partial x}(L/2,t) = c_2^2 \frac{\partial u_2}{\partial x}(L/2,t), \tag{5.69}$$

$$u_i(x,0) = u_{i0}(x), \quad \frac{\partial u_i}{\partial t}(x,0) = u_{i1}(x), \quad i = 1,2, \tag{5.70}$$

where k is a positive constant. Define

$$E(t) = \frac{1}{2} \int_0^{L/2} \left(\left| \frac{\partial u_1}{\partial t}(x,t) \right|^2 + c_1^2 \left| \frac{\partial u_1}{\partial x}(x,t) \right|^2 \right) dx$$

$$+ \frac{1}{2} \int_{L/2}^L \left(\left| \frac{\partial u_2}{\partial t}(x,t) \right|^2 + c_2^2 \left| \frac{\partial u_2}{\partial x}(x,t) \right|^2 \right) dx,$$

$$F(t) = 2\int_0^{L/2} x\frac{\partial u_1}{\partial t}(x,t)\frac{\partial u_1}{\partial x}(x,t)dx + 2\int_{L/2}^{L} x\frac{\partial u_2}{\partial t}(x,t)\frac{\partial u_2}{\partial x}(x,t)dx,$$

$$E_\varepsilon = E(t) + \varepsilon F(t).$$

a. Show that

$$\frac{dE}{dt} = -k\left|\frac{\partial u_2}{\partial t}(L,t)\right|^2.$$

b. Show that if ε is small enough, then there exist positive constants C_1 and C_2 such that

$$C_1 E(t) \le E_\varepsilon(t) \le C_2 E(t).$$

c. Show that if $c_2 \le c_1$ and ε is small enough, then there exists a positive constant δ such that

$$\frac{dE_\varepsilon(t)}{dt} \le -\delta E_\varepsilon(t).$$

d. Solve the above differential equation and then show that

$$E(t) \le ME(0)e^{-\delta t},$$

where M is a positive constant.

5.4 References and Notes

This chapter is mainly based on the references [14, 15, 16, 17, 18, 22, 23, 24, 47, 70, 71, 80, 93, 110, 111]. The interior feedback control in Section 5.2 is a simplified version of the work by Castro and Zuazua [12, 13], Freitas and Zuazuan [36], and Cox and Zuazua [22]. In their work, these authors considered the following feedback control problem

$$\frac{\partial^2 u}{\partial t^2} = c^2\frac{\partial^2 u}{\partial x^2} + a(x)\phi \quad \text{in } (0,L) \times (0,\infty),$$

$$u(0,t) = 0, \quad u(L,t) = 0, \quad t \ge 0,$$

$$u(x,0) = u_0(x), \quad \frac{\partial u}{\partial t}(x,0) = u_1(x), \quad x \in (0,L),$$

where $a(x)$ is a function supported on a subset $\omega = (x_0 - \varepsilon, x_0 + \varepsilon) \subset [0,L]$, that is, $a(x) = 0$ if $x \notin \omega$. The effect of time delay in the boundary feedback control was investigated by Datko, Lagnese, and Polis [25]; by Datko [26, 27, 28, 29]; by Li and Liu [68]. We mention more references for further studies: Dehman, Lebeau, and Zuazua [33]; Zhang and Zuazua [107, 108].

Chapter 6
Higher-dimensional Wave Equation

In this chapter, we study the control problem of the linear wave equation

$$\frac{\partial^2 u}{\partial t^2} = c^2 \nabla^2 u.$$

This equation can serve as a mathematical model for many physical problems, such as the vibration of a membrane. In the membrane problem, $u = u(x,y,t)$ represents the vertical displacement of the membrane from its equilibrium and the positive constant c (m/s) is a wave speed.

In this chapter, we use the following notation. Let Ω be a bounded open set in $\mathbb{R}^n$ and $\mathbf{x}^0 \in \mathbb{R}^n$. Set (see Figure 6.1)

$$\Gamma = \partial \Omega, \tag{6.1}$$

$$\mathbf{m}(\mathbf{x}) = \mathbf{x} - \mathbf{x}^0 = (x_1 - x_1^0, \cdots, x_n - x_n^0), \tag{6.2}$$

$$\Gamma(\mathbf{x}^0) = \{\mathbf{x} \in \Gamma \ : \ \mathbf{m}(\mathbf{x}) \cdot \mathbf{n}(\mathbf{x}) > 0\}, \tag{6.3}$$

$$\Gamma_*(\mathbf{x}^0) = \Gamma - \Gamma(\mathbf{x}^0) = \{\mathbf{x} \in \Gamma \ : \ \mathbf{m}(\mathbf{x}) \cdot \mathbf{n}(\mathbf{x}) \leq 0\}, \tag{6.4}$$

where $\mathbf{n}$ denotes the unit normal pointing towards the outside of Ω.

6.1 Stability

Consider the wave equation

$$\frac{\partial^2 u}{\partial t^2} = c^2 \nabla^2 u \quad \text{in } \Omega \times (0, \infty), \tag{6.5}$$

$$u = 0 \quad \text{on } \Gamma_*(\mathbf{x}^0) \times (0, \infty), \tag{6.6}$$

$$\frac{\partial u}{\partial \mathbf{n}} = 0 \quad \text{on } \Gamma(\mathbf{x}^0) \times (0, \infty), \tag{6.7}$$

W. Liu, *Elementary Feedback Stabilization of the Linear Reaction-Convection-Diffusion Equation and the Wave Equation*, Mathématiques et Applications 66, DOI 10.1007/978-3-642-04613-1_6, © Springer-Verlag Berlin Heidelberg 2010

Fig. 6.1 Parts $\Gamma(\mathbf{x}^0)$ and $\Gamma_*(\mathbf{x}^0)$ of the boundary of the domain Ω.

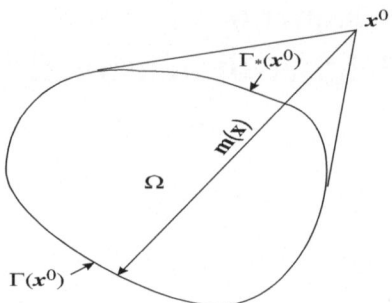

$$u(\mathbf{x},0) = u_0(\mathbf{x}), \quad \frac{\partial u}{\partial t}(\mathbf{x},0) = u_1(\mathbf{x}) \quad \text{in } \Omega, \tag{6.8}$$

where u_0 and u_1 are initial conditions. We define the energy of the system (6.5)-(6.8) by

$$E(t) = \frac{1}{2} \int_\Omega \left(\left| \frac{\partial u}{\partial t}(\mathbf{x},t) \right|^2 + c^2 |\nabla u(\mathbf{x},t)|^2 \right) dV. \tag{6.9}$$

The following theorem shows that the equilibrium 0 is stable, but not exponentially stable.

Theorem 6.1. *The energy of the system (6.5)-(6.8) satisfies the following identity*

$$E(t) = E(0) \quad \text{for all } t \geq 0. \tag{6.10}$$

Proof. Using the equation (6.5) and Green's identity (2.26), we derive that

$$\begin{aligned}
\frac{dE}{dt} &= \int_\Omega \left(\frac{\partial u}{\partial t}(\mathbf{x},t) \frac{\partial^2 u}{\partial t^2}(\mathbf{x},t) + c^2 \nabla u(\mathbf{x},t) \cdot \nabla \frac{\partial u}{\partial t}(\mathbf{x},t) \right) dV \\
&= \int_\Omega \left(c^2 \frac{\partial u}{\partial t}(\mathbf{x},t) \nabla^2 u + c^2 \nabla u(\mathbf{x},t) \cdot \nabla \frac{\partial u}{\partial t}(\mathbf{x},t) \right) dV \\
&= \int_\Gamma c^2 \frac{\partial u}{\partial t}(\mathbf{x},t) \frac{\partial u}{\partial \mathbf{n}}(\mathbf{x},t) dS - \int_\Omega c^2 \nabla u(\mathbf{x},t) \cdot \nabla \frac{\partial u}{\partial t}(\mathbf{x},t) dV \\
&\quad + \int_\Omega c^2 \nabla u(\mathbf{x},t) \cdot \nabla \frac{\partial u}{\partial t}(\mathbf{x},t) dV \\
&= 0.
\end{aligned}$$

So $E(t) = E(0)$ for all $t \geq 0$.

We can easily show that Theorem 5.2 also holds for the higher-dimensional wave equation.

6.2 Linear Boundary Feedback Stabilization

Consider the wave equation with a boundary control

$$\frac{\partial^2 u}{\partial t^2} = c^2 \nabla^2 u \quad \text{in } \Omega \times (0,\infty), \tag{6.11}$$

$$u = 0 \quad \text{on } \Gamma_*(\mathbf{x}^0) \times (0,\infty), \tag{6.12}$$

$$c^2 \frac{\partial u}{\partial \mathbf{n}} = \phi \quad \text{on } \Gamma(\mathbf{x}^0) \times (0,\infty), \tag{6.13}$$

$$u(\mathbf{x},0) = u_0(\mathbf{x}), \quad \frac{\partial u}{\partial t}(\mathbf{x},0) = u_1(\mathbf{x}) \quad \text{in } \Omega, \tag{6.14}$$

where $\phi = \phi(\mathbf{x},t)$ is a control to be designed. In this control system, the state variables are u and $\frac{\partial u}{\partial t}$, and the vertical component of the tensile force is controlled. Mathematically, a boundary control of this kind is called a Neumann boundary control. Since the equilibrium 0 is stable, but not exponentially stable, we are interested in stabilizing it exponentially. If there is a feedback control $\phi = F(u, u_t)$ such that the above closed-loop system is exponentially stable, we say that the wave equation is *exponentially stabilizable* by boundary feedback.

To design a feedback control, we examine the derivative of the energy

$$\frac{dE}{dt} = \int_\Omega \left(\frac{\partial u}{\partial t} \frac{\partial^2 u}{\partial t^2} + c^2 \nabla u \cdot \nabla \frac{\partial u}{\partial t} \right) dV$$

$$= \int_\Omega \left(\frac{\partial u}{\partial t} c^2 \nabla^2 u + c^2 \nabla u \cdot \nabla \frac{\partial u}{\partial t} \right) dV$$

$$= \int_\Gamma \frac{\partial u}{\partial t} c^2 \frac{\partial u}{\partial \mathbf{n}} d\Gamma$$

$$= \int_\Gamma \frac{\partial u}{\partial t} \phi \chi_{\Gamma(\mathbf{x}^0)} d\Gamma.$$

This leads us to take

$$\phi = -k \mathbf{m} \cdot \mathbf{n} \frac{\partial u}{\partial t} \tag{6.15}$$

so that

$$\frac{dE}{dt} = -k \int_{\Gamma(\mathbf{x}^0)} \mathbf{m} \cdot \mathbf{n} \left| \frac{\partial u}{\partial t} \right|^2 d\Gamma \leq 0, \tag{6.16}$$

where k is a positive constant. Then we have designed the feedback control $\phi = F(u, u_t) = -k \mathbf{m} \cdot \mathbf{n} \frac{\partial u}{\partial t} \big|_{\Gamma(\mathbf{x}^0)}$. With this feedback control, the problem (6.11)-(6.14) becomes a closed-loop system:

$$\frac{\partial^2 u}{\partial t^2} = c^2 \nabla^2 u \quad \text{in } \Omega \times (0,\infty), \tag{6.17}$$

$$u = 0 \quad \text{on } \Gamma_*(\mathbf{x}^0) \times (0,\infty), \tag{6.18}$$

$$c^2 \frac{\partial u}{\partial \mathbf{n}} = -k\mathbf{m} \cdot \mathbf{n} \frac{\partial u}{\partial t} \quad \text{on } \Gamma(\mathbf{x}^0) \times (0, \infty), \tag{6.19}$$

$$u(\mathbf{x}, 0) = u_0(\mathbf{x}), \quad \frac{\partial u}{\partial t}(\mathbf{x}, 0) = u_1(\mathbf{x}) \quad \text{in } \Omega. \tag{6.20}$$

This feedback control can be regarded as an output feedback control since only the velocity on the part of boundary is used for feedback.

Evidently, the above energy functional E is a Lyapunov functional. The above procedure of designing a feedback controller to make the energy functional become Lyapunov is called a Lyapunov design. We now show that the system is exponentially stable. To this end, we need Poincaré's inequality.

Lemma 6.1. *(Poincaré's inequality) There exists a positive constant $M > 0$ such that*

$$\int_\Omega |u|^2 dV \le M \int_\Omega |\nabla u|^2 dV \tag{6.21}$$

holds for any $u(x) \in \{u \in H^1(\Omega) \mid u = 0 \text{ on } \Gamma_*(\mathbf{x}^0)\}$.

The proof of this lemma is referred to [30].

Using Poincaré's inequality, we can readily prove the energy norm is equivalent to the usual norm in $\{u \in H^1(\Omega) \mid u = 0 \text{ on } \Gamma_*(\mathbf{x}^0)\}$.

Lemma 6.2. *There exists a positive constant $M > 0$ such that*

$$M \int_\Omega (u^2 + c^2 |\nabla u|^2) dV \le \int_\Omega c^2 |\nabla u|^2 dx$$

$$\le \int_\Omega (u^2 + c^2 |\nabla u|^2) dV \tag{6.22}$$

holds for any $u \in \{u \in H^1(\Omega) \mid u = 0 \text{ on } \Gamma_*(\mathbf{x}^0)\}$.

We use the perturbed energy method to establish the exponential stability of (6.17)-(6.20). So we construct the following perturbed energy functional E_δ

$$F(t) = \int_\Omega [2u_t m \cdot \nabla u + (n-1)u_t u] dV, \tag{6.23}$$

$$E_\delta(t) = E(t) + \delta F(t), \tag{6.24}$$

where δ is a positive constant. How is this perturbed energy constructed? Here is the idea. To achieve the exponential stability, the perturbed energy should be constructed such that

$$E_\delta'(t) = E'(t) + \delta F'(t) \le -\delta E(t).$$

Then F should be constructed such that F' contains $-E(t)$. In the proof below, we will see that

$$F'(t) \le -E(t) + C \int_{\Gamma(\mathbf{x}^0)} \mathbf{m} \cdot \mathbf{n} |u_t|^2 dS.$$

where C is a positive constant. The positive term $C \int_{\Gamma(\mathbf{x}^0)} \mathbf{m} \cdot \mathbf{n} |u_t|^2 dS$ can be cancelled by $E'(t)$ by making δ sufficiently small and then we obtain the desired differential inequality. The reason why $m \cdot \nabla u$ is chosen in F will be further explained in the proof of Theorem 6.3 below.

We first show that, by choosing δ sufficiently small, E_δ and E are equivalent.

Lemma 6.3. *There exists a constant $C > 0$ such that*

$$(1 - \delta C)E(t) \leq E_\delta(t) \leq (1 + \delta C)E(t). \tag{6.25}$$

Proof. Using Poincaré's inequality, we deduce that there exists a constant $C > 0$ such that

$$\left| \int_\Omega [2u_t m \cdot \nabla u + (n-1)u_t u] dV \right| \leq C \int_\Omega [|u_t|^2 + |\nabla u|^2 + |u|^2] dV$$

$$\leq C \int_\Omega [|u_t|^2 + c^2 |\nabla u|^2] dV$$

$$\leq CE(t), \tag{6.26}$$

where the constant C changes from line to line. Then (6.25) follows from (6.26).

To prove the exponential stability, we need the trace theorem.

Theorem 6.2. *Let Ω be a bounded domain in $\mathbb{R}^n$ with C^1 boundary. Then the operator $\gamma : H^1(\Omega) \to L^2(\Gamma)$ defined by*

$$\gamma(u) = u|_\Gamma$$

is a linear continuous operator.

The proof of this theorem is referred to [73, p. 41]. This theorem implies that there exists a constant $C > 0$ such that

$$\int_\Gamma |u(\mathbf{x})|^2 dS \leq C \int_\Omega \left(|u(\mathbf{x})|^2 + |\nabla u(\mathbf{x})|^2 \right) dV \quad \text{for all } u \in H^1(\Omega). \tag{6.27}$$

To better understand this inequality, we consider the case $\Omega = (0,1)$. Since

$$f(0) = f(x) - \int_0^x f'(t)dt, \quad f(1) = f(x) + \int_x^1 f'(t)dt,$$

it follows that

$$\int_\Gamma |f(\mathbf{x})|^2 dS = f^2(0) + f^2(1)$$

$$= \int_0^1 \left(f^2(0) + f^2(1) \right) dx$$

$$= \int_0^1 \left[\left(f(x) - \int_0^x f'(t)dt \right)^2 + \left(f(x) + \int_x^1 f'(t)dt \right)^2 \right] dx$$

$$\leq C \int_0^1 \left[f^2(x) + (f'(x))^2 \right] dx.$$

Theorem 6.3. *Assume that* $\Gamma_*(\mathbf{x}^0)$ *has a non-empty interior and*

$$\Gamma_*(\mathbf{x}^0) \cap \bar{\Gamma}(\mathbf{x}^0) = \emptyset. \tag{6.28}$$

Then there exist constants $M, \sigma > 0$ *such that the solution of (6.17)-(6.20) satisfies*

$$E(t) \leq ME(0)e^{-\sigma t} \quad \text{for } t \geq 0. \tag{6.29}$$

Proof. By (6.25), it suffices to show

$$E'_\delta(t) = E'(t) + \delta F'(t) \leq -\delta E(t)$$

to prove (6.29). Thus we need to estimate F'. By (6.23), we have

$$F'(t) = \int_\Omega 2u_{tt}\mathbf{m} \cdot \nabla u dV + \int_\Omega 2u_t \mathbf{m} \cdot \nabla u_t dV$$
$$+ \int_\Omega (n-1)uu_{tt}dV + (n-1)\int_\Omega |u_t|^2 dV. \tag{6.30}$$

We now estimate every integral in (6.30) as follows. Since $u = 0$ on $\Gamma_*(\mathbf{x}^0)$, we have

$$\mathbf{n} = \frac{\nabla u}{\|\nabla u\|}, \quad \frac{\partial u}{\partial \mathbf{n}} = \nabla u \cdot \mathbf{n} = \|\nabla u\|.$$

It then follows that

$$\frac{\partial u}{\partial x_k} = \|\nabla u\| n_k = \frac{\partial u}{\partial \mathbf{n}} n_k \quad \text{on} \quad \Gamma_*(\mathbf{x}^0), \tag{6.31}$$

where $\mathbf{n} = (n_1 \cdots, n_n)$. Using Green's identity (2.26) and integration by parts (2.25), we derive that

$$2\int_\Omega u_{tt}\mathbf{m} \cdot \nabla u dV \quad \text{(use (6.17))}$$

$$= 2\int_\Omega c^2 \nabla^2 u \mathbf{m} \cdot \nabla u dV \quad \text{(use (6.21))}$$

$$= \int_\Gamma 2c^2 \frac{\partial u}{\partial \mathbf{n}}\mathbf{m} \cdot \nabla u dS - 2\int_\Omega c^2 \nabla u \cdot \nabla (\mathbf{m} \cdot \nabla u) dV$$

$$= \int_\Gamma 2c^2 \frac{\partial u}{\partial \mathbf{n}}\mathbf{m} \cdot \nabla u dS - 2\int_\Omega c^2 \sum_{i=1}^n \sum_{j=1}^n \frac{\partial u}{\partial x_i} \frac{\partial}{\partial x_i}\left((x_j - x_j^0) \frac{\partial u}{\partial x_j} \right) dV$$

$$= \int_\Gamma 2c^2 \frac{\partial u}{\partial \mathbf{n}}\mathbf{m} \cdot \nabla u dS - \int_\Omega c^2 \left(2|\nabla u|^2 + \mathbf{m} \cdot \nabla(|\nabla u|^2) \right) dV \quad \text{(use (2.25))}$$

$$= c^2 \int_\Gamma \left[2\frac{\partial u}{\partial \mathbf{n}} \mathbf{m} \cdot \nabla u - \mathbf{m} \cdot \mathbf{n} |\nabla u|^2 \right] dS + (n-2)c^2 \int_\Omega |\nabla u|^2 dV$$

$$= c^2 \int_{\Gamma(\mathbf{x}^0)} \left(2\frac{\partial u}{\partial \mathbf{n}} \mathbf{m} \cdot \nabla u - \mathbf{m} \cdot \mathbf{n} |\nabla u|^2 \right) dS$$

$$+ c^2 \int_{\Gamma_*(\mathbf{x}^0)} \mathbf{m} \cdot \mathbf{n} \left| \frac{\partial u}{\partial \mathbf{n}} \right|^2 dS + (n-2)c^2 \int_\Omega |\nabla u|^2 dV. \tag{6.32}$$

This equation can be obtained by multiplying (6.17) by the multiplier $\mathbf{m} \cdot \nabla u$ and then integrating over Ω. Thus the construction of $F(t)$ is motivated by the multiplier technique.

Since $\mathbf{m} \cdot \mathbf{n} \leq 0$ on $\Gamma_*(\mathbf{x}^0)$, we have

$$c^2 \int_{\Gamma_*(\mathbf{x}^0)} \mathbf{m} \cdot \mathbf{n} \left| \frac{\partial u}{\partial \mathbf{n}} \right|^2 dS \leq 0. \tag{6.33}$$

It therefore follows from (6.32) that

$$2 \int_\Omega u_{tt} \mathbf{m} \cdot \nabla u \, dV \leq c^2 \int_{\Gamma(\mathbf{x}^0)} \left(2\frac{\partial u}{\partial \mathbf{n}} \mathbf{m} \cdot \nabla u - \mathbf{m} \cdot \mathbf{n} |\nabla u|^2 \right) dS$$

$$+ (n-2)c^2 \int_\Omega |\nabla u|^2 dV. \tag{6.34}$$

For the second integral in (6.30), we derive from (2.25) that

$$2 \int_\Omega u_t \mathbf{m} \cdot \nabla u_t \, dV = \int_\Omega \mathbf{m} \cdot \nabla (|u_t|^2) dV \tag{6.35}$$

$$= -n \int_\Omega |u_t(t)|^2 dV + \int_{\Gamma(\mathbf{x}^0)} \mathbf{m} \cdot \mathbf{n} |u_t|^2 dS. \tag{6.36}$$

From this equation we can see how the multiplier $\mathbf{m} \cdot \nabla u$ is critical for producing the negative term $-n \int_\Omega |u_t(t)|^2 dV$, which is necessary for producing $-E(t)$.

Using (6.17), we deduce

$$(n-1) \int_\Omega u u_{tt} \, dV = (n-1) \int_\Omega u c^2 \nabla^2 u \, dV$$

$$= (n-1) \int_{\Gamma(\mathbf{x}^0)} c^2 \frac{\partial u}{\partial \mathbf{n}} u \, dS$$

$$- (n-1)c^2 \int_\Omega |\nabla u|^2 dV. \tag{6.37}$$

This equation tells why the term $u u_t$ is included in F to produce the negative term $-(n-1)c^2 \int_\Omega |\nabla u|^2 dV$, which is necessary for producing $-E(t)$. It therefore follows from (6.30) and (6.34)-(6.37) that

$$F'(t) \leq c^2 \int_{\Gamma(\mathbf{x}^0)} \left(2\frac{\partial u}{\partial \mathbf{n}} \mathbf{m} \cdot \nabla u - \mathbf{m} \cdot \mathbf{n} |\nabla u|^2 \right) dS$$

$$+ \int_{\Gamma(\mathbf{x}^0)} \mathbf{m} \cdot \mathbf{n} |u_t|^2 dS + (n-1) \int_{\Gamma(\mathbf{x}^0)} c^2 \frac{\partial u}{\partial \mathbf{n}} u \, dS$$

$$- c^2 \int_{\Omega} |\nabla u|^2 dV - \int_{\Omega} |u_t|^2 dV. \tag{6.38}$$

We now estimate the boundary integrals in the above inequality. Using Young's inequality (2.4) and the boundary control (6.19), we deduce that

$$c^2 \int_{\Gamma(\mathbf{x}^0)} \left(2\frac{\partial u}{\partial \mathbf{n}} \mathbf{m} \cdot \nabla u - \mathbf{m} \cdot \mathbf{n} |\nabla u|^2 \right) dS \quad \text{(use (6.19))}$$

$$= \int_{\Gamma(\mathbf{x}^0)} \left(-2k\mathbf{m} \cdot \mathbf{n} u_t \mathbf{m} \cdot \nabla u - \mathbf{m} \cdot \mathbf{n} |\nabla u|^2 \right) dS$$

$$\text{(use Young's inequality (2.4))} \tag{6.39}$$

$$\leq C \int_{\Gamma(\mathbf{x}^0)} \mathbf{m} \cdot \mathbf{n} |u_t|^2 dS + \int_{\Gamma(\mathbf{x}^0)} \mathbf{m} \cdot \mathbf{n} |\nabla u|^2 dS$$

$$- \int_{\Gamma(\mathbf{x}^0)} \mathbf{m} \cdot \mathbf{n} |\nabla u|^2 dS \tag{6.40}$$

$$= C \int_{\Gamma(\mathbf{x}^0)} \mathbf{m} \cdot \mathbf{n} |u_t|^2 dS. \tag{6.41}$$

Hereafter C denotes a generic positive constant that may change from line to line. Moreover, by Young's inequality, the boundary control (6.19), and the trace theorem (the trace inequality (6.27)), we deduce that

$$(n-1) \int_{\Gamma(\mathbf{x}^0)} c^2 \frac{\partial u}{\partial \mathbf{n}} u \, dS$$

$$= -(n-1) \int_{\Gamma(\mathbf{x}^0)} k\mathbf{m} \cdot \mathbf{n} u_t u \, dS$$

$$\leq C \int_{\Gamma(\mathbf{x}^0)} \mathbf{m} \cdot \mathbf{n} |u_t|^2 dS + \omega \int_{\Gamma(\mathbf{x}^0)} |u|^2 dS$$

$$\leq C \int_{\Gamma(\mathbf{x}^0)} \mathbf{m} \cdot \mathbf{n} |u_t|^2 dS + \omega M \int_{\Omega} |\nabla u|^2 dV$$

$$\leq C \int_{\Gamma(\mathbf{x}^0)} \mathbf{m} \cdot \mathbf{n} |u_t|^2 dS + \frac{c^2}{2} \int_{\Omega} |\nabla u|^2 dV, \tag{6.42}$$

where ω is chosen such that $\omega M = \frac{c^2}{2}$. It therefore follows from (6.38)-(6.42) that

$$F'(t) \leq -E(t) + C \int_{\Gamma(\mathbf{x}^0)} \mathbf{m} \cdot \mathbf{n} |u_t|^2 dS. \tag{6.43}$$

We then derive from (6.16) that

$$
\begin{aligned}
E_\delta'(t) &= E'(t) + \delta F'(t) \\
&\leq -\delta E(t) + (\delta C - k) \int_{\Gamma(\mathbf{x}^0)} \mathbf{m} \cdot \mathbf{n} |u_t|^2 dS \\
&\leq -\delta E(t)
\end{aligned}
$$

by taking $\delta < k/C$. It then follows from (6.25) that

$$
E_\delta'(t) \leq -\delta(1 - \delta C) E_\delta(t). \tag{6.44}
$$

Using the differential inequality (4.13), we then derive that

$$
E_\delta(t) \leq E_\delta(0) \exp[-\delta(1 - \delta C)t]. \tag{6.45}
$$

Then (6.29) follows from (6.25).

If the position of a string or membrane can be controlled, we can rewrite the controller (6.19) as follows. Integrating (6.19) from 0 to t, we obtain

$$
u(\mathbf{x},t) = u_0(\mathbf{x}) - \frac{c^2}{k\mathbf{m} \cdot \mathbf{n}} \frac{\partial}{\partial \mathbf{n}} \int_0^t u(\mathbf{x},s) ds. \tag{6.46}
$$

This is an integral controller since it involves the integral $\int_0^t u(\mathbf{x},s) ds$. The boundary control of this kind is called Dirichlet boundary control.

If a term au with $a \geq 0$ is added to the equation (6.17), Theorem 6.3 still holds. For details, we refer to [47, Chapter 8]. The geometrical condition (6.28) can be removed by using microlocal analysis. This is beyond the scope of the text and is referred to [5, 63].

Exercises 6.2

1. Define the operator A on the Hilbert space $\mathscr{H} = \{(u,v) \in H^1((0,1) \times (0,1)) \times L^2((0,1) \times (0,1)) \mid u(0,y) = u(x,0) = u(x,1) = 0\}$ by

$$
A = \begin{pmatrix} 0 & I \\ c^2 \nabla^2 & 0 \end{pmatrix}
$$

with the domain $D(A) = \{(u,v) \in H^2((0,1) \times (0,1)) \times H^1((0,1) \times (0,1)) \mid u(0,y) = u(x,0) = u(x,1) = v(0,y) = v(x,0) = v(x,1) = 0, \ c^2 \frac{\partial u}{\partial x}(1,y) = -kv(1,y)\}$, where I denotes the identity operator. Find eigenvalues and eigenfunctions of A.

6.3 Nonlinear Boundary Feedback Stabilization

We now design a nonlinear boundary feedback control for the wave equation. As in
the case of linear control, we look for the velocity feedback control

$$\frac{\partial^2 u}{\partial t^2} = c^2 \nabla^2 u \quad \text{in } \Omega \times (0, \infty), \tag{6.47}$$

$$u = 0 \quad \text{on } \Gamma_*(\mathbf{x}^0) \times (0, \infty), \tag{6.48}$$

$$c^2 \frac{\partial u}{\partial \mathbf{n}} = -\mathbf{m} \cdot \mathbf{n} f \left(\frac{\partial u}{\partial t} \right) \quad \text{on } \Gamma(\mathbf{x}^0) \times (0, \infty), \tag{6.49}$$

$$u(\mathbf{x}, 0) = u_0(\mathbf{x}), \quad \frac{\partial u}{\partial t}(\mathbf{x}, 0) = u_1(\mathbf{x}) \quad \text{in } \Omega, \tag{6.50}$$

where f is a function to be designed.

Theorem 6.4. *Suppose that the function f is continuous and satisfies the following
conditions:*

$$f(0) = 0, \tag{6.51}$$

$$|f(u_1) - f(u_2)| \le k_1 |u_1 - u_2|^q \text{ for } |u_1 - u_2| \le 1, \tag{6.52}$$

$$|f(u_1) - f(u_2)| \le k_1 |u_1 - u_2| \text{ for } |u_1 - u_2| \ge 1, \tag{6.53}$$

$$f(u) \cdot u \ge k_2 |u|^{p+1} \text{ for } |u| \le 1, \tag{6.54}$$

$$f(u) \cdot u \ge k_2 |u|^2 \text{ for } |u| \ge 1, \tag{6.55}$$

$$0 \le [f(u_1) - f(u_2)](u_1 - u_2) \text{ for } u_1, u_2 \in \mathbb{R}, \tag{6.56}$$

for some constants $k_1, k_2 > 0$ and p, q with $0 < q \le 1$.

(1) If $p = q = 1$, there exist some constants $M \ge 1$ and $\omega > 0$ such that

$$E(t) \le ME(0)e^{-\omega t}, \quad \forall t \ge 0. \tag{6.57}$$

(2) If $p + 1 > 2q$, then

$$E(t) \le \frac{E(0) + CE^{\sigma+1}(0)}{[1 + F(E(0))t]^{1/\sigma}}, \quad \forall t \ge 0, \tag{6.58}$$

*where C is a positive constant, F is a positive function with $F(0) = 0$, and σ is
given by*

$$\sigma = \frac{p + 1 - 2q}{2q}. \tag{6.59}$$

Proof. As in the case of the proof of Theorem 6.3, we define a perturbed energy
function by

$$E_{\delta,\sigma}(t) = E(t) + \delta[E(t)]^\sigma F(t), \tag{6.60}$$

where F is defined by (6.23). We now show that $E_{\delta,\sigma}$ satisfies

$$E'_{\delta,\sigma}(t) \leq -\alpha E_{\delta,\sigma}^{\sigma+1}(t), \tag{6.61}$$

where α is a positive constant.

Note that calculations of F' (defined by (6.23)) in Theorem 6.3 are still valid. We then have the estimate for F'

$$F'(t) \leq c^2 \int_{\Gamma(\mathbf{x}^0)} \left(2\frac{\partial u}{\partial \mathbf{n}}\mathbf{m}\cdot\nabla u - \mathbf{m}\cdot\mathbf{n}|\nabla u|^2\right) dS$$
$$+ \int_{\Gamma(\mathbf{x}^0)} \mathbf{m}\cdot\mathbf{n}|u_t|^2 dS + (n-1)\int_{\Gamma(\mathbf{x}^0)} c^2\frac{\partial u}{\partial \mathbf{n}} u dS$$
$$- c^2 \int_\Omega |\nabla u|^2 dV - \int_\Omega |u_t|^2 dV. \tag{6.62}$$

We now estimate the boundary integrals in the above inequality. Using Young's inequality and the boundary control (6.49), we deduce that

$$c^2 \int_{\Gamma(\mathbf{x}^0)} \left(2\frac{\partial u}{\partial \mathbf{n}}\mathbf{m}\cdot\nabla u - \mathbf{m}\cdot\mathbf{n}|\nabla u|^2\right) dS \quad \text{(use (6.49))}$$
$$= \int_{\Gamma(\mathbf{x}^0)} \left(-2\mathbf{m}\cdot\mathbf{n}f(u_t)\mathbf{m}\cdot\nabla u - \mathbf{m}\cdot\mathbf{n}|\nabla u|^2\right) dS$$
$$\text{(use Young's inequality (2.4))}$$
$$\leq C \int_{\Gamma(\mathbf{x}^0)} \mathbf{m}\cdot\mathbf{n}f^2(u_t) dS + \int_{\Gamma(\mathbf{x}^0)} \mathbf{m}\cdot\mathbf{n}|\nabla u|^2 dS$$
$$- \int_{\Gamma(\mathbf{x}^0)} \mathbf{m}\cdot\mathbf{n}|\nabla u|^2 dS$$
$$= C \int_{\Gamma(\mathbf{x}^0)} \mathbf{m}\cdot\mathbf{n}f^2(u_t) dS. \tag{6.63}$$

Hereafter C denotes a generic positive constant that may change from line to line. Moreover, by Young's inequality, the boundary control (6.49), and the trace theorem (the trace inequality (6.27)), we deduce that

$$(n-1)\int_{\Gamma(\mathbf{x}^0)} c^2\frac{\partial u}{\partial \mathbf{n}} u dS$$
$$= -(n-1)\int_{\Gamma(\mathbf{x}^0)} \mathbf{m}\cdot\mathbf{n}f(u_t) u dS$$
$$\leq C \int_{\Gamma(\mathbf{x}^0)} \mathbf{m}\cdot\mathbf{n}f^2(u_t) dS + \omega \int_{\Gamma(\mathbf{x}^0)} |u|^2 dS$$
$$\leq C \int_{\Gamma(\mathbf{x}^0)} \mathbf{m}\cdot\mathbf{n}f^2(u_t) dS + \omega M \int_\Omega |\nabla u|^2 dV$$
$$\leq C \int_{\Gamma(\mathbf{x}^0)} \mathbf{m}\cdot\mathbf{n}f^2(u_t) dS + \frac{c^2}{2}\int_\Omega |\nabla u|^2 dV, \tag{6.64}$$

where ω is chosen such that $\omega M = \frac{c^2}{2}$. It therefore follows from (6.62)-(6.64) that

$$F'(t) \le -E(t) + C \int_{\Gamma(x^0)} \mathbf{m} \cdot \mathbf{n}(|u_t|^2 + f^2(u_t))dS. \tag{6.65}$$

As in the proof of (6.16), we can show that

$$E'(t) = -\int_{\Gamma(x^0)} \mathbf{m} \cdot \mathbf{n} f(u_t) u_t dS. \tag{6.66}$$

It then follows that

$$
\begin{aligned}
E'_{\delta,\sigma}(t) &= E'(t) + \delta \frac{d}{dt}(E^\sigma(t)F(t)) \\
&= E'(t) + \delta \sigma E^{\sigma-1}(t)E'(t)F(t) + \delta E^\sigma(t)F'(t) \\
&\quad \text{(use (6.26) for the estimate of } F) \\
&\le [1 - \delta\sigma C E^\sigma(t)]E'(t) + \delta E^\sigma(t)F'(t) \\
&\quad \text{(use (6.65) and (6.66))} \\
&\le -\delta E^{\sigma+1}(t) - [1 - \delta\sigma C E^\sigma(t)]\int_{\Gamma(x^0)} \mathbf{m} \cdot \mathbf{n} f(u_t) u_t dS \\
&\quad + \delta C E^\sigma(t) \int_{\Gamma(x^0)} \mathbf{m} \cdot \mathbf{n}(|u_t|^2 + f^2(u_t))dS. \tag{6.67}
\end{aligned}
$$

We now distinguish the cases $p = q = 1$ and $p + 1 > 2q$. If $p = q = 1$, we take $\sigma = 0$. It follows from (6.67) that

$$
\begin{aligned}
E'_{\delta,\sigma}(t) &\le -\delta E(t) - \int_{\Gamma(x^0)} \mathbf{m} \cdot \mathbf{n} f(u_t) u_t dS + \delta C \int_{\Gamma(x^0)} \mathbf{m} \cdot \mathbf{n}(|u_t|^2 + f^2(u_t))dS \\
&\quad \text{(use (6.51)-(6.56))} \\
&\le -\delta E(t) + [\delta C - k_2] \int_{\Gamma(x^0)} \mathbf{m} \cdot \mathbf{n}|u_t|^2 dS \\
&\le -\delta E(t) \quad (\text{use (6.25)}) \\
&\le -\frac{\delta}{1 + \delta C} E_{\delta,\sigma}(t) \tag{6.68}
\end{aligned}
$$

if δ is sufficiently small. Solving this differential inequality and using (6.25), we obtain

$$
\begin{aligned}
E(t) &\le \frac{1}{1 - \delta C} E_{\delta,\sigma}(t) \\
&\le \frac{1}{1 - \delta C} E_{\delta,\sigma}(0) e^{-\delta t/(1+\delta C)} \\
&\le \frac{1 + \delta C}{1 - \delta C} E(0) e^{-\delta t/(1+\delta C)}, \tag{6.69}
\end{aligned}
$$

which proves (6.57).

We then consider the case where $p + 1 > 2q$. In the following, C denotes a generic constant that may change from line to line. By (6.52), we deduce that

$$CE^\sigma(t) \int_{\Gamma(\mathbf{x}^0) \cap [|u_t| \leq 1]} \mathbf{m} \cdot \mathbf{n}(|u_t|^2 + f^2(u_t)) dS$$

$$\leq CE^\sigma(t) \int_{\Gamma(\mathbf{x}^0) \cap [|u_t| \leq 1]} \mathbf{m} \cdot \mathbf{n}(|u_t|^2 + k_1^2|u_t|^{2q}) dS$$

(note that $|u_t|^2 \leq |u_t|^{2q}$ since $q \leq 1$)

$$\leq CE^\sigma(t) \int_{\Gamma(\mathbf{x}^0) \cap [|u_t| \leq 1]} \mathbf{m} \cdot \mathbf{n}|u_t|^{2q} dS$$

(use Young's inequality (2.4) with $b > 0$)

$$\leq \frac{p + 1 - 2q}{p + 1} b^{(p+1)/(p+1-2q)} E^{\sigma(p+1)/(p+1-2q)}$$

$$+ \frac{2q}{(p+1)b^{(p+1)/2q}} \left(C \int_{\Gamma(\mathbf{x}^0) \cap [|u_t| \leq 1]} \mathbf{m} \cdot \mathbf{n}|u_t|^{2q} dS \right)^{(p+1)/(2q)}$$

(use Hölder's inequality (2.5))

$$\leq \frac{p + 1 - 2q}{p + 1} b^{(p+1)/(p+1-2q)} E^{\sigma(p+1)/(p+1-2q)}$$

$$+ \frac{2qC}{(p+1)b^{(p+1)/(2q)}} \int_{\Gamma(\mathbf{x}^0) \cap [|u_t| \leq 1]} \mathbf{m} \cdot \mathbf{n}|u_t|^{p+1} dS$$

(use (6.54))

$$\leq \frac{p + 1 - 2q}{p + 1} b^{(p+1)/(p+1-2q)} E^{\sigma(p+1)/(p+1-2q)}$$

$$+ \frac{2qC}{(p+1)b^{(p+1)/(2q)}} \int_{\Gamma(\mathbf{x}^0) \cap [|u_t| \leq 1]} \mathbf{m} \cdot \mathbf{n} f(u_t) u_t dS$$

$$\leq \frac{1}{2} E^{\sigma(p+1)/(p+1-2q)} + C \int_{\Gamma(\mathbf{x}^0) \cap [|u_t| \leq 1]} \mathbf{m} \cdot \mathbf{n} f(u_t) u_t dS \qquad (6.70)$$

if we take

$$b = \left(\frac{p + 1}{2(p + 1 - 2q)} \right)^{(p+1-2q)/(p+1)}.$$

It therefore follows from (6.67) and (6.70) that

$$E'_{\delta,\sigma}(t) \leq -\delta E^{\sigma+1}(t) - [1 - \delta\sigma CE^\sigma(t)] \int_{\Gamma(\mathbf{x}^0)} \mathbf{m} \cdot \mathbf{n} f(u_t) u_t dS$$

$$+ \delta CE^\sigma(t) \int_{\Gamma(\mathbf{x}^0)} \mathbf{m} \cdot \mathbf{n}(|u_t|^2 + f^2(u_t)) dS$$

$$= -\delta E^{\sigma+1}(t) - [1 - \delta\sigma CE^{\sigma}(t)] \int_{\Gamma(\mathbf{x}^0)\cap[|u_t|\le 1]} \mathbf{m}\cdot\mathbf{n} f(u_t)u_t dS$$

$$- [1 - \delta\sigma CE^{\sigma}(t)] \int_{\Gamma(\mathbf{x}^0)\cap[|u_t|\ge 1]} \mathbf{m}\cdot\mathbf{n} f(u_t)u_t dS$$

$$+ \delta CE^{\sigma}(t) \int_{\Gamma(\mathbf{x}^0)\cap[|u_t|\le 1]} \mathbf{m}\cdot\mathbf{n}(|u_t|^2 + f^2(u_t))dS$$

$$+ \delta CE^{\sigma}(t) \int_{\Gamma(\mathbf{x}^0)\cap[|u_t|\ge 1]} \mathbf{m}\cdot\mathbf{n}(|u_t|^2 + f^2(u_t))dS$$

(use (6.53) and (6.55) to derive

$|u_t|^2 + f^2(u_t) \le Cf(u_t)u_t$ for $|u_t| \ge 1$)

$$\le -\delta E^{\sigma+1}(t) - [1 - \delta\sigma CE^{\sigma}(t)] \int_{\Gamma(\mathbf{x}^0)\cap[|u_t|\le 1]} \mathbf{m}\cdot\mathbf{n} f(u_t)u_t dS$$

$$- [1 - \delta CE^{\sigma}(t)] \int_{\Gamma(\mathbf{x}^0)\cap[|u_t|\ge 1]} \mathbf{m}\cdot\mathbf{n} f(u_t)u_t dS$$

$$+ \delta CE^{\sigma}(t) \int_{\Gamma(\mathbf{x}^0)\cap[|u_t|\le 1]} \mathbf{m}\cdot\mathbf{n}(|u_t|^2 + f^2(u_t))dS$$

(use(6.70))

$$\le -\delta E^{\sigma+1}(t) + \frac{\delta}{2} E^{\sigma(p+1)/(p+1-2q)}$$

$$- [1 - \delta CE^{\sigma}(t)] \int_{\Gamma(\mathbf{x}^0)\cap[|u_t|\le 1]} \mathbf{m}\cdot\mathbf{n} f(u_t)u_t dS$$

$$- [1 - \delta CE^{\sigma}(t)] \int_{\Gamma(\mathbf{x}^0)\cap[|u_t|\ge 1]} \mathbf{m}\cdot\mathbf{n} f(u_t)u_t dS. \tag{6.71}$$

We now choose σ so that

$$\sigma + 1 = \frac{\sigma(p+1)}{p+1-2q}.$$

Then

$$\sigma = \frac{p+1-2q}{2q}.$$

Taking δ small enough, we derive from (6.71) that

$$E'_{\delta,\sigma}(t) \le -\frac{\delta}{2} E^{\sigma+1}(t). \tag{6.72}$$

By (6.26), noting that $E(t) \le E(0)$, we derive that

$$E_{\delta,\sigma}(t)(t) \le [1 + \delta CE^{\sigma}(0)]E(t). \tag{6.73}$$

It therefore follows from (6.72) that

$$E'_{\delta,\sigma}(t) \le -\frac{\delta_2}{2[1 + \delta CE^{\sigma}(0)]^{\sigma+1}} E^{\sigma+1}_{\delta,\sigma}(t). \tag{6.74}$$

Solving this differential inequality, we obtain

$$E_{\delta,\sigma}(t) \leq \left[(E_{\delta,\sigma}(0))^{-\sigma} + \frac{\sigma \delta t}{2[1 + \delta C E^{\sigma}(0)]^{\sigma+1}} \right]^{-1/\sigma}$$

$$\leq \left[(E(0) + \delta C E^{\sigma+1}(0))^{-\sigma} + \frac{\sigma \delta t}{2[1 + \delta C E^{\sigma}(0)]^{\sigma+1}} \right]^{-1/\sigma},$$

which, combined with (6.25), implies (6.58).

Example 6.1. Let $f(u) = ku$, where k is a positive constant. This is a linear feedback control and all the assumptions of Theorem 6.4 are satisfied.

Example 6.2. We can easily verify that the functions

$$f_1(u) = \frac{u^3}{1+u^2}, \quad f_2(u) = \begin{cases} u^3, & |u| \leq 1, \\ u, & |u| \geq 1. \end{cases}$$

satisfy all the assumptions of Theorem 6.4 with $p = 3$ and $q = 1$.

Exercises 6.3

1. Show that the function

$$f(u) = \frac{u^3}{1+u^2}$$

   satisfies all the conditions of Theorem 6.4 with $p = 3$ and $q = 1$.
2. Let $E : [0,\infty) \to [0,\infty)$ be a non-increasing function and assume that there exist two constants $\alpha > 0$ and $T > 0$ such that

$$\int_t^\infty E^{\alpha+1}(s)ds \leq TE(0)^\alpha E(t), \quad t \geq 0.$$

Show that

$$E(t) \leq E(0) \left(\frac{T + \alpha t}{T + \alpha T} \right)^{-1/\alpha}, \quad t \geq T.$$

   (Hint: Define $f(t) = \int_t^\infty E^{\alpha+1}(s)ds$ and differentiate f to establish a differential inequality for f. Then solve the inequality and use $f(t) \geq \int_t^{T+(\alpha+1)t} E^{\alpha+1}(s)ds \geq (T + \alpha t)E(T + (\alpha+1)t)^{\alpha+1}$.)
3. Consider the wave equation with a nonlinear boundary velocity feedback control

$$u_{tt} = c^2 u_{xx} \quad \text{in } (0,L) \times (0,\infty), \tag{6.75}$$

$$u(0,t) = 0, \quad c^2 u_x(L,t) = -\frac{ku_t^3(L,t)}{1 + u_t^2(L,t)}, \quad t \geq 0, \tag{6.76}$$

$$u(x,0) = u_0(x), \quad u_t(x,0) = u_1(x), \quad x \in (0,L), \tag{6.77}$$

where k is a positive constant. Define the energy $E(t)$ by

$$E(t) = \frac{1}{2} \int_0^L \left(|u_t(x,t)|^2 + c^2 |u_x(x,t)|^2 \right) dx.$$

a. Show that

$$\frac{dE}{dt} = -\frac{ku_t^4(L,t)}{1 + u_t^2(L,t)}.$$

b. Multiply the equation (6.75) by $2xu_x E(t)$ and integrate over $(0,L) \times (t,T))$ to show that

$$2\int_t^T E^2(s)ds = -2\left[E(s) \int_0^L xu_x u_s dx \right]_t^T + 2\int_t^T E'(s) \int_0^L xu_x u_s dxds$$

$$+ L\int_t^T E(s) \left[c^2 u_x^2(L,s) + u_s^2(L,s) \right] ds.$$

c. Assume that $E(0) \leq 1$. Prove that there exists a constant $C > 0$ such that

$$\left| \left[E(s) \int_0^L xu_x u_s dx \right]_t^T \right| \leq CE(t),$$

$$\left| \int_t^T E'(s) \int_0^L xu_x u_s dxds \right| \leq CE(t).$$

d. Assume that $E(0) \leq 1$ and let $I_1 = \{ s \in [t,T] : |u_s(L,s)| > 1 \}$. Prove that there exists a constant $C > 0$ such that

$$\left| \int_{I_1} E(s) \left[c^2 u_x^2(L,s) + u_s^2(L,s) \right] ds \right| \leq CE(t).$$

e. Assume that $E(0) \leq 1$ and let $I_2 = \{ s \in [t,T] : |u_s(L,s)| \leq 1 \}$. Prove that for every $\varepsilon > 0$ there exists a constant $C(\varepsilon) > 0$ such that

$$\left| \int_{I_2} E(s) \left[c^2 u_x^2(L,s) + u_s^2(L,s) \right] ds \right| \leq C(\varepsilon)E(t) + \varepsilon \int_t^T E^2(s)ds.$$

f. Show that there exists a positive constant M such that

$$E(t) \leq \frac{M}{1+t}.$$

6.4 Observability Inequalities

To address the problems of interior feedback stabilization, we need observability inequalities of the wave equation

$$\frac{\partial^2 v}{\partial t^2} = c^2 \nabla^2 v \quad \text{in } \Omega \times (0,T), \tag{6.78}$$

$$v = 0 \quad \text{on } \partial\Omega \times (0,T), \tag{6.79}$$

$$v(\mathbf{x},0) = v_0(\mathbf{x}), \quad \frac{\partial v}{\partial t}(\mathbf{x},0) = v_1(\mathbf{x}) \quad \text{in } \Omega. \tag{6.80}$$

We start by establishing identities. In what follows, we denote $Q = \Omega \times (0,T)$ and $\Sigma = \Gamma \times (0,T)$ for some $T > 0$.

Lemma 6.4. *The solution v of (6.78)-(6.80) satisfies*

$$\left(v(t), \frac{\partial v}{\partial t}(t) \right) \Big|_0^T = \int_Q \left(\left| \frac{\partial v}{\partial t} \right|^2 - c^2 |\nabla v|^2 \right) dV dt, \tag{6.81}$$

where

$$\left(v(t), \frac{\partial v}{\partial t}(t) \right) = \int_\Omega v(t) \frac{\partial v}{\partial t}(t) dV.$$

Proof. Multiplying the equation (6.78) by v and then integrating over Q by parts, we obtain

$$\left(v(t), \frac{\partial v}{\partial t}(t) \right) \Big|_0^T - \int_Q \left| \frac{\partial v}{\partial t} \right|^2 dV dt = - \int_Q c^2 |\nabla v|^2 dV dt,$$

which gives (6.81). $\qquad\blacksquare$

Lemma 6.5. *Let $\mathbf{q} = (q_k)$ be a vector field in $[C^1(\bar{\Omega})]^n$. Suppose v is the solution of (6.78)-(6.80). Then the following identity holds:*

$$\frac{1}{2} \int_\Sigma c^2 \mathbf{q} \cdot \mathbf{n} \left| \frac{\partial v}{\partial \mathbf{n}} \right|^2 d\Sigma$$

$$= \left(\frac{\partial v}{\partial t}(t), \mathbf{q} \cdot \nabla v(t) \right) \Big|_0^T + \sum_{i=1}^n \sum_{j=1}^n \int_Q c^2 \frac{\partial v}{\partial x_i} \frac{\partial q_j}{\partial x_i} \frac{\partial v}{\partial x_j} dV dt$$

$$+ \frac{1}{2} \int_Q \text{div}(\mathbf{q}) \left(\left| \frac{\partial v}{\partial t} \right|^2 - c^2 |\nabla v|^2 \right) dV dt, \tag{6.82}$$

where

$$\left(\frac{\partial v}{\partial t}(t), \mathbf{q} \cdot \nabla v(t) \right) = \int_\Omega \frac{\partial v}{\partial t}(t) \mathbf{q} \cdot \nabla v(t) dV.$$

Proof. Multiplying (6.78) by $\mathbf{q} \cdot \nabla v$ and integrating over Q yields

$$\int_Q \frac{\partial^2 v}{\partial t^2} \mathbf{q} \cdot \nabla v dV dt = \int_Q c^2 \nabla^2 v \mathbf{q} \cdot \nabla v dV dt. \tag{6.83}$$

Integration by parts gives

$$\int_Q \frac{\partial^2 v}{\partial t^2} \mathbf{q} \cdot \nabla v dV dt = \left(\frac{\partial v}{\partial t}(t), \mathbf{q} \cdot \nabla v \right) \Big|_0^T + \frac{1}{2} \int_Q \text{div}(\mathbf{q}) \left| \frac{\partial v}{\partial t} \right|^2 dV dt \tag{6.84}$$

and

$$\int_Q c^2 \nabla^2 v \mathbf{q} \cdot \nabla v \, dV \, dt$$

$$= \int_\Sigma c^2 \frac{\partial v}{\partial \mathbf{n}} \mathbf{q} \cdot \nabla v \, d\Sigma - \sum_{i=1}^n \sum_{j=1}^n \int_Q c^2 \frac{\partial v}{\partial x_i} \frac{\partial q_j}{\partial x_i} \frac{\partial v}{\partial x_j} V \, dt$$

$$- \sum_{i=1}^n \sum_{j=1}^n \int_Q c^2 \frac{\partial v}{\partial x_i} q_j \frac{\partial^2 v}{\partial x_i \partial x_j} V \, dt$$

$$= \int_\Sigma c^2 \frac{\partial v}{\partial \mathbf{n}} \mathbf{q} \cdot \nabla v \, d\Sigma - \sum_{i=1}^n \sum_{j=1}^n \int_Q c^2 \frac{\partial v}{\partial x_i} \frac{\partial q_j}{\partial x_i} \frac{\partial v}{\partial x_j} V \, dt$$

$$- \frac{1}{2} \int_\Sigma c^2 \mathbf{q} \cdot \mathbf{n} |\nabla v|^2 \, d\Sigma + \frac{1}{2} \int_Q c^2 \operatorname{div}(\mathbf{q}) |\nabla v|^2 V \, dt.$$

Since $v = 0$ on Γ, we have $\mathbf{n} = \nabla v / |\nabla v|$ and then $|\nabla v| = \frac{\partial v}{\partial \mathbf{n}}$. Using this result, we derive from the above equation that

$$\int_Q c^2 \nabla^2 v \mathbf{q} \cdot \nabla v \, dV \, dt$$

$$= \frac{1}{2} \int_\Sigma c^2 \mathbf{q} \cdot \mathbf{n} \left| \frac{\partial v}{\partial \mathbf{n}} \right|^2 d\Sigma - \sum_{i=1}^n \sum_{j=1}^n \int_Q c^2 \frac{\partial v}{\partial x_i} \frac{\partial q_j}{\partial x_i} \frac{\partial v}{\partial x_j} V \, dt$$

$$+ \frac{1}{2} \int_Q c^2 \operatorname{div}(\mathbf{q}) |\nabla v|^2 V \, dt. \tag{6.85}$$

Then (6.82) follows from (6.84) and (6.85). $\quad\blacksquare$

From (6.10) and (6.82) we can easily derive the following lemma.

Lemma 6.6. *The solution v of (6.78)-(6.80) satisfies*

$$\int_\Sigma c^2 \left| \frac{\partial v}{\partial \mathbf{n}} \right|^2 d\Sigma \leq CE(0), \tag{6.86}$$

where E is the energy function defined in (6.9) and C is a positive constant independent of initial conditions.

Proof. Taking $\mathbf{q} = \mathbf{n}$ in (6.82), we obtain

$$\frac{1}{2} \int_\Sigma c^2 \left| \frac{\partial v}{\partial \mathbf{n}} \right|^2 d\Sigma$$

$$= \left(\frac{\partial v}{\partial t}(t), \mathbf{n} \cdot \nabla v(t) \right) \Big|_0^T + \sum_{i=1}^n \sum_{j=1}^n \int_Q c^2 \frac{\partial v}{\partial x_i} \frac{\partial n_j}{\partial x_i} \frac{\partial v}{\partial x_j} dV \, dt$$

$$+ \frac{1}{2} \int_Q \operatorname{div}(\mathbf{n}) \left(\left| \frac{\partial v}{\partial t} \right|^2 - c^2 |\nabla v|^2 \right) dV \, dt$$

$$\leq C(E(0)+E(T))+C\int_0^T E(t)dt$$

$$\leq CE(0).$$

We introduce an important constant used in the control theory of the wave equation:

$$R(\mathbf{x}^0) = \max_{\mathbf{x}\in\bar{\Omega}}|\mathbf{m}(\mathbf{x})| = \max_{\mathbf{x}\in\bar{\Omega}}\left|\sum_{k=1}^n (x_k-x_k^0)^2\right|^{1/2}.$$

Lemma 6.7. *If $T > \frac{2R(\mathbf{x}^0)}{c}$, then there exists a positive constant C such that any solutions of (6.78)-(6.80) satisfy*

$$C\|(v_0,v_1)\|_{H_0^1(\Omega)\times L^2(\Omega)}^2 \leq \int_0^T\int_{\Gamma(\mathbf{x}^0)}\left|\frac{\partial v}{\partial\mathbf{n}}(\mathbf{x},t)\right|^2 d\Sigma \qquad (6.87)$$

for any initial conditions $(v_0,v_1)\in H_0^1(\Omega)\times L^2(\Omega)$.

Proof. Taking $\mathbf{q}=\mathbf{m}(\mathbf{x})$ in (6.82) and using (6.10) and (6.81), we obtain

$$\frac{1}{2}\int_\Sigma c^2\mathbf{m}\cdot\mathbf{n}\left|\frac{\partial v}{\partial\mathbf{n}}\right|^2 d\Sigma = \left(\frac{\partial v}{\partial t}(t),\mathbf{m}\cdot\nabla v(t)\right)\Big|_0^T + \int_Q c^2|\nabla v|^2 dVdt$$

$$+\frac{n}{2}\int_Q\left(\left|\frac{\partial v}{\partial t}\right|^2 - c^2|\nabla v|^2\right)dVdt$$

$$= \left(\frac{\partial v}{\partial t}(t),\mathbf{m}\cdot\nabla v(t)\right)\Big|_0^T$$

$$+\frac{1}{2}\int_Q\left(\left|\frac{\partial v}{\partial t}\right|^2 + c^2|\nabla v|^2\right)dVdt$$

$$+\frac{n-1}{2}\int_Q\left(\left|\frac{\partial v}{\partial t}\right|^2 - c^2|\nabla v|^2\right)dVdt$$

$$= \left(\frac{\partial v}{\partial t}(t),\mathbf{m}\cdot\nabla v(t)+\frac{n-1}{2}v(t)\right)\Big|_0^T$$

$$+TE(0). \qquad (6.88)$$

Furthermore,

$$\left(\frac{\partial v}{\partial t}(t),\mathbf{m}\cdot\nabla v(t)+\frac{n-1}{2}v(t)\right)$$

$$\leq \frac{R(\mathbf{x}^0)}{2}\int_\Omega\left|\frac{\partial v}{\partial t}(t)\right|^2 dV$$

$$+\frac{1}{2R(\mathbf{x}^0)}\int_\Omega\left|\mathbf{m}\cdot\nabla v(t)+\frac{n-1}{2}v(t)\right|^2 dV$$

$$= \frac{R(\mathbf{x}^0)}{2c} \int_\Omega \left| \frac{\partial v}{\partial t}(t) \right|^2 dV$$

$$+ \frac{c}{2R(\mathbf{x}^0)} \int_\Omega \left(|\mathbf{m} \cdot \nabla v(t)|^2 + \left(\frac{n-1}{2} \right)^2 |v(t)|^2 + (n-1)v(t)\mathbf{m} \cdot \nabla v(t) \right) dV$$

$$= \frac{R(\mathbf{x}^0)}{2c} \int_\Omega \left| \frac{\partial v}{\partial t}(t) \right|^2 dV$$

$$+ \frac{c}{2R(\mathbf{x}^0)} \int_\Omega \left(|\mathbf{m} \cdot \nabla v(t)|^2 + \left(\frac{n-1}{2} \right)^2 |v(t)|^2 - \frac{n(n-1)}{2} |v(t)|^2 \right) dV$$

$$\leq \frac{R(\mathbf{x}^0)}{2c} \int_\Omega \left| \frac{\partial v}{\partial t}(t) \right|^2 dV + \frac{c}{2R(\mathbf{x}^0)} \int_\Omega |\mathbf{m} \cdot \nabla v(t)|^2 dV$$

$$\leq \frac{R(\mathbf{x}^0)}{2c} \int_\Omega \left| \frac{\partial v}{\partial t}(t) \right|^2 dV + \frac{cR(\mathbf{x}^0)}{2} \int_\Omega |\nabla v(t)|^2 dV$$

$$\leq \frac{R(\mathbf{x}^0)}{c} E(0).$$

It then follows from (6.88) that

$$\frac{1}{2} \int_{\Sigma(\mathbf{x}^0)} c^2 \mathbf{m} \cdot \mathbf{n} \left| \frac{\partial v}{\partial \mathbf{n}} \right|^2 d\Sigma \geq \frac{1}{2} \int_\Sigma c^2 \mathbf{m} \cdot \mathbf{n} \left| \frac{\partial v}{\partial \mathbf{n}} \right|^2 d\Sigma$$

$$\geq TE(0) - \left| \left(\frac{\partial v}{\partial t}(t), \mathbf{m} \cdot \nabla v(t) \right) \right|_0^T \right|$$

$$\geq \left(T - \frac{2R(\mathbf{x}^0)}{c} \right) E(0).$$

The inequality (6.87) is called a boundary *observability inequality*. The inequality implies that if $\frac{\partial v}{\partial \mathbf{n}}(\mathbf{x},t) = 0$ on $\Gamma(\mathbf{x}^0)$ during the period of time T, then the initial condition $(v_0, v_1) = (0,0)$. Thus initial states of the system are uniquely determined by its output $\frac{\partial v}{\partial \mathbf{n}}$ on the boundary $\Gamma(\mathbf{x}^0)$. That is, the system (6.78)-(6.80) is observable from the boundary $\Gamma(\mathbf{x}^0)$ in time T.

Lemma 6.8. *Let ω be a neighborhood of $\Gamma(\mathbf{x}^0)$. If $T > \frac{2R(\mathbf{x}^0)}{c}$, then there exists a constant M such that the solution of the system (6.78)-(6.80) satisfies that*

$$E(0) \leq M \int_0^T \int_\omega [|v_t|^2 + |\nabla v|^2] dV dt. \tag{6.89}$$

Proof. Let $\alpha > 0$ be such that $T - 2\alpha > \frac{2R(\mathbf{x}^0)}{c}$. By Lemma 6.7, there exists $M > 0$ such that

$$E(0) \leq M \int_\alpha^{T-\alpha} \int_{\Gamma(\mathbf{x}^0)} \left| \frac{\partial v}{\partial \mathbf{n}}(\mathbf{x},t) \right|^2 dS dt.$$

It then suffices to prove that

$$\int_\alpha^{T-\alpha} \int_{\Gamma(\mathbf{x}^0)} \left| \frac{\partial v}{\partial \mathbf{n}}(\mathbf{x},t) \right|^2 dSdt \leq M \int_0^T \int_\omega [|v_t|^2 + |\nabla v|^2] dVdt.$$

This can be proved by using identity (6.82) with

$$\mathbf{q} = t(T-t)\mathbf{h}(\mathbf{x}),$$

where $\mathbf{h} \in [C^1(\bar{\Omega})]^n$ satisfies

$$\mathbf{h} \cdot \mathbf{n} = 1 \text{ on } \Gamma(\mathbf{x}^0), \ \mathbf{h} \cdot \mathbf{n} \geq 0 \text{ on } \Gamma, \ \mathrm{supp}(\mathbf{h}) \subset \omega.$$

In fact, it follows from (6.82) that

$$\int_\alpha^{T-\alpha} \int_{\Gamma(\mathbf{x}^0)} \left| \frac{\partial v}{\partial \mathbf{n}}(\mathbf{x},t) \right|^2 dSdt \leq M \int_0^T \int_{\Gamma(\mathbf{x}^0)} \mathbf{h} \cdot \mathbf{n} \left| \frac{\partial v}{\partial \mathbf{n}}(\mathbf{x},t) \right|^2 dSdt$$

$$\leq M \int_0^T \int_\omega [|v_t|^2 + |\nabla v|^2] dVdt.$$

Lemma 6.9. *If* $T > \frac{2R(\mathbf{x}^0)}{c}$, *then there exists a constant M such that the solution of the system (6.78)-(6.80) satisfies that*

$$E(0) \leq M \int_0^T \int_\omega [|v_t|^2 + |v|^2] dVdt. \tag{6.90}$$

Proof. For any $\varepsilon > 0$, we construct the neighborhood

$$O_\varepsilon = \cup_{\mathbf{x} \in \Gamma(\mathbf{x}^0)} B(\mathbf{x},\varepsilon), \quad \omega_\varepsilon = O_\varepsilon \cap \Omega,$$

where $B(\mathbf{x},\varepsilon)$ denotes the ball of the radius ε at the center $\mathbf{x}$. Let $\alpha > 0$ be such that $T - 2\alpha > \frac{2R(\mathbf{x}^0)}{c}$. By (6.89), there exists $M > 0$ such that

$$E(0) \leq M \int_\alpha^{T-\alpha} \int_{\omega_{\varepsilon/2}} [|v_t|^2 + |\nabla v|^2] dVdt. \tag{6.91}$$

It suffices to prove that

$$\int_\alpha^{T-\alpha} \int_{\omega_{\varepsilon/2}} |\nabla v|^2 dVdt \leq M \int_0^T \int_{\omega_\varepsilon} [|v_t|^2 + |v|^2] dVdt. \tag{6.92}$$

For this, we construct $\phi \in W^{1,\infty}(O_\varepsilon)$ such that

$$0 \leq \phi \leq 1 \text{ in } O_\varepsilon, \ \phi = 1 \text{ in } \omega_{\varepsilon/2}, \ \left\| \frac{|\nabla \phi|^2}{\phi} \right\|_{L^\infty(O_\varepsilon)} < \infty.$$

Multiplying (6.5) by $t(T-t)\phi(\mathbf{x})v(\mathbf{x},t)$ and integrating over $\Omega \times (0,T)$, we obtain

$$\int_0^T \int_\Omega [t(T-t)\phi(\mathbf{x})|v_t(\mathbf{x},t)|^2 + (T-2t)\phi(\mathbf{x})v(\mathbf{x},t)v_t(\mathbf{x},t)]dVdt$$

$$= \int_0^T \int_\Omega c^2[t(T-t)\phi(\mathbf{x})|\nabla v|^2 + t(T-t)v\nabla\phi(\mathbf{x})\cdot\nabla v]dVdt$$

$$\geq \int_0^T \int_\Omega c^2t(T-t)\phi(\mathbf{x})|\nabla v|^2 dVdt$$

$$-\frac{1}{2}\int_0^T \int_\Omega c^2t(T-t)\phi(\mathbf{x})|\nabla v|^2 dVdt$$

$$-\frac{1}{2}\int_0^T \int_\Omega c^2t(T-t)\frac{|\nabla\phi(\mathbf{x})|}{\phi}|v|^2 dVdt$$

$$\geq \frac{1}{2}\int_0^T \int_\Omega c^2t(T-t)\phi(\mathbf{x})|\nabla v|^2 dVdt$$

$$-\frac{1}{2}\int_0^T \int_\Omega c^2t(T-t)\frac{|\nabla\phi(\mathbf{x})|}{\phi}|v|^2 dVdt$$

$$\geq \frac{1}{2}\int_\alpha^{T-\alpha} \int_\Omega c^2t(T-t)\phi(\mathbf{x})|\nabla v|^2 dVdt$$

$$-\frac{1}{2}\int_0^T \int_\Omega c^2t(T-t)\frac{|\nabla\phi(\mathbf{x})|}{\phi}|v|^2 dVdt,$$

which implies (6.92).

The inequalities (6.89) and (6.90) are called *interior observability inequalities*.

Exercises 6.4

1. Consider the wave equation

$$\frac{\partial^2 v}{\partial t^2} = c^2\frac{\partial^2 v}{\partial x^2} \quad \text{in } \Omega \times (0,T),$$

$$v(0,t) = v(L,t) = 0 \quad t \in (0,T),$$

$$v(x,0) = v_0(x), \quad \frac{\partial v}{\partial t}(x,0) = v_1(x) \quad \text{in } \Omega.$$

Use the series solution

$$v(x,t) = \sum_{n=1}^\infty \left[a_n\cos\left(\frac{nc\pi t}{L}\right) + b_n\sin\left(\frac{nc\pi t}{L}\right)\right]\sin\left(\frac{n\pi x}{L}\right)$$

to prove:

a. If $\omega = [x_0 - \varepsilon, x_0 + \varepsilon] \subset [0, L]$ and $T \geq 2L/c$, then

$$CE(0) \leq \int_0^T \int_\omega \left| \frac{\partial v}{\partial t}(x,t) \right|^2 dx dt,$$

where C is a positive constant.

b. If $T \geq 2L/c$, then

$$C_1 E(0) \leq \int_0^T \left| \frac{\partial v}{\partial x}(L,t) \right|^2 dt \leq C_2 E(0),$$

where C_1 and C_2 are positive constants.

6.5 Linear Interior Feedback Stabilization

Consider the wave equation with an interior control

$$\frac{\partial^2 u}{\partial t^2} = c^2 \nabla^2 u + \phi \chi_\omega \quad \text{in } \Omega \times (0, \infty), \tag{6.93}$$

$$u = 0 \quad \text{on } \Gamma \times (0, \infty), \tag{6.94}$$

$$u(\mathbf{x}, 0) = u_0(\mathbf{x}), \quad \frac{\partial u}{\partial t}(\mathbf{x}, 0) = u_1(\mathbf{x}) \quad \text{in } \Omega, \tag{6.95}$$

where ω is a nonempty open subset of Ω, χ_ω denotes the characteristic function of ω (equal to 1 if $\mathbf{x} \in \omega$ and equal to 0 if $\mathbf{x} \notin \omega$) and $\phi = \phi(\mathbf{x}, t)$ is a control to be found. Physically, the control ϕ represents an external force exerted on the part ω of the domain. To find a feedback control to regulate the solution to the equilibrium zero, we compute the derivative of the energy

$$\frac{dE}{dt} = \int_\Omega \left(\frac{\partial u}{\partial t} \frac{\partial^2 u}{\partial t^2} + c^2 \nabla u \cdot \nabla \frac{\partial u}{\partial t} \right) dV$$

$$= \int_\Omega \left(\frac{\partial u}{\partial t} c^2 \nabla^2 u + \frac{\partial u}{\partial t} \phi \chi_\omega + c^2 \nabla u \cdot \nabla \frac{\partial u}{\partial t} \right) dV$$

$$= \int_\Omega \frac{\partial u}{\partial t} \phi \chi_\omega dV \quad \text{(use Green's identity (2.26)).}$$

This leads us to take

$$\phi = -k \frac{\partial u}{\partial t} \tag{6.96}$$

so that

$$\frac{dE}{dt} = -k \int_\omega \left| \frac{\partial u}{\partial t} \right|^2 dV \leq 0, \tag{6.97}$$

where k is a positive constant. With this feedback control, the equilibrium 0 of the system (6.11)-(6.14) is exponentially stable.

To prove this result, we construct the following perturbed energy functional E_δ defined by

$$E_\delta(t) = E(t) + 2\delta \int_\Omega u(x,t)u_t(x,t)dV, \tag{6.98}$$

where δ is a positive constant. We first show that, by choosing δ sufficiently small, E_δ and E are equivalent.

Lemma 6.10. *The perturbed energy functional satisfies*

$$\left(1 - \frac{2\delta M}{c}\right) E(t) \le E_\delta(t) \le \left(1 + \frac{2\delta M}{c}\right) E(t), \tag{6.99}$$

where M is the constant in Poincaré's inequality.

Proof. Using Young's inequality and Poincaré's inequality, we deduce that

$$\begin{aligned}
\left| \int_\Omega 2u_t u\, dV \right| &\le \frac{M}{c} \int_\Omega 2|u_t| \left| \frac{c}{M} u \right| dV \\
&\le \frac{M}{c} \int_\Omega \left(|u_t|^2 + \left| \frac{c}{M} u \right|^2 \right) dV \\
&\le \frac{M}{c} \int_\Omega \left(|u_t|^2 + |c\nabla u|^2 \right) dV \\
&\le \frac{2M}{c} E(t), \tag{6.100}
\end{aligned}$$

which implies (6.99).

Theorem 6.5. *Assume $\omega = \Omega$. Then there exist constants $M, \sigma > 0$ such that the solution of (6.93)-(6.95) with the feedback control (6.96) satisfies*

$$E(t) \le ME(0)e^{-\sigma t} \quad \text{for } t \ge 0. \tag{6.101}$$

Proof. It follows from (6.98), Young's inequality, and Poincaré's inequality that

$$\begin{aligned}
E_\delta'(t) &= E'(t) + \delta \int_\Omega 2u_{tt}u\, dV + \delta \int_\Omega 2u_t^2 dV \quad \text{(use (6.93) and (6.96))} \\
&= -k \int_\Omega \left| \frac{\partial u}{\partial t} \right|^2 dV + \delta \int_\Omega 2c^2 \nabla^2 u u\, dV - k\delta \int_\Omega 2u_t u\, dV + \delta \int_\Omega 2u_t^2 dV \\
&\quad \text{(use Green's identity (2.26))} \\
&= -k \int_\Omega \left| \frac{\partial u}{\partial t} \right|^2 dV - \delta \int_\Omega 2c^2 |\nabla u|^2 dV - \delta \int_\Omega 2 \frac{kMu_t}{c} \frac{cu}{M} dV \\
&\quad + \delta \int_\Omega 2u_t^2 dV \\
&\quad \text{(use Young's inequality (2.4) and Poincaré's inequality (6.21))}
\end{aligned}$$

$$\leq -k \int_\Omega \left|\frac{\partial u}{\partial t}\right|^2 dV - \delta \int_\Omega 2c^2 |\nabla u|^2 dV$$

$$+ \delta \int_\Omega \left(\frac{k^2 M^2 u_t^2}{c^2} + c^2 |\nabla u|^2\right) dV + \delta \int_\Omega 2u_t^2 dV.$$

We then derive that

$$E_\delta'(t) \leq -2\delta E(t) + \left(\frac{\delta k^2 M^2}{c^2} + 3\delta - k\right) \int_\Omega u_t^2 dV$$

$$\leq -2\delta E(t)$$

when δ is small enough. It then follows from (6.99) that

$$E_\delta'(t) \leq -2\delta \left(1 - \frac{2\delta M}{c}\right) E_\delta(t). \tag{6.102}$$

We then deduce from the differential inequality (4.13) that

$$E_\delta(t) \leq E_\delta(0)e^{-\sigma t}. \tag{6.103}$$

Thus (6.101) follows from (6.99).

If $\omega \subset \Omega$ is a neighborhood of $\Gamma(\mathbf{x}^0)$, the solution of (6.93)-(6.95) with the feedback control (6.96) still decays exponentially. The proof of this result is technically sophisticated and readers may skip it in the first reading.

Lemma 6.11. *Let ω be a neighborhood of $\Gamma(\mathbf{x}^0)$. If $T > \frac{2R(\mathbf{x}^0)}{c}$, there exists a constant $M > 0$ such that the solution of (6.93)-(6.95) with the feedback control (6.96) satisfies*

$$E(0) \leq M \int_0^T \int_\omega (|u|^2 + |u_t|^2) dV dt. \tag{6.104}$$

Proof. We write the solution u as

$$u = v + w$$

where v and w are solutions of

$$\frac{\partial^2 v}{\partial t^2} = c^2 \nabla^2 v \quad \text{in } \Omega \times (0,T), \tag{6.105}$$

$$v = 0 \quad \text{on } \partial\Omega \times (0,T), \tag{6.106}$$

$$v(\mathbf{x},0) = u_0(\mathbf{x}), \quad \frac{\partial v}{\partial t}(\mathbf{x},0) = u_1(\mathbf{x}) \quad \text{in } \Omega, \tag{6.107}$$

and

$$\frac{\partial^2 w}{\partial t^2} = c^2 \nabla^2 w - k\frac{\partial u}{\partial t}\chi_\omega \quad \text{in } \Omega \times (0,T), \tag{6.108}$$

$$w = 0 \quad \text{on } \partial\Omega \times (0,T), \tag{6.109}$$

$$w(\mathbf{x},0) = 0, \quad \frac{\partial w}{\partial t}(\mathbf{x},0) = 0 \quad \text{in } \Omega. \tag{6.110}$$

Energy estimates give

$$\int_\Omega (|w_t|^2 + c^2|\nabla w|^2)dV \leq M \int_0^T \int_\omega |u_t|^2 dV dt.$$

The observability inequality (6.90) yields

$$\begin{aligned}
E(0) &\leq M \int_0^T \int_\omega (|v|^2 + |v_t|^2)dV dt \\
&= M \int_0^T \int_\omega (|u-w|^2 + |u_t - w_t|^2)dV dt \\
&\leq M \int_0^T \int_\omega (|u|^2 + |u_t|^2)dV dt + M \int_0^T \int_\omega (|w|^2 + |w_t|^2)dV dt \\
&\leq M \int_0^T \int_\omega (|u|^2 + |u_t|^2)dV dt.
\end{aligned}$$

Lemma 6.12. *There exists a constant $M > 0$ such that the solution of (6.93)-(6.95) with the feedback control (6.96) satisfies*

$$\int_0^T \int_\Omega |u|^2 dV dt \leq M \int_0^T \int_\omega |u_t|^2 dV dt. \tag{6.111}$$

Proof. We argue by contradiction. If (6.111) is not true, there exists a sequence of solutions $\{u_n\}$ of

$$\frac{\partial^2 u_n}{\partial t^2} = c^2 \nabla^2 u_n - k\frac{\partial u_n}{\partial t}\chi_\omega \quad \text{in } \Omega \times (0,T), \tag{6.112}$$

$$u_n = 0 \quad \text{on } \partial\Omega \times (0,T), \tag{6.113}$$

$$u_n(\mathbf{x},0) = u_{n0}(\mathbf{x}), \quad \frac{\partial u_n}{\partial t}(\mathbf{x},0) = u_{n1}(\mathbf{x}) \quad \text{in } \Omega, \tag{6.114}$$

satisfying

$$\int_0^T \int_\Omega |u_n|^2 dV dt = 1, \tag{6.115}$$

$$\int_0^T \int_\Omega |u_n|^2 dV dt \geq n \int_0^T \int_\omega |u_{nt}|^2 dV dt. \tag{6.116}$$

It then follows from (6.104) that for $0 \leq t \leq T$

$$\begin{aligned}
E(u_n,t) &\leq E(u_n,0) \\
&\leq M \int_0^T \int_\omega (|u_n|^2 + |u_{nt}|^2)dV dt
\end{aligned}$$

$$\leq \frac{M(1+n)}{n} \int_0^T \int_\omega |u_n|^2 dV dt$$
$$= \frac{M(1+n)}{n}.$$

We then can extract a subsequence (still denoted by $\{u_n\}$) such that

$$
\begin{aligned}
u_n &\to v &&\text{star-weakly in}\quad L^\infty([0,T];H_0^1(\Omega)),\\
u_{nt} &\to v_t &&\text{star-weakly in}\quad L^\infty([0,T];L^2(\Omega)),\\
u_n &\to v &&\text{strongly in}\quad L^2([0,T];L^2(\Omega)).
\end{aligned}
$$

Passing to the limit in (6.115) and (6.116) gives

$$\int_0^T \int_\Omega |v|^2 dV dt = 1, \quad v_t = 0 \quad \text{in}\quad \omega \times (0,T). \tag{6.117}$$

Passing to the limit in (6.112)-(6.114) gives

$$\frac{\partial^2 v}{\partial t^2} = c^2 \nabla^2 v \quad \text{in}\ \Omega \times (0,T), \tag{6.118}$$
$$v = 0 \quad \text{on}\ \partial\Omega \times (0,T). \tag{6.119}$$

Differentiating the above equation gives

$$\frac{\partial^2 v_t}{\partial t^2} = c^2 \nabla^2 v_t \quad \text{in}\ \Omega \times (0,T), \tag{6.120}$$
$$v_t = 0 \quad \text{on}\ \partial\Omega \times (0,T). \tag{6.121}$$

It then follows from (6.104) that

$$
\begin{aligned}
E(v_t,t) &= E(v_t,0)\\
&\leq M \int_0^T \int_\omega (|v_t|^2 + |v_{tt}|^2) dV dt\\
&= 0.
\end{aligned}
$$

So we have $v_t = 0$ in $\Omega \times (0,T)$. Then equation (6.120) reduces to

$$c^2 \nabla^2 v = 0 \quad \text{in}\ \Omega \times (0,T), \tag{6.122}$$
$$v = 0 \quad \text{on}\ \partial\Omega \times (0,T), \tag{6.123}$$

which implies that $v = 0$. This is a contradiction of (6.117).

Theorem 6.6. *Let ω be a neighborhood of $\Gamma(\mathbf{x}^0)$. Then there exist constants $M, \sigma > 0$ such that the solution of (6.93)-(6.95) with the feedback control (6.96) satisfies*

$$E(t) \leq ME(0)e^{-\sigma t} \quad \text{for}\ t \geq 0. \tag{6.124}$$

Proof. Let $T > \frac{2R(\mathbf{x}^0)}{c}$. By (6.97), we first have

$$E(0) - E(T) = k \int_0^T \int_\omega \left| \frac{\partial u}{\partial t} \right|^2 dV dt.$$

By (6.104) and (6.111), we have

$$M \int_0^T \int_\omega \left| \frac{\partial u}{\partial t} \right|^2 dV dt \geq E(T).$$

It then follows that

$$E(0) - E(T) \geq ME(T),$$

and then

$$E(T) \leq \frac{1}{1+M} E(0). \tag{6.125}$$

Let $u(t; u_0, u_1)$ denote the solution of (6.93)-(6.95) corresponding to the initial condition (u_0, u_1). Then we have $u(s+t; u_0, u_1) = u(s; u(t; u_0, u_1), u_t(t; u_0, u_1))$ due to the uniqueness of the solution. For any $t > 0$, there exists an integer N such that $t = NT + r, 0 \leq r < T$. It then follows from (6.125) that

$$
\begin{aligned}
E(t) &= E(u(NT + r; u_0, u_1)) \\
&= E(u(T; u((N-1)T + r; u_0, u_1), u_t((N-1)T + r; u_0, u_1))) \\
&\leq \frac{1}{1+M} E(u((N-1)T + r; u_0, u_1)) \\
&\ \vdots \\
&\leq \frac{1}{(1+M)^N} E(u(r; u_0, u_1)) \\
&\leq \frac{1}{(1+M)^N} E(0) \\
&= e^{\ln \frac{1}{(1+M)^N}} E(0) \\
&= e^{-N \ln(1+M)} E(0) \\
&= e^{-\frac{t-r}{T} \ln(1+M)} E(0) \\
&= e^{\frac{r}{T} \ln(1+M)} E(0) e^{-\frac{t}{T} \ln(1+M)} \\
&\leq (1+M) E(0) e^{-\frac{t}{T} \ln(1+M)}.
\end{aligned}
$$

As in the case of the reaction-convection-diffusion equations, the stabilization theory for the wave equation has been extended to the abstract dynamical control system (4.347)-(4.348), where the operator A generates a C_0 semigroup. For details, we refer to [24].

Fig. 6.2 Legendre transformation.

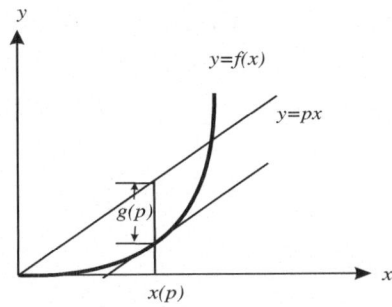

6.6 Nonlinear Interior Feedback Stabilization

This section is technically involved and may be skipped in the first reading.

Consider the wave equation with a nonlinear interior control

$$\frac{\partial^2 u}{\partial t^2} = c^2 \nabla^2 u + f\left(\frac{\partial u}{\partial t}\right) \quad \text{in } \Omega \times (0, \infty), \tag{6.126}$$

$$u = 0 \quad \text{on } \Gamma \times (0, \infty), \tag{6.127}$$

$$u(\mathbf{x}, 0) = u_0(\mathbf{x}), \quad \frac{\partial u}{\partial t}(\mathbf{x}, 0) = u_1(\mathbf{x}) \quad \text{in } \Omega. \tag{6.128}$$

To design a feedback controller f, we need to introduce a generalized Young's inequality and Jensen's inequality.

Let $y = f(x)$ be a convex function and $f''(x) > 0$. The Legendre transformation of the function f is a new function g of a new variable p. Consider the straight line $y = px$ (Figure 6.2). We take the point $x = x(p)$ at which the curve of $f(x)$ is farthest from the straight line in the vertical direction: for each p the function $F(p, x) = px - f(x)$ has a maximum with respect to x at the point $x(p)$. Now we define

$$g(p) = F(p, x) = px - f(x).$$

Definition 6.1. Two functions f and g, which are the Legendre transformation of one another, are called dual in the sense of Young.

Generalized Young's inequality. Let $y = f(x)$ be a convex function and $f''(x) > 0$. Let $g(p)$ be the dual function of $f(x)$ in the sense of Young. By definition of the Legendre transformation, we have

$$px \le f(x) + g(p). \tag{6.129}$$

Example 6.3. If $f(x) = \frac{1}{2}x^2$, then $F(p, x) = px - \frac{1}{2}x^2$ and $\frac{\partial}{\partial x}F(p, x) = p - x = 0$. So $x = p$ and $g(p) = F(p, p) = \frac{1}{2}p^2$. Hence we have

$$px \le f(x) + g(p) = \frac{1}{2}(x^2 + p^2).$$

Example 6.4. If $f(x) = \frac{x^\alpha}{\alpha}$ with $\alpha > 1$, then $F(p,x) = px - \frac{x^\alpha}{\alpha}$ and $\frac{\partial}{\partial x}F(p,x) = p - x^{\alpha-1} = 0$. So $x = p^{1/(\alpha-1)}$ and $g(p) = F(p,p) = pp^{1/(\alpha-1)} - \frac{p^{\alpha/(\alpha-1)}}{\alpha} = \frac{\alpha-1}{\alpha}p^{\alpha/(\alpha-1)} = \frac{p^\beta}{\beta}$, where $\beta = \alpha/(\alpha-1)$. Hence we have

$$px \le f(x) + g(p) = \frac{x^\alpha}{\alpha} + \frac{p^\beta}{\beta},$$

where $\frac{1}{\alpha} + \frac{1}{\beta} = 1$.

Lemma 6.13. *(Jensen's inequality) Let $\varphi(x)$ be a convex function. Then*

$$\varphi\left(\frac{1}{\mathrm{mes}(\Omega)}\int_\Omega f(\mathbf{x})dV\right) \le \frac{1}{\mathrm{mes}(\Omega)}\int_\Omega \varphi[f(\mathbf{x})]dV, \qquad (6.130)$$

where $\mathrm{mes}(\Omega)$ denotes the volume of Ω.

Proof. Since φ is convex, we deduce that

$$\varphi\left(\frac{1}{\mathrm{mes}(\Omega)}\int_\Omega f(\mathbf{x})dV\right) = \varphi\left(\frac{1}{\mathrm{mes}(\Omega)}\lim_{n\to\infty}\sum_{i=1}^n f(\mathbf{x}_i)\Delta V_i\right)$$

$$= \lim_{n\to\infty}\varphi\left(\frac{1}{\mathrm{mes}(\Omega)}\sum_{i=1}^n f(\mathbf{x}_i)\Delta V_i\right)$$

$$\left(\text{since } \sum_{i=1}^n \frac{\Delta V_i}{\mathrm{mes}(\Omega)} = 1\right)$$

$$\le \lim_{n\to\infty}\frac{1}{\mathrm{mes}(\Omega)}\sum_{i=1}^n \varphi[f(\mathbf{x}_i)]\Delta V_i$$

$$= \frac{1}{\mathrm{mes}(\Omega)}\int_\Omega \varphi[f(\mathbf{x})]dV.$$

We also need the following Sobolev embedding theorem from [1, p. 97].

Theorem 6.7. *Let Ω be a bounded smooth domain in $\mathbb{R}^n$.*

(1) If $k < n/2$, then

$$H^k(\Omega) \subset L^{2n/(n-2k)}(\Omega)$$

and

$$\|u\|_{L^{2n/(n-2k)}} \le C\|u\|_{H^k}, \quad u \in H^k(\Omega).$$

(2) If $k = n/2$, then for any $1 \le p < \infty$

$$H^k(\Omega) \subset L^p(\Omega)$$

and

$$\|u\|_{L^p} \le C\|u\|_{H^k}, \quad u \in H^k(\Omega).$$

(3) If $k > j + n/2$, then

$$H^k(\Omega) \subset C^j(\bar{\Omega})$$

and

$$\|u\|_{C^j} \le C\|u\|_{H^k}, \quad u \in H^k(\Omega).$$

Choose p to satisfy

$$1 \le p \le \frac{n+2}{n-2}, \quad \text{if } n > 2, \tag{6.131}$$

$$1 \le p < \infty, \quad \text{if } n \le 2. \tag{6.132}$$

Then, by Sobolev's embedding theorem, $H_0^1(\Omega)$ is embedded into $L^{p+1}(\Omega)$, and consequently, $L^{(p+1)/p}(\Omega)$ is continuously embedded into $H^{-1}(\Omega)$. Let $\alpha = \alpha(p) > 0$ denote the best constant such that

$$\|u\|_{H^{-1}(\Omega)} \le \alpha \left(\int_\Omega |u|^{(p+1)/p} dx \right)^{p/(p+1)}, \quad \forall u \in L^{(p+1)/p}(\Omega). \tag{6.133}$$

Theorem 6.8. *Assume that $f \in C(\mathbb{R})$ satisfies the following conditions:*

(1) $f(0) = 0$;
(2) f is increasing on $\mathbb{R}$;
(3) there are constants $c_1, c_2 > 0$ and $p \ge 1$ satisfying (6.131)-(6.132) such that

$$c_1|s| \le |f(s)| \le c_2|s|^p, \quad \text{for } |s| \ge 1; \tag{6.134}$$

(4) there exist a strictly increasing positive function $h(s)$ of class C^2 defined on $[0,\infty)$ and constants $c_3, c_4 > 0$ such that

$$c_3 h(|s|) \le |f(s)| \le c_4 h^{-1}(|s|), \quad \text{for } |s| \le 1, \tag{6.135}$$

where h^{-1} denotes the inverse of h;
(5) there exists an increasing, positive and convex function $\varphi = \varphi(s)$ defined on $[0,\infty)$ and twice differentiable outside $s = 0$ such that $\varphi(|s|^{(p+1)/p}) \le h(|s|)|s|$ and $\varphi''(s)s$ is increasing on $[0,\infty)$.

Then the energy $E(t)$ of solutions of (6.126)-(6.128) with $(u^0, u^1) \in H_0^1(\Omega) \times L^2(\Omega)$ satisfies

$$E(t) \le 2V(t), \quad \text{for } t \ge 0, \tag{6.136}$$

where $V(t)$ is the solution of the following differential equation:

$$V'(t) = -\frac{\varepsilon V(t)}{b} \varphi' \left(\frac{aV(t)}{b} \right) - \varepsilon m_1 \varphi \left(\frac{aV(t)}{b} \right)$$

$$+ m_2 \varepsilon \lambda^{p+1} \varphi' \left(\frac{aV(t)}{c} \right) \left(\frac{V(t)}{c} \right)^{(p+1)/2}, \tag{6.137}$$

and $a, b, c, m_1, m_2, \lambda, \varepsilon$ are positive constants. Furthermore, we have

$$\lim_{t \to \infty} E(t) = \lim_{t \to \infty} V(t) = 0. \qquad (6.138)$$

Proof. By a straightforward calculation, we obtain

$$E'(t) = -\int_{\Omega} u_t f(u_t) dV \le 0. \qquad (6.139)$$

If $E(t_0) = 0$ for some $t_0 \ge 0$, then, by (6.139), we have $E(t) \equiv 0$ for $t \ge t_0$ and the theorem holds. Therefore, we may assume that $E(t) > 0$ for $t \ge 0$. This assumption ensures that, in the following proof, $\varphi''(aE(t))$ is well defined as we have assumed that $\varphi(s)$ is twice differentiable outside $s = 0$.

Define the perturbed energy

$$V(t) = E(t) + \varepsilon \psi(E(t)) \int_{\Omega} u u_t dV, \qquad (6.140)$$

where $\psi(s)$ and $\psi'(s)s$ are positive and increasing functions on $(0, +\infty)$ that will be determined in the proof. Using (6.139) and Poincaré inequality, we deduce

$$V'(t) = E'(t) + \varepsilon \psi'(E(t)) E'(t) \int_{\Omega} u u_t dV$$

$$+ \varepsilon \psi(E(t)) \int_{\Omega} [|u_t|^2 - |\nabla u|^2 - u f(u_t)] dV$$

(use (6.139) for $E'(t)$)

$$= -\int_{\Omega} u_t f(u_t) dV - \varepsilon \psi'(E(t)) \int_{\Omega} u u_t dV \int_{\Omega} u_t f(u_t) dV$$

$$+ \varepsilon \psi(E(t)) \int_{\Omega} [-|u_t|^2 - |\nabla u|^2] dV \quad \text{(subtract } |u_t|^2 \text{ and then add } |u_t|^2\text{)}$$

$$+ 2\varepsilon \psi(E(t)) \int_{\Omega} |u_t|^2 dV - \varepsilon \psi(E(t)) \int_{\Omega} u f(u_t) dV$$

(use Young inequality (2.4) for $u u_t$)

$$\le -\int_{\Omega} u_t f(u_t) dV + \frac{1}{2} k \varepsilon \psi'(E(t)) \int_{\Omega} \left(\frac{|u|^2}{k^2} + |u_t|^2 \right) dV \int_{\Omega} u_t f(u_t) dV$$

$$- 2\varepsilon \psi(E(t)) E(t) + 2\varepsilon \psi(E(t)) \int_{\Omega} |u_t|^2 dV - \varepsilon \psi(E(t)) \int_{\Omega} u f(u_t) dV$$

(use Poincaré inequality (2.29) for u^2/k^2)

$$\le -\int_{\Omega} u_t f(u_t) dV + \frac{1}{2} k \varepsilon \psi'(E(t)) \int_{\Omega} \left(|\nabla u|^2 + |u_t|^2 \right) dV \int_{\Omega} u_t f(u_t) dV$$

$$- 2\varepsilon \psi(E(t)) E(t) + 2\varepsilon \psi(E(t)) \int_{\Omega} |u_t|^2 dV - \varepsilon \psi(E(t)) \int_{\Omega} u f(u_t) dV$$

$$= -\int_{\Omega} u_t f(u_t) dV + k \varepsilon \psi'(E(t)) E(t) \int_{\Omega} u_t f(u_t) dV$$

$$- 2\varepsilon \psi(E(t)) E(t) + 2\varepsilon \psi(E(t)) \int_{\Omega} |u_t|^2 dV - \varepsilon \psi(E(t)) \int_{\Omega} u f(u_t) dV,$$

where k denotes the constant in Poincaré inequality. Moreover, by (6.134), we have $|u_t|^2 \leq c_1^{-1} u_t f(u_t)$ for $|u_t| \geq 1$. Therefore, taking into account the fact that $\psi'(s)s$ is non-decreasing, we deduce that

$$
\begin{aligned}
V'(t) \leq{} & -2\varepsilon\psi(E(t))E(t) + [\varepsilon k\psi'(E(0))E(0) - 1]\int_\Omega u_t f(u_t)dV \\
& +2\varepsilon c_1^{-1}\psi(E(t))\int_{[|u_t|\geq 1]} u_t f(u_t)dV + 2\varepsilon\psi(E(t))\int_{[|u_t|\leq 1]} |u_t|^2 dV \\
& -\varepsilon\psi(E(t))\int_\Omega u f(u_t)dV \\
\leq{} & -2\varepsilon\psi(E(t))E(t) \\
& +[\varepsilon k\psi'(E(0))E(0) + 2\varepsilon c_1^{-1}\psi(E(0)) - 1]\int_{[|u_t|\geq 1]} u_t f(u_t)dV \\
& +[\varepsilon k\psi'(E(0))E(0) - 1]\int_{[|u_t|\leq 1]} u_t f(u_t)dV \\
& +2\varepsilon\psi(E(t))\int_{[|u_t|\leq 1]} |u_t|^2 dV \quad (= I_1) \\
& -\varepsilon\psi(E(t))\int_\Omega u f(u_t)dV \quad (= I_2).
\end{aligned}
\tag{6.141}
$$

We now estimate I_1 and I_2 in terms of $\int_\Omega u_t f(u_t)dV$ as follows. Let φ^* denote the dual of φ in the sense of Young. Then, by the generalized Young's inequality (6.129) and Jensen's inequality (6.130), we deduce

$$
\begin{aligned}
I_1 ={} & 2\varepsilon\mathrm{mes}([|u_t| \leq 1])\psi(E)\frac{1}{\mathrm{mes}([|u_t| \leq 1])}\int_{[|u_t|\leq 1]} |u_t|^2 dV \\
& \text{(note that } |u_t| \leq 1 \text{ and } p \geq 1) \\
\leq{} & 2\varepsilon\mathrm{mes}([|u_t| \leq 1])\psi(E)\frac{1}{\mathrm{mes}([|u_t| \leq 1])}\int_{[|u_t|\leq 1]} |u_t|^{(p+1)/p} dV \\
& \text{(Use the generalized Young's inequality (6.129))} \\
\leq{} & 2\varepsilon\mathrm{mes}([|u_t| \leq 1])\left[\varphi^*(\psi(E)) + \varphi\left(\frac{1}{\mathrm{mes}([|u_t| \leq 1])}\int_{[|u_t|\leq 1]} |u_t|^{(p+1)/p} dV\right)\right] \\
& \text{(Use Jensen's inequality (6.130))} \\
\leq{} & 2\varepsilon\mathrm{mes}([|u_t| \leq 1])\varphi^*(\psi(E)) + 2\varepsilon\int_{[|u_t|\leq 1]} \varphi(|u_t|^{(p+1)/p})dV \\
\leq{} & 2\varepsilon\mathrm{mes}([|u_t| \leq 1])\varphi^*(\psi(E)) + 2\varepsilon\int_{[|u_t|\leq 1]} |u_t| h(|u_t|)dV \quad \text{(use (6.135))} \\
\leq{} & 2\varepsilon\mathrm{mes}(\Omega)\varphi^*(\psi(E)) + 2\varepsilon c_3^{-1}\int_{[|u_t|\leq 1]} u_t f(u_t)dV.
\end{aligned}
\tag{6.142}
$$

Using the embedding inequality (6.133), the condition (6.134), and Young's inequality (2.4), we deduce that

$$|I_2| = \varepsilon \psi(E(t)) \left| \int_{\Omega} u f(u_t) dV \right| \tag{6.143}$$

$$\leq \varepsilon \psi(E(t)) \|u\|_{H_0^1(\Omega)} \|f(u_t)\|_{H^{-1}(\Omega)}$$

(use the embedding inequality (6.133))

$$\leq \alpha \varepsilon \psi(E(t)) \|u\|_{H_0^1(\Omega)} \left(\int_{\Omega} |f(u_t)|^{(p+1)/p} dV \right)^{p/(p+1)}$$

$$\leq \alpha \varepsilon \psi(E(t)) \|u\|_{H_0^1(\Omega)} \left(\int_{[|u_t| \leq 1]} |f(u_t)|^{(p+1)/p} dV \right)^{p/(p+1)}$$

(use condition (6.134):

$$|f(u_t)|^{(p+1)} = |f(u_t)||f(u_t)|^p \leq c_2 |u_t||f(u_t)|^p = c_2 [u_t f(u_t)]^p)$$

$$+ \alpha \varepsilon c_2^{1/(p+1)} \psi(E(t)) \|u\|_{H_0^1(\Omega)} \left(\int_{[|u_t| \geq 1]} u_t f(u_t) dV \right)^{p/(p+1)}$$

(use Young inequality (2.4))

$$\leq \alpha \varepsilon \psi(E(t)) \left[\frac{p}{p+1} \lambda^{-(p+1)/p} \int_{[|u_t| \leq 1]} |f(u_t)|^{(p+1)/p} dV \right.$$

$$\left. + \frac{1}{p+1} \lambda^{p+1} \|u\|_{H_0^1(\Omega)}^{p+1} \right] + \alpha \varepsilon c_2^{1/(p+1)} \psi(E(t)) \left[\frac{p}{p+1} \lambda^{-(p+1)/p} \right.$$

$$\left. \int_{[|u_t| \geq 1]} u_t f(u_t) dV + \frac{1}{p+1} \lambda^{p+1} \|u\|_{H_0^1(\Omega)}^{p+1} \right],$$

where λ is any positive number. We now use the generalized Young's inequality (6.129) and Jensen's inequality (6.130) to estimate $\psi(E) \int_{[|u_t| \leq 1]} |f(u_t)|^{(p+1)/p} dV$ as follows:

$$\psi(E) \int_{[|u_t| \leq 1]} |f(u_t)|^{(p+1)/p} dV$$

$$= c_4^{(p+1)/p} \text{mes}([|u_t| \leq 1]) \psi(E) \frac{1}{\text{mes}([|u_t| \leq 1])} \int_{[|u_t| \leq 1]} |c_4^{-1} f(u_t)|^{(p+1)/p} dV$$

(Use the generalized Young's inequality (6.129))

$$\leq c_4^{(p+1)/p} \text{mes}([|u_t| \leq 1]) \left[\varphi^*(\psi(E)) + \varphi \left(\frac{1}{\text{mes}([|u_t| \leq 1])} \right. \right.$$

$$\left. \left. \int_{[|u_t| \leq 1]} |c_4^{-1} f(u_t)|^{(p+1)/p} \right) \right]$$

(Use Jensen's inequality (6.130))

$$\leq c_4^{(p+1)/p} \text{mes}([|u_t| \leq 1) \varphi^*(\psi(E)) + c_4^{(p+1)/p} \int_{[|u_t| \leq 1]} \varphi \left(|c_4^{-1} f(u_t)|^{(p+1)/p} \right) dV$$

(use condition (v))

$$\leq c_4^{(p+1)/p} \text{mes}(\Omega) \varphi^*(\psi(E)) + c_4^{(p+1)/p} \int_{[|u_t| \leq 1]} |c_4^{-1} f(u_t)| h(c_4^{-1} |f(u_t)|) dV$$

(use (6.135))

$$\leq c_4^{(p+1)/p}\mathrm{mes}(\Omega)\varphi^*(\psi(E))+c_4^{1/p}\int_{[|u_t|\leq1]}u_tf(u_t)dV. \tag{6.144}$$

Thus it follows from (6.143) that

$$
\begin{aligned}
|I_2| &\leq \alpha\varepsilon\psi(E(t))\left[\frac{p}{p+1}\lambda^{-(p+1)/p}\int_{[|u_t|\leq1]}|f(u_t)|^{(p+1)/p}dV\right.\\
&\quad\left.+\frac{1}{p+1}\lambda^{p+1}\|u\|_{H_0^1(\Omega)}^{p+1}\right]+\alpha\varepsilon c_2^{1/(p+1)}\psi(E(t))\left[\frac{p}{p+1}\lambda^{-(p+1)/p}\right.\\
&\quad\left.\int_{[|u_t|\geq1]}u_tf(u_t)dV+\frac{1}{p+1}\lambda^{p+1}\|u\|_{H_0^1(\Omega)}^{p+1}\right]\\
&\leq\frac{\alpha\varepsilon p}{p+1}c_4^{(p+1)/p}\mathrm{mes}(\Omega)\lambda^{-(p+1)/p}\varphi^*(\psi(E)) \quad(\text{use }(6.144))\\
&\quad(\text{use }\|u\|_{H_0^1(\Omega)}^{p+1}=\left(\frac{2}{c^2}\frac{1}{2}\int_\Omega c^2|\nabla u|^2dV\right)^{(p+1)/2}\leq\left(\frac{2}{c^2}\right)^{(p+1)/2}E^{(p+1)/2}(t))\\
&\quad+\frac{\alpha\varepsilon 2^{(p+1)/2}}{(p+1)c^{p+1}}(1+c_2^{1/(p+1)})\lambda^{p+1}\psi(E(t))E^{(p+1)/2}(t)\\
&\quad+\frac{\alpha\varepsilon p}{p+1}c_4^{1/p}\lambda^{-(p+1)/p}\int_{[|u_t|\leq1]}u_tf(u_t)dV\\
&\quad+\frac{\alpha\varepsilon p}{p+1}c_2^{1/(p+1)}\lambda^{-(p+1)/p}\psi(E(0))\int_{[|u_t|\geq1]}u_tf(u_t)dV. \tag{6.145}
\end{aligned}
$$

It therefore follows from (6.141), (6.142) and (6.145) that

$$
\begin{aligned}
V'(t) &\leq -2\varepsilon\psi(E(t))E(t)+2\varepsilon\mathrm{mes}(\Omega)\varphi^*(\psi(E))\\
&\quad+\frac{\alpha\varepsilon p}{p+1}c_4^{(p+1)/p}\mathrm{mes}(\Omega)\lambda^{-(p+1)/p}\varphi^*(\psi(E))\\
&\quad+\frac{\alpha\varepsilon 2^{(p+1)/2}}{(p+1)c^{p+1}}(1+c_2^{1/(p+1)})\lambda^{p+1}\psi(E(t))E^{(p+1)/2}(t)\\
&\quad+(\varepsilon k_1-1)\int_{[|u_t|\geq1]}u_tf(u_t)dV+(\varepsilon k_2-1)\int_{[|u_t|\leq1]}u_tf(u_t)dV. \tag{6.146}
\end{aligned}
$$

where k_1 and k_2 are positive constants depending on $E(0)$. By the definition of the dual function in the sense of Young, $\varphi^*(t)$ is the Legendre transform of $\varphi(s)$, which is given by

$$\varphi^*(t)=t[\varphi']^{-1}(t)-\varphi[[\varphi]'^{-1}(t)]. \tag{6.147}$$

Thus, we have

$$\varphi^*(\psi(E))=\psi(E(t))[\varphi]'^{-1}(\psi(E(t)))-\varphi[[\varphi]'^{-1}(\psi(E(t))). \tag{6.148}$$

This motivates us to make the choice

$$\psi(s)=\varphi'(as) \tag{6.149}$$

so that

$$\varphi^*(\psi(E)) = \varphi'(aE)aE - \varphi(aE)$$

where the constant a will be determined later. By condition (v), $\psi(s)$ satisfies the requirement we set at the beginning of the proof, that is, ψ and $\psi'(s)s$ are positive and increasing on $(0, +\infty)$. Taking

$$a = \left(2\text{mes}(\Omega) + \frac{\alpha p}{p+1} c_4^{(p+1)/p} \text{mes}(\Omega)\lambda^{-(p+1)/p} \right)^{-1}, \qquad (6.150)$$

we deduce from (6.146) that

$$V'(t) \leq -\varepsilon\varphi'(aE(t))E(t) - \varepsilon m_1 \varphi(aE(t)) + m_2\varepsilon\lambda^{p+1}\varphi'(aE(t))E^{(p+1)/2}(t), \qquad (6.151)$$

where m_1, m_2, and ε are positive constants.

On the other hand, since $\varphi(s)$ and $\varphi'(s)$ are positive and increasing on $(0, \infty)$, it follows from Poincaré's inequality that

$$[1 - \varepsilon k\varphi'(aE(0))]E(t) \leq V(t) \leq [1 + \varepsilon k\varphi'(aE(0))]E(t). \qquad (6.152)$$

Therefore, we deduce from (6.151) and (6.152) that

$$
\begin{aligned}
V'(t) \leq &-\frac{\varepsilon V(t)}{1 + \varepsilon k\varphi'(aE(0))} \varphi'\left(\frac{aV(t)}{1 + \varepsilon k\varphi'(aE(0))} \right) \\
&-\varepsilon m_1 \varphi\left(\frac{aV(t)}{1 + \varepsilon k\varphi'(aE(0))} \right) \\
&+m_2\varepsilon\lambda^{p+1}\varphi'\left(\frac{aV(t)}{1 - \varepsilon k\varphi'(aE(0))} \right) \left[\frac{V(t)}{(1 - \varepsilon k\varphi'(aE(0)))} \right]^{(p+1)/2}
\end{aligned}
\qquad (6.153)
$$

which proves (6.137).

It remains to prove (6.138). We argue by contradiction. Suppose that $E(t)$ doesn't tend to zero as $t \to \infty$. Since $E(t)$ is decreasing on $[0, \infty)$, we have

$$E(0) \geq E(t) \geq \sigma > 0, \quad \forall t \geq 0, \qquad (6.154)$$

and by (6.152), we have

$$bE(0) \geq V(t) \geq \beta > 0, \quad \forall t \geq 0. \qquad (6.155)$$

Thus we have

$$\varphi'(aE(0)) \geq \varphi'\left(\frac{aV(t)}{b} \right) \geq \gamma > 0, \quad \forall t \geq 0. \qquad (6.156)$$

Let $\lambda > 0$ be so small that

$$m_2\lambda^{p+1}\varphi'\left(\frac{aV(t)}{c} \right) \left(\frac{V(t)}{c} \right)^{(p+1)/2} \leq m_1\varphi(a\beta/b), \quad \forall t \geq 0. \qquad (6.157)$$

It therefore follows from (6.137) that

$$V'(t) \leq -\frac{\varepsilon\gamma}{b}V(t), \quad \forall t \geq 0, \tag{6.158}$$

which is in contradiction of (6.155). This completes the proof.

Corollary 6.1. *Assume that $f \in C(\mathbb{R})$ satisfies all the conditions of Theorem 6.8. Suppose $\varphi(s) = s^{p/(p+1)}h(s^{p/(p+1)})$ is convex and twice continuously differentiable. Then the energy $E(t)$ of (6.126)-(6.128) satisfies the following decay rate:*

$$E(t) \leq 2V(t), \quad \text{for } t \geq 0, \tag{6.159}$$

where $V(t)$ satisfies the following differential equation:

$$V'(t) = -\frac{\varepsilon(2p+1)a^{\frac{-1}{p+1}}}{(p+1)b^{\frac{p}{p+1}}}V^{\frac{p}{p+1}}h\left(\left(\frac{aV}{b}\right)^{\frac{p}{p+1}}\right) - \frac{\varepsilon p}{(p+1)b}\left(\frac{a}{b}\right)^{\frac{p-1}{p+1}}V^{\frac{2p}{p+1}}h'\left(\left(\frac{aV}{b}\right)^{\frac{p}{p+1}}\right)$$
$$+\frac{pm_2\varepsilon l^{(p+1)}}{p+1}\left[\left(\frac{aV}{c}\right)^{\frac{-1}{p+1}}h\left(\left(\frac{aV}{c}\right)^{\frac{p}{p+1}}\right) + \left(\frac{aV}{c}\right)^{\frac{p-1}{p+1}}h'\left(\left(\frac{aV}{c}\right)^{\frac{p}{p+1}}\right)\right]\left(\frac{V}{c}\right)^{\frac{p+1}{2}}. \tag{6.160}$$

Proof. Since

$$\varphi(s) = s^{p/(p+1)}h(s^{p/(p+1)}), \tag{6.161}$$

$$\varphi'(s) = \frac{p}{p+1}[s^{-1/(p+1)}h(s^{p/(p+1)}) + s^{(p-1)/(p+1)}h'(s^{p/(p+1)})], \tag{6.162}$$

by substituting (6.161) and (6.162) into (6.137), we obtain (6.160).

We give three examples to illustrate how to derive the usual exponential or polynomial decay rate and the logarithmic decay rate for the exponentially degenerate damping from our general result. In what follows, ω denotes various positive constants that may vary from line to line.

Example 6.5. Exponential decay rate. Let $f(s) = ks$ and $p = 1$, where k is a positive constant. Then $h(s) = ks$. In this case, all the assumptions of Corollary 6.1 are satisfied and (6.160) becomes

$$V'(t) = -\omega V(t), \tag{6.163}$$

where ω is a positive constant. Thus, as usual, we obtain an exponential decay rate.

Example 6.6. Polynomial decay rate. Assume $f(s) = k|s|^{q-1}s$ with $q > 1$ and $k > 0$. Then $h(s) = ks^q$ and $p = 1$, $q > 1$. Then (6.160) becomes

$$V'(t) = -\omega[V(t)]^{(q+1)/2}, \tag{6.164}$$

which, as usual, implies the polynomial decay rate

$$E(t) \leq C(E(0))t^{-2/(q-1)}, \quad \forall t > 0. \tag{6.165}$$

Example 6.7. Logarithmic decay rate. Let $p = 1$ and $f(s) = s^3 e^{-\frac{1}{s^2}}$ near the origin. Let

$$h(s) = s^3 e^{-\frac{1}{s^2}}, \quad s > 0. \tag{6.166}$$

Then, by (6.160), V satisfies

$$V'(t) \leq -\omega V^2 e^{-\frac{b}{aV}}, \tag{6.167}$$

which is the same as

$$\left(e^{\frac{b}{aV}}\right)' \geq \frac{b\omega}{a}. \tag{6.168}$$

Solving the inequality, we obtain the logarithmic decay rate

$$V(t) \leq \frac{b}{a} \left[\ln\left(\frac{b\omega}{a}t + e^{\frac{b}{aV(0)}}\right)\right]^{-1}. \tag{6.169}$$

Exercises 6.6

1. Consider the beam equation with a nonlinear velocity feedback control

$$u_{tt} = -u_{xxxx} - \frac{ku_t^3}{1+u_t^2} \quad \text{in } (0,L) \times (0,\infty),$$
$$u(0,t) = u(L,t) = u_x(0,t) = u_x(L,t) = 0, \quad t \geq 0,$$
$$u(x,0) = u_0(x), \quad u_t(x,0) = u_1(x), \quad x \in (0,L),$$

where k is a positive constant. Define the energy $E(t)$ by

$$E(t) = \frac{1}{2} \int_0^L \left(|u_t(x,t)|^2 + |u_{xx}(x,t)|^2\right) dx.$$

a. Show that

$$\frac{dE}{dt} = -\int_0^L \frac{ku_t^4}{1+u_t^2} dV.$$

b. Use the perturbed energy

$$E_\varepsilon(t) = E(t) + \varepsilon E(t) \int_0^L uu_t dV$$

to show that there exists a positive constant M such that

$$E(t) \leq \frac{M}{1+t}.$$

6.7 Optimal Boundary Control

Consider the boundary control problem

$$\frac{\partial^2 u}{\partial t^2} = c^2 \nabla^2 u, \tag{6.170}$$

$$u|_\Gamma = \phi, \tag{6.171}$$

$$u(\mathbf{x}, s) = u_0(\mathbf{x}), \quad \frac{\partial u}{\partial t}(\mathbf{x}, s) = u_1(\mathbf{x}), \tag{6.172}$$

where ϕ is a control input. For $T > s$, we define the quadratic performance functional by

$$J(\phi; u_0, u_1, s, T) = \alpha \int_s^T \int_\Omega |u|^2 dV dt + \beta \int_s^T \int_\Gamma |\phi|^2 dS dt + \gamma \int_\Omega |u(T)|^2 dV, \tag{6.173}$$

where $\alpha, \beta, \gamma \geq 0$ are weight constants. The optimal control problem is to minimize the functional J over $L^2(\Gamma \times (s, T))$ with respect to ϕ, where u_0, u_1, and s are parameters. This problem can be solved in the same way as for the reaction-convection-diffusion equation. We first assume that an optimal control exists and prove the existence at the end of this section by using solutions of Riccati equations.

We denote the solution of (6.170)-(6.172) corresponding to the control ϕ and the initial condition u_0, u_1 by $u(\mathbf{x}, t; u_0, u_1, \phi)$. Since the equation is linear, we can readily see that

$$u(\mathbf{x}, t; u_0, u_1, \phi + \lambda \psi) = u(\mathbf{x}, t; u_0, u_1, \phi) + \lambda u(\mathbf{x}, t; 0, 0, \psi) \tag{6.174}$$

for any control functions ϕ, ψ and any number λ. We now calculate the Gâteaux-differential of J.

Theorem 6.9. *The Gâteaux-differential of J defined by (6.173) at ϕ is given by*

$$\int_s^T \int_\Gamma J'(\phi; u_0, u_1, s, T) \psi dS dt = 2\alpha \int_s^T \int_\Omega u(\mathbf{x}, t; u_0, u_1, \phi) u(\mathbf{x}, t; 0, 0, \psi) dV dt$$

$$+ 2\gamma \int_\Omega u(\mathbf{x}, T; u_0, u_1, \phi) u(\mathbf{x}, T; 0, 0, \psi) dV$$

$$+ 2\beta \int_s^T \int_\Gamma \phi \psi dS dt, \tag{6.175}$$

where $u(\mathbf{x}, t; 0, 0, \psi)$ is the solution of

$$\frac{\partial^2 u}{\partial t^2} = c^2 \nabla^2 u, \tag{6.176}$$

$$u|_\Gamma = \psi, \tag{6.177}$$

$$u(\mathbf{x}, s) = 0, \quad \frac{\partial u}{\partial t}(\mathbf{x}, s) = 0. \tag{6.178}$$

Proof. Using (6.173) and (6.174), we derive that

$$
\int_s^T \int_\Gamma J'(\phi; u_0, u_1, s) \psi \, dS \, dt
$$

$$
= \lim_{\lambda \to 0^+} \frac{J(\phi + \lambda \psi; u_0, u_1, s) - J(\phi; u_0, u_1, s)}{\lambda}
$$

$$
= \lim_{\lambda \to 0^+} \frac{\alpha}{\lambda} \int_s^T \int_\Omega \left(|u(\mathbf{x}, t; u_0, u_1, \phi + \lambda \psi)|^2 - |u(\mathbf{x}, t; u_0, u_1, \phi)|^2 \right) dV \, dt
$$

$$
+ \lim_{\lambda \to 0^+} \frac{\beta}{\lambda} \int_s^T \int_\Gamma \left(|\phi + \lambda \psi|^2 - |\phi|^2 \right) dS \, dt
$$

$$
+ \lim_{\lambda \to 0^+} \frac{\gamma}{\lambda} \int_\Omega \left(|u(\mathbf{x}, T; u_0, u_1, \phi + \lambda \psi)|^2 - |u(\mathbf{x}, T; u_0, u_1, \phi)|^2 \right) dV
$$

$$
= \lim_{\lambda \to 0^+} \frac{\alpha}{\lambda} \int_s^T \int_\Omega \left(2\lambda u(\mathbf{x}, t; u_0, u_1, \phi) u(\mathbf{x}, t; 0, 0, \psi) + \lambda^2 |u(\mathbf{x}, t; 0, 0, \psi)|^2 \right) dV \, dt
$$

$$
+ \lim_{\lambda \to 0^+} \frac{\beta}{\lambda} \int_s^T \int_\Gamma \left(2\lambda \phi \psi + \lambda^2 |\psi|^2 \right) dS \, dt
$$

$$
+ \lim_{\lambda \to 0^+} \frac{\gamma}{\lambda} \int_\Omega \left(2\lambda u(\mathbf{x}, T; u_0, u_1, \phi) u(\mathbf{x}, T; 0, 0, \psi) + \lambda^2 |u(\mathbf{x}, T; 0, 0, \psi)|^2 \right) dV
$$

$$
= 2\alpha \int_s^T \int_\Omega u(\mathbf{x}, t; u_0, u_1, \phi) u(\mathbf{x}, t; 0, 0, \psi) dV \, dt
$$

$$
+ 2\beta \int_s^T \int_\Gamma \phi \psi \, dS \, dt + 2\gamma \int_\Omega u(\mathbf{x}, T; u_0, u_1, \phi) u(\mathbf{x}, T; 0, 0, \psi) dV.
$$

In a manner analogous to Theorem 4.35, we have the optimality (also called Hamiltonian) system for the optimal control.

Theorem 6.10. *The optimal control* ϕ^* *satisfies the following system*

$$
\frac{\partial^2 u}{\partial t^2} = c^2 \nabla^2 u, \tag{6.179}
$$

$$
\frac{\partial^2 v}{\partial t^2} = c^2 \nabla^2 v + \alpha u, \tag{6.180}
$$

$$
u|_\Gamma = \phi^*, \quad v|_\Gamma = 0, \tag{6.181}
$$

$$
\phi^* = \frac{c^2}{\beta} \frac{\partial v}{\partial \mathbf{n}} \bigg|_\Gamma, \tag{6.182}
$$

$$
u(\mathbf{x}, s) = u_0(\mathbf{x}), \quad \frac{\partial u}{\partial t}(\mathbf{x}, s) = u_1(\mathbf{x}),
$$

$$
v(\mathbf{x}, T) = 0, \quad \frac{\partial v}{\partial t}(\mathbf{x}, T) = -\gamma u(\mathbf{x}, T). \tag{6.183}
$$

Proof. Multiplying (6.176) by the solution v of (6.180) and integrating over $\Omega \times (s, T)$ by parts, we have

$$\int_s^T \int_\Omega v \frac{\partial^2 u}{\partial t^2}(\mathbf{x},t;0,0,\psi)dVdt = c^2 \int_s^T \int_\Omega v\nabla^2 u(\mathbf{x},t;0,0,\psi)dVdt \quad (6.184)$$

$$= -c^2 \int_s^T \int_\Omega \nabla v \cdot \nabla u(\mathbf{x},t;0,0,\psi)dVdt.$$

Multiplying (6.180) by the solution $u(\mathbf{x},t;0,0,\psi)$ of (6.176) and integrating over $\Omega \times (s,T)$ by parts, we have

$$\int_s^T \int_\Omega u(\mathbf{x},t;0,0,\psi)\frac{\partial^2 v}{\partial t^2}dVdt \quad (6.185)$$

$$= c^2 \int_s^T \int_\Omega u(\mathbf{x},t;0,0,\psi)\nabla^2 v dVdt + \alpha \int_s^T \int_\Omega u(\mathbf{x},t;0,0,\psi)u(\mathbf{x},t;u_0,u_1,\phi^*)dVdt$$

$$= c^2 \int_s^T \int_\Gamma \psi \frac{\partial v}{\partial \mathbf{n}}dSdt - c^2 \int_s^T \int_\Omega \nabla u(\mathbf{x},t;0,0,\psi) \cdot \nabla v dVdt$$

$$+ \alpha \int_s^T \int_\Omega u(\mathbf{x},t;0,0,\psi)u(\mathbf{x},t;u_0,u_1,\phi^*)dVdt.$$

Substituting (6.184) into (6.185) gives

$$c^2 \int_s^T \int_\Gamma \psi \frac{\partial v}{\partial \mathbf{n}}dSdt = \int_s^T \int_\Omega u(\mathbf{x},t;0,0,\psi)\frac{\partial^2 v}{\partial t^2}dVdt \quad (6.186)$$

$$- \int_s^T \int_\Omega v \frac{\partial^2 u}{\partial t^2}(\mathbf{x},t;0,0,\psi)dVdt$$

$$- \alpha \int_s^T \int_\Omega u(\mathbf{x},t;0,0,\psi)u(\mathbf{x},t;u_0,u_1,\phi^*)dVdt$$

$$= -\gamma \int_\Omega u(\mathbf{x},T;0,0,\psi)u(\mathbf{x},T;u_0,u_1,\phi^*)dVdt$$

$$- \alpha \int_s^T \int_\Omega u(\mathbf{x},t;0,0,\psi)u(\mathbf{x},t;u_0,u_1,\phi^*)dVdt.$$

$$(6.187)$$

On the other hand, it follows from Theorems 4.34 and 6.9 that

$$\alpha \int_s^T \int_\Omega u(\mathbf{x},t;u_0,u_1,\phi^*)u(\mathbf{x},t;0,0,\psi)dVdt + \beta \int_s^T \int_\Gamma \phi^* \psi dSdt$$

$$+ \gamma \int_\Omega u(\mathbf{x},T;u_0,u_1,\phi^*)u(\mathbf{x},T;0,0,\psi)dV = 0$$

for all functions $\psi \in L^2(\Gamma \times (s,T))$. Substituting (6.187) into this equation, we obtain

$$\int_s^T \int_\Gamma \left(\beta \phi^* - c^2 \frac{\partial v}{\partial \mathbf{n}} \right) \psi dVdt = 0$$

for all functions $\psi \in L^2(\Gamma \times (s,T))$. Thus $\beta \phi^* = c^2 \frac{\partial v}{\partial \mathbf{n}}\big|_\Gamma$.

To decouple the optimality system (6.179)-(6.183), we define a family of linear operators $P(s)$ by

$$P(s)(u_0, u_1) = \left(-\frac{\partial v}{\partial t}(s; u_0, u_1), v(s; u_0, u_1)\right) \tag{6.188}$$

for any given function $(u_0, u_1) \in L^2(\Omega) \times H^{-1}(\Omega)$, where $v(s; u_0, u_1)$ is the solution of (6.179)-(6.183).

Theorem 6.11. *The operator $P(s)$ defined by (6.188) has the following properties:*

(1) $P(s)$ is symmetric, that is,

$$\langle P(s)(u_0, u_1), (p_0, p_1) \rangle = \langle (u_0, u_1), P(s)(p_0, p_1) \rangle \tag{6.189}$$

for $(u_0, u_1), (p_0, p_1) \in L^2(\Omega) \times H^{-1}(\Omega)$, where $\langle \cdot, \cdot \rangle$ denotes the dual product between $L^2(\Omega) \times H^{-1}(\Omega)$ and $L^2(\Omega) \times H_0^1(\Omega)$.
(2) If ϕ^* is the optimal control of (6.170)-(6.172), then

$$\langle P(s)(u_0, u_1), (u_0, u_1) \rangle = \alpha \int_s^T \int_\Omega |u|^2 dV dt + \beta \int_s^T \int_\Gamma |\phi^*|^2 dS dt$$
$$+ \gamma \int_\Omega |u(T)|^2) dV. \tag{6.190}$$

Hence $P(s)$ is nonnegative.
(3) $P(T)(u_0, u_1) = \gamma(u_0, 0)$ for all $(u_0, u_1) \in L^2(\Omega) \times H^{-1}(\Omega)$.
(4) $P(s)$ is a linear bounded operator from $L^2(\Omega) \times H^{-1}(\Omega)$ to $(L^2(\Omega) \times H_0^1(\Omega)$.

Proof. (1) Consider the system

$$\frac{\partial^2 p}{\partial t^2} = c^2 \nabla^2 p, \tag{6.191}$$

$$\frac{\partial^2 q}{\partial t^2} = c^2 \nabla^2 q + \alpha p, \tag{6.192}$$

$$p|_\Gamma = \psi^*, \quad q|_\Gamma = 0, \tag{6.193}$$

$$\psi^* = \frac{c^2}{\beta} \frac{\partial q}{\partial \mathbf{n}}\bigg|_\Gamma, \tag{6.194}$$

$$p(\mathbf{x}, s) = p_0(\mathbf{x}), \quad \frac{\partial p}{\partial t}(\mathbf{x}, s) = p_1(\mathbf{x}),$$

$$q(\mathbf{x}, T) = 0, \quad \frac{\partial q}{\partial t}(\mathbf{x}, T) = -\gamma p(\mathbf{x}, T). \tag{6.195}$$

According to the definition of P in the equation (6.188), we have

$$P(s)(p_0, p_1) = \left(-\frac{\partial p}{\partial t}(s; p_0, p_1), p(s; p_0, p_1)\right).$$

Multiplying (6.179) by the solution q of (6.192) and integrating over $\Omega \times (s, T)$ by parts, we have

$$\int_s^T \int_\Omega q\frac{\partial^2 u}{\partial t^2} dV dt = c^2 \int_s^T \int_\Omega q\nabla^2 u dV dt$$

$$= -c^2 \int_s^T \int_\Omega \nabla q \cdot \nabla u dV dt. \tag{6.196}$$

Multiplying (6.192) by the solution u of (6.179) and integrating over $\Omega \times (s, T)$ by parts, we have

$$\int_s^T \int_\Omega u\frac{\partial^2 q}{\partial t^2} dV dt = c^2 \int_s^T \int_\Omega u\nabla^2 q dV dt + \alpha \int_s^T \int_\Omega u p dV dt$$

$$= c^2 \int_s^T \int_\Gamma \phi^* \frac{\partial q}{\partial \mathbf{n}} dS dt - c^2 \int_s^T \int_\Omega \nabla q \cdot \nabla u dV dt$$

$$+ \alpha \int_s^T \int_\Omega u p dV dt. \tag{6.197}$$

Subtracting (6.196) from (6.197) gives

$$\langle (u_0, u_1), P(s)(p_0, p_1) \rangle$$

$$= \int_\Omega \left(u_1 q(s) - u_0 \frac{\partial q(s)}{\partial t} \right) dV$$

$$= \alpha \int_s^T \int_\Omega u p dV dt + c^2 \int_s^T \int_\Gamma \phi^* \frac{\partial q}{\partial \mathbf{n}} dS dt + \gamma \int_\Omega u(T) p(T) dV$$

$$= \alpha \int_s^T \int_\Omega u p dV dt + \beta \int_s^T \int_\Gamma \phi^* \psi^* dS dt + \gamma \int_\Omega u(T) p(T) dV. \tag{6.198}$$

Repeating the above procedure by multiplying (6.191) by the solution v of (6.180) and multiplying (6.180) by the solution p of (6.191), we can obtain

$$\langle P(s)(u_0, u_1), (p_0, p_1) \rangle$$

$$= \int_\Omega \left(p_1 v(s) - p_0 \frac{\partial v(s)}{\partial t} \right) dV$$

$$= \alpha \int_s^T \int_\Omega u p dV dt + \beta \int_s^T \int_\Gamma \phi^* \psi^* dS dt + \gamma \int_\Omega u(T) p(T) dV. \tag{6.199}$$

Then (6.189) follows from (6.198) and (6.199).

(2) If $p_0 = u_0, p_1 = u_1$, then $u = p, v = q$, and $\phi^* = \psi^*$. It then follows from (6.199) that

$$\langle P(s)(u_0, u_1), (u_0, u_1) \rangle = \alpha \int_s^T \int_\Omega u^2 dV dt + \beta \int_s^T \int_\Gamma |\phi^*|^2 dS dt + \gamma \int_\Omega u(T)^2 dV.$$

(3) Letting $s \to T$ in (6.199), we derive that

$$\langle P(s)(u_0, u_1), (p_0, p_1) \rangle = \int_\Omega (\gamma u_0 p_0 + 0 \cdot p_1) dV \text{ for all } (p_0, p_1) \in L^2(\Omega) \times H^{-1}(\Omega).$$

Thus $P(T)(u_0, u_1) = (\gamma u_0, 0)$ for all $(u_0, u_1) \in L^2(\Omega) \times H^{-1}(\Omega)$.

(4) By (6.199), we deduce from the Hölder's inequality and the Cauchy-Schwarz inequality that

$$|\langle P(s)(u_0, u_1), (p_0, p_1)\rangle|$$

$$\leq \gamma \left(\int_\Omega |u(T)|^2 dV \right)^{1/2} \left(\int_\Omega |p(T)|^2 dV \right)^{1/2}$$

$$+ \beta \left(\int_s^T \int_\Gamma |\phi^*|^2 dSdt \right)^{1/2} \left(\int_s^T \int_\Gamma |\psi^*|^2 dSdt \right)^{1/2}$$

$$+ \alpha \left(\int_s^T \int_\Omega |u|^2 dVdt \right)^{1/2} \left(\int_s^T \int_\Omega |p|^2 dVdt \right)^{1/2}$$

$$\leq \left(\gamma \int_\Omega |u(T)|^2 dV + \beta \int_s^T \int_\Gamma |\phi^*|^2 dSdt + \alpha \int_s^T \int_\Omega |u|^2 dVdt \right)^{1/2}$$

$$\times \left(\gamma \int_\Omega |p(T)|^2 dV + \beta \int_s^T \int_\Gamma |\psi^*|^2 dSdt + \alpha \int_s^T \int_\Omega |p|^2 dVdt \right)^{1/2}$$

$$= (J(\phi^*; u_0, u_1, s, T))^{1/2} (J(\psi^*; p_0, p_1, s, T))^{1/2}. \tag{6.200}$$

To estimate J in terms of (u_0, u_1), we introduce the change of variable:

$$v = \int_s^t u(r)dr + \psi,$$

where u is the solution of (6.170)-(6.172) with $\phi = 0$ and ψ is the solution of

$$c^2 \nabla^2 \psi = u_1, \quad \psi|_\Gamma = 0.$$

Multiplying this equation by ψ and integrating over Ω, we obtain

$$c^2 \int_\Omega |\nabla \psi|^2 dV = - \int_\Omega u_1 \psi dV \leq \|u_1\|_{H^{-1}} \|\psi\|_{H^1},$$

and so

$$\|\psi\|_{H^1} \leq \frac{1}{c^2} \|u_1\|_{H^{-1}}.$$

On the other hand, we can easily verify that v satisfies the wave equation

$$\frac{\partial^2 v}{\partial t^2} = c^2 \nabla^2 v,$$

$$v|_\Gamma = 0,$$

$$v(\mathbf{x}, s) = \psi(\mathbf{x}), \quad \frac{\partial v}{\partial t}(\mathbf{x}, s) = u_0(\mathbf{x}).$$

Evidently, the energy identity (6.10) also holds for this equation. It therefore follows that

$$\int_\Omega \left(|u|^2 + \left| \int_s^t \nabla u(s)ds + \nabla \psi \right|^2 \right) dV = \int_\Omega \left(|v_t|^2 + |\nabla v|^2 \right) dV$$

$$= \int_\Omega \left(|u_0|^2 + |\nabla \psi|^2 \right) dV$$

$$\leq C \left(\|u_0\|_{L^2}^2 + |u_1|_{H^{-1}}^2 \right)^2.$$

We then deduce that there exists a constant $C > 0$ such that

$$J(\phi^*; u_0, u_1, s, T) \leq J(0; u_0, u_1, s, T)$$

$$= \gamma \int_\Omega |u(T)|^2 dV + \alpha \int_s^T \int_\Omega |u|^2 dV dt$$

$$\leq C \left(\|u_0\|_{L^2}^2 + |u_1|_{H^{-1}}^2 \right)$$

and

$$J(\psi^*; p_0, p_1, s, T) \leq J(0; p_0, p_1, s, T)$$

$$= \gamma \int_\Omega |p(T)|^2 dV + \alpha \int_s^T \int_\Omega |p|^2 dV dt$$

$$\leq C \left(\|p_0\|_{L^2}^2 + |p_1|_{H^{-1}}^2 \right).$$

It therefore follows from (6.200) that

$$|\langle P(s)(u_0, u_1), (p_0, p_1) \rangle_{L^2}| \leq C \left(\left(\|u_0\|_{L^2}^2 + |u_1|_{H^{-1}}^2 \right) \left(\|p_0\|_{L^2}^2 + |p_1|_{H^{-1}}^2 \right) \right)^{1/2}$$

for all $(u_0, u_1) \in L^2(\Omega) \times H^{-1}(\Omega))$. This implies that

$$\|P(s)(u_0, u_1)\|_{L^2 \times H^1} \leq C \|(u_0, u_1)\|_{L^2 \times H^{-1}}.$$

Thus $P(s)$ is bounded. Since the optimality system (6.179)-(6.183) is linear, $P(s)$ is linear.

Since the optimality system (6.179)-(6.183) has a solution for any initial condition $(u_0, u_1) \in L^2(\Omega) \times H^{-1}(\Omega)$, the state space is $L^2(\Omega) \times H^{-1}(\Omega)$.

Consider the optimal control problem

$$\frac{\partial^2 u}{\partial t^2} = c^2 \nabla^2 u, \tag{6.201}$$

$$u|_\Gamma = \phi, \tag{6.202}$$

$$u(\mathbf{x}, 0) = w_0(\mathbf{x}), \quad \frac{\partial u}{\partial t}(\mathbf{x}, 0) = w_1(\mathbf{x}). \tag{6.203}$$

By Theorem 6.10, we derive that the optimal control ϕ^* satisfies the following system:

$$\frac{\partial^2 u}{\partial t^2} = c^2 \nabla^2 u, \tag{6.204}$$

$$\frac{\partial^2 v}{\partial t^2} = c^2 \nabla^2 v + \alpha u, \tag{6.205}$$

$$u|_\Gamma = \phi^*, \quad v|_\Gamma = 0, \tag{6.206}$$

$$\phi^* = \frac{c^2}{\beta} \left.\frac{\partial v}{\partial \mathbf{n}}\right|_\Gamma, \tag{6.207}$$

$$u(\mathbf{x},0) = w_0(\mathbf{x}), \ \frac{\partial u}{\partial t}(\mathbf{x},0) = w_1(\mathbf{x}),$$

$$v(\mathbf{x},T) = 0, \ \frac{\partial v}{\partial t}(\mathbf{x},T) = -\gamma u(\mathbf{x},T). \tag{6.208}$$

If we restrict the system (6.204)-(6.208) on $[t,T]$ and compare it with the system (6.179)-(6.183), we find that

$$P(t)(u(t),u_t(t)) = (-v_t(t),v(t)). \tag{6.209}$$

Let

$$P(t)[u(t),u_t(t)] = (P_1(t)[u(t),u_t(t)]), P_2(t)[u(t),u_t(t)]). \tag{6.210}$$

Then we have

$$v(t) = P_2(t)[u(t),u_t(t)],$$

$$-\frac{\partial v}{\partial t}(t) = P_1(t)[u(t),u_t(t)].$$

We then obtain the optimal state feedback control

$$u|_\Gamma = \phi^* = \frac{c^2}{\beta} \frac{\partial}{\partial \mathbf{n}} \left.(P_2(t)[u(t),u_t(t)])\right|_\Gamma. \tag{6.211}$$

We now formally derive an equation for the operator P. For this, we define the operator

$$A = \begin{bmatrix} 0 & I \\ c^2 \nabla^2 & 0 \end{bmatrix}, B = \begin{bmatrix} I & 0 \\ 0 & 0 \end{bmatrix}.$$

The adjoint of A is given by

$$A^* = \begin{bmatrix} 0 & c^2 \nabla^2 \\ I & 0 \end{bmatrix}.$$

Since the state space is $L^2(\Omega) \times H^{-1}(\Omega)$, the domains of the operators A and A^* are $H_0^1(\Omega) \times L^2(\Omega)$ and $L^2(\Omega) \times H_0^1(\Omega)$, respectively. Using the operators A and A^*, we can rewrite the equations (6.204) and (6.205) as follows:

$$\frac{\partial}{\partial t}\begin{bmatrix} u \\ u_t \end{bmatrix} = A \begin{bmatrix} u \\ u_t \end{bmatrix}, \tag{6.212}$$

$$\frac{\partial}{\partial t}\begin{bmatrix} -v_t \\ v \end{bmatrix} = -A^*\begin{bmatrix} -v_t \\ v \end{bmatrix} - \alpha B\begin{bmatrix} u \\ u_t \end{bmatrix}. \tag{6.213}$$

Substituting (6.209) into (6.213) gives

$$P'(t)\begin{bmatrix} u \\ u_t \end{bmatrix} + P(t)\left(\frac{\partial}{\partial t}\begin{bmatrix} u \\ u_t \end{bmatrix}\right) = -A^*P(t)\begin{bmatrix} u \\ u_t \end{bmatrix} - \alpha B\begin{bmatrix} u \\ u_t \end{bmatrix}.$$

Taking the dual product of this equation with $(\xi, \eta) \in H_0^1(\Omega) \times L^2(\Omega)$, we obtain

$$\left\langle P'(t)\begin{bmatrix} u \\ u_t \end{bmatrix}, \begin{bmatrix} \xi \\ \eta \end{bmatrix}\right\rangle + \left\langle P(t)\left(\frac{\partial}{\partial t}\begin{bmatrix} u \\ u_t \end{bmatrix}\right), \begin{bmatrix} \xi \\ \eta \end{bmatrix}\right\rangle$$

$$= \left\langle -A^*P(t)\begin{bmatrix} u \\ u_t \end{bmatrix}, \begin{bmatrix} \xi \\ \eta \end{bmatrix}\right\rangle - \alpha\left\langle B\begin{bmatrix} u \\ u_t \end{bmatrix}, \begin{bmatrix} \xi \\ \eta \end{bmatrix}\right\rangle.$$

Substituting (6.212) into this equation for $\frac{\partial}{\partial t}\begin{bmatrix} u \\ u_t \end{bmatrix}$, we deduce

$$\left\langle P'\begin{bmatrix} u \\ u_t \end{bmatrix}, \begin{bmatrix} \xi \\ \eta \end{bmatrix}\right\rangle + \left\langle PA\begin{bmatrix} u \\ u_t \end{bmatrix}, \begin{bmatrix} \xi \\ \eta \end{bmatrix}\right\rangle$$

$$= \left\langle -A^*P\begin{bmatrix} u \\ u_t \end{bmatrix}, \begin{bmatrix} \xi \\ \eta \end{bmatrix}\right\rangle - \alpha\left\langle B\begin{bmatrix} u \\ u_t \end{bmatrix}, \begin{bmatrix} \xi \\ \eta \end{bmatrix}\right\rangle. \tag{6.214}$$

Moreover,

$$\left\langle PA\begin{bmatrix} u \\ u_t \end{bmatrix}, \begin{bmatrix} \xi \\ \eta \end{bmatrix}\right\rangle \tag{6.215}$$

$$= \left\langle \begin{bmatrix} u_t \\ c^2\nabla^2 u \end{bmatrix}, P\begin{bmatrix} \xi \\ \eta \end{bmatrix}\right\rangle$$

$$= \int_\Omega u_t P_1\begin{bmatrix} \xi \\ \eta \end{bmatrix} dV + \int_\Omega c^2\nabla^2 u P_2\begin{bmatrix} \xi \\ \eta \end{bmatrix} dV$$

(integrate by parts twice and note that $P_2\begin{bmatrix} \xi \\ \eta \end{bmatrix}_\Gamma = 0$)

$$= \int_\Omega u_t P_1\begin{bmatrix} \xi \\ \eta \end{bmatrix} dV + \int_\Omega c^2 u \nabla^2 P_2\begin{bmatrix} \xi \\ \eta \end{bmatrix} dV$$

$$- \int_\Gamma c^2 u \frac{\partial}{\partial \mathbf{n}} P_2\begin{bmatrix} \xi \\ \eta \end{bmatrix} dS$$

$$= \int_\Omega u_t P_1\begin{bmatrix} \xi \\ \eta \end{bmatrix} dV + \int_\Omega c^2 u \nabla^2 P_2\begin{bmatrix} \xi \\ \eta \end{bmatrix} dV$$

$$- \int_\Gamma \frac{c^4}{\beta}\frac{\partial}{\partial \mathbf{n}} P_2\begin{bmatrix} u \\ u_t \end{bmatrix}\frac{\partial}{\partial \mathbf{n}} P_2\begin{bmatrix} \xi \\ \eta \end{bmatrix} dS$$

$$= \left\langle \begin{bmatrix} u \\ u_t \end{bmatrix}, A^*P\begin{bmatrix} \xi \\ \eta \end{bmatrix}\right\rangle - \int_\Gamma \frac{c^4}{\beta}\frac{\partial}{\partial \mathbf{n}} P_2\begin{bmatrix} u \\ u_t \end{bmatrix}\frac{\partial}{\partial \mathbf{n}} P_2\begin{bmatrix} \xi \\ \eta \end{bmatrix} dS.$$

Substituting this equation into (6.214) gives

$$\left\langle P' \begin{bmatrix} u \\ u_t \end{bmatrix}, \begin{bmatrix} \xi \\ \eta \end{bmatrix} \right\rangle$$

$$= -\left\langle \begin{bmatrix} u \\ u_t \end{bmatrix}, A^*P \begin{bmatrix} \xi \\ \eta \end{bmatrix} \right\rangle - \left\langle A^*P \begin{bmatrix} u \\ u_t \end{bmatrix}, \begin{bmatrix} \xi \\ \eta \end{bmatrix} \right\rangle$$

$$-\alpha \left\langle B \begin{bmatrix} u \\ u_t \end{bmatrix}, \begin{bmatrix} \xi \\ \eta \end{bmatrix} \right\rangle + \int_\Gamma \frac{c^4}{\beta} \frac{\partial}{\partial \mathbf{n}} P_2 \begin{bmatrix} u \\ u_t \end{bmatrix} \frac{\partial}{\partial \mathbf{n}} P_2 \begin{bmatrix} \xi \\ \eta \end{bmatrix} dS.$$

Since u is arbitrary, it follows from the property (3) of Theorem 6.11 that

$$\left\langle P' \begin{bmatrix} \varphi \\ \psi \end{bmatrix}, \begin{bmatrix} \xi \\ \eta \end{bmatrix} \right\rangle$$

$$= -\left\langle \begin{bmatrix} \varphi \\ \psi \end{bmatrix}, A^*P \begin{bmatrix} \xi \\ \eta \end{bmatrix} \right\rangle - \left\langle A^*P \begin{bmatrix} \varphi \\ \psi \end{bmatrix}, \begin{bmatrix} \xi \\ \eta \end{bmatrix} \right\rangle$$

$$-\alpha \left\langle B \begin{bmatrix} \varphi \\ \psi \end{bmatrix}, \begin{bmatrix} \xi \\ \eta \end{bmatrix} \right\rangle + \int_\Gamma \frac{c^4}{\beta} \frac{\partial}{\partial \mathbf{n}} P_2 \begin{bmatrix} \varphi \\ \psi \end{bmatrix} \frac{\partial}{\partial \mathbf{n}} P_2 \begin{bmatrix} \xi \\ \eta \end{bmatrix} dS, \quad (6.216)$$

$$\left\langle P(T) \begin{bmatrix} \varphi \\ \psi \end{bmatrix}, \begin{bmatrix} \xi \\ \eta \end{bmatrix} \right\rangle = \left\langle \begin{bmatrix} \gamma\varphi \\ 0 \end{bmatrix}, \begin{bmatrix} \xi \\ \eta \end{bmatrix} \right\rangle \quad (6.217)$$

for all $(\varphi, \psi), (\xi, \eta) \in H_0^1(\Omega) \times L^2(\Omega)$. This equation is called a differential Riccati equation.

The study of the Riccati equation is beyond the scope of this text and is referred to one of the advanced control books [6, 24, 64, 67, 69, 99]. We state a result from [6, p.463, Theorem 2.1] without proof.

Theorem 6.12. *The differential Riccati equation (6.216)-(6.217) has a unique, nonnegative, symmetric, strongly continuous solution.*

Using the solutions of the Riccati equation, we can prove the existence of an optimal control. For this we define the Riccati wave energy of (6.170)-(6.172) by

$$V(u, u_t) = \left\langle P(t) \begin{bmatrix} u \\ u_t \end{bmatrix}, \begin{bmatrix} u \\ u_t \end{bmatrix} \right\rangle. \quad (6.218)$$

Lemma 6.14. *Let P be the nonnegative symmetric solution of (6.216)-(6.217). Then*

$$\frac{dV}{dt} = -\alpha \int_\Omega u^2 dV - \beta \int_\Gamma \phi^2 dS + \frac{1}{\beta} \int_\Gamma \left[c^2 \frac{\partial}{\partial \mathbf{n}} P_2 \begin{bmatrix} u \\ u_t \end{bmatrix} - \beta\phi \right]^2 dS. \quad (6.219)$$

Proof. Using the equation (6.212), we compute

$$\frac{dV}{dt} = \left\langle P'(t) \begin{bmatrix} u \\ u_t \end{bmatrix}, \begin{bmatrix} u \\ u_t \end{bmatrix} \right\rangle + \left\langle P(t) \frac{\partial}{\partial t} \begin{bmatrix} u \\ u_t \end{bmatrix}, \begin{bmatrix} u \\ u_t \end{bmatrix} \right\rangle$$

$$+ \left\langle P(t) \begin{bmatrix} u \\ u_t \end{bmatrix}, \frac{\partial}{\partial t} \begin{bmatrix} u \\ u_t \end{bmatrix} \right\rangle$$

$$
= \left\langle P'(t) \begin{bmatrix} u \\ u_t \end{bmatrix}, \begin{bmatrix} u \\ u_t \end{bmatrix} \right\rangle + \left\langle P(t)A \begin{bmatrix} u \\ u_t \end{bmatrix}, \begin{bmatrix} u \\ u_t \end{bmatrix} \right\rangle
$$

$$
+ \left\langle P(t) \begin{bmatrix} u \\ u_t \end{bmatrix}, A \begin{bmatrix} u \\ u_t \end{bmatrix} \right\rangle
$$

(use the symmetry of P)

$$
= \left\langle P'(t) \begin{bmatrix} u \\ u_t \end{bmatrix}, \begin{bmatrix} u \\ u_t \end{bmatrix} \right\rangle + 2 \left\langle P(t) \begin{bmatrix} u \\ u_t \end{bmatrix}, A \begin{bmatrix} u \\ u_t \end{bmatrix} \right\rangle
$$

$$
= \left\langle P'(t) \begin{bmatrix} u \\ u_t \end{bmatrix}, \begin{bmatrix} u \\ u_t \end{bmatrix} \right\rangle + 2 \left\langle \begin{bmatrix} u \\ u_t \end{bmatrix}, A^*P \begin{bmatrix} u \\ u_t \end{bmatrix} \right\rangle \quad \text{(use (6.215))}
$$

$$
- 2 \int_\Gamma c^2 \phi \frac{\partial}{\partial \mathbf{n}} P_2 \begin{bmatrix} u \\ u_t \end{bmatrix} dS
$$

$$
= -\alpha \left\langle B \begin{bmatrix} u \\ u_t \end{bmatrix}, \begin{bmatrix} u \\ u_t \end{bmatrix} \right\rangle + \int_\Gamma \frac{c^4}{\beta} \frac{\partial}{\partial \mathbf{n}} P_2 \begin{bmatrix} u \\ u_t \end{bmatrix} \frac{\partial}{\partial \mathbf{n}} P_2 \begin{bmatrix} u \\ u_t \end{bmatrix} dS
$$

$$
- 2 \int_\Gamma c^2 \phi \frac{\partial}{\partial \mathbf{n}} P_2 \begin{bmatrix} u \\ u_t \end{bmatrix} dS \quad \text{(use (6.216))}
$$

$$
= -\alpha \int_\Omega u^2 dV - \beta \int_\Gamma \phi^2 dS + \int_\Gamma \frac{c^4}{\beta} \frac{\partial}{\partial \mathbf{n}} P_2 \begin{bmatrix} u \\ u_t \end{bmatrix} \frac{\partial}{\partial \mathbf{n}} P_2 \begin{bmatrix} u \\ u_t \end{bmatrix} dS
$$

$$
- 2 \int_\Gamma c^2 \phi \frac{\partial}{\partial \mathbf{n}} P_2 \begin{bmatrix} u \\ u_t \end{bmatrix} dS + \beta \int_\Gamma \phi^2 dS
$$

$$
= -\alpha \int_\Omega u^2 dV - \beta \int_\Gamma \phi^2 dS + \frac{1}{\beta} \int_\Gamma \left[c^2 \frac{\partial}{\partial \mathbf{n}} P_2 \begin{bmatrix} u \\ u_t \end{bmatrix} - \beta \phi \right]^2 dS.
$$

Integrating the identity (6.219) from s to T, we obtain the following lemma.

Lemma 6.15. *Let P be the nonnegative symmetric solution of (6.216)-(6.217). Then*

$$
J(\phi, u_0, u_1, s, T) = \left\langle P(s) \begin{bmatrix} u_0 \\ u_1 \end{bmatrix}, \begin{bmatrix} u_0 \\ u_1 \end{bmatrix} \right\rangle
$$

$$
+ \frac{1}{\beta} \int_s^T \int_\Gamma \left[c^2 \frac{\partial}{\partial \mathbf{n}} P_2 \begin{bmatrix} u \\ u_t \end{bmatrix} - \beta \phi \right]^2 dSdt. \quad (6.220)
$$

Proof. Integrating the identity (6.219) from s to T, we obtain

$$
V(T) - V(s) = -\alpha \int_s^T \int_\Omega u^2 dV dt - \beta \int_s^T \int_\Gamma \phi^2 dS dt
$$

$$
+ \frac{1}{\beta} \int_s^T \int_\Gamma \left[c^2 \frac{\partial}{\partial \mathbf{n}} P_2 \begin{bmatrix} u \\ u_t \end{bmatrix} - \beta \phi \right]^2 dS dt.
$$

This implies (6.220) since $V(T) = \gamma \int_\Omega u(T)^2 dV$ and

$$
V(s) = \left\langle P(s) \begin{bmatrix} u_0 \\ u_1 \end{bmatrix}, \begin{bmatrix} u_0 \\ u_1 \end{bmatrix} \right\rangle.
$$

Theorem 6.13. *Let J be defined by (6.173). Then there exists a unique optimal control ϕ^* such that*

$$\min_{\phi \in L^2(\Gamma \times (s,T))} J(\phi; u_0, u_1, s, T) = J(\phi^*; u_0, u_1, s, T) = \left\langle P(s) \begin{bmatrix} u_0 \\ u_1 \end{bmatrix}, \begin{bmatrix} u_0 \\ u_1 \end{bmatrix} \right\rangle,$$

(6.221)

where P is the nonnegative symmetric solution of (6.216)-(6.217) and

$$\phi^* = c^2 \frac{\partial}{\partial \mathbf{n}} P_2 \begin{bmatrix} u \\ u_t \end{bmatrix} \bigg|_{\Gamma}.$$

(6.222)

Proof. It follows from (6.220) that

$$J(\phi^*; u_0, u_1, s, T) \geq \min_{\phi \in L^2(\Gamma \times (s,T))} J(\phi; u_0, u_1, s, T)$$

$$\geq \left\langle P(s) \begin{bmatrix} u_0 \\ u_1 \end{bmatrix}, \begin{bmatrix} u_0 \\ u_1 \end{bmatrix} \right\rangle$$

$$= J(\phi^*; u_0, u_1, s, T)$$

if we take

$$\phi = \phi^* = \frac{c^2}{\beta} \frac{\partial}{\partial \mathbf{n}} P_2 \begin{bmatrix} u \\ u_t \end{bmatrix} \bigg|_{\Gamma}.$$

In addition, we can readily show that J is strictly convex and then the optimal control is unique.

To design an optimal boundary state feedback controller, we consider the boundary control problem

$$\frac{\partial^2 u}{\partial t^2} = c^2 \nabla^2 u,$$

(6.223)

$$u|_\Gamma = \phi,$$

(6.224)

$$u(\mathbf{x}, 0) = u_0(\mathbf{x}), \quad \frac{\partial u}{\partial t}(\mathbf{x}, 0) = u_1(\mathbf{x})$$

(6.225)

with the quadratic performance functional over an infinite time interval

$$J(\phi; u_0, u_1) = \alpha \int_0^\infty \int_\Omega |u|^2 dv dt + \beta \int_0^\infty \int_\Gamma |\phi|^2 dS dt,$$

(6.226)

where $\alpha, \beta \geq 0$ are weight constants. The optimal control problem is to minimize the functional J over $L^2(\Gamma \times (0, \infty))$. Since the wave equation can be stabilized by the feedback controller (6.46), it is well posed.

The following optimality theorem is the infinite time version of Theorem 6.10. Its proof is beyond the scope of the text and is referred to [69].

Theorem 6.14. *The optimal control ϕ^* satisfies the following system*

$$\frac{\partial^2 u}{\partial t^2} = c^2 \nabla^2 u, \tag{6.227}$$

$$\frac{\partial^2 v}{\partial t^2} = c^2 \nabla^2 v + \alpha u, \tag{6.228}$$

$$u|_\Gamma = \phi^*, \quad v|_\Gamma = 0, \tag{6.229}$$

$$\phi^* = \frac{c^2}{\beta} \frac{\partial v}{\partial \mathbf{n}}\bigg|_\Gamma, \tag{6.230}$$

$$u(\mathbf{x},0) = u_0(\mathbf{x}), \quad \frac{\partial u}{\partial t}(\mathbf{x},0) = u_1(\mathbf{x}), \tag{6.231}$$

$$v(\mathbf{x},\infty) = 0, \quad \frac{\partial v}{\partial t}(\mathbf{x},\infty) = 0. \tag{6.232}$$

We define a linear operator Π by

$$\Pi(u_0,u_1) = \left(-\frac{\partial v}{\partial t}(0;u_0,u_1), v(0;u_0,u_1) \right) \tag{6.233}$$

for any given function $(u_0,u_1) \in L^2(\Omega) \times H^{-1}(\Omega)$, where $v(t;u_0,u_1)$ is the solution of (6.227)-(6.232).

Theorem 6.15. *The operator Π defined by (6.233) has the following properties:*

(1) Π is symmetric, that is,

$$\langle \Pi(u_0,u_1),(p_0,p_1)\rangle = \langle (u_0,u_1), \Pi(p_0,p_1)\rangle \tag{6.234}$$

for $(u_0,u_1),(p_0,p_1) \in L^2(\Omega) \times H^{-1}(\Omega)$.
(2) If ϕ^* is the optimal control of (6.223)-(6.225), then

$$\langle \Pi(u_0,u_1),(u_0,u_1)\rangle = \alpha \int_0^\infty \int_\Omega |u|^2 dV dt + \beta \int_0^\infty \int_\Gamma |\phi^*|^2 dS dt. \tag{6.235}$$

Hence Π is nonnegative.

The proof is similar to that of Theorem 6.11 and is left as an exercise. It can be seen from the definition (6.233) of the operator Π that

$$\Pi\left(u(t), \frac{\partial u}{\partial t}(t) \right) = \left(-\frac{\partial v}{\partial t}(t), v(t) \right). \tag{6.236}$$

Let

$$\Pi = (\Pi_1, \Pi_2). \tag{6.237}$$

We then obtain the optimal state feedback control

$$\phi^* = \frac{c^2}{\beta} \frac{\partial}{\partial \mathbf{n}} (\Pi_2(u(t),u_t(t)))\bigg|_\Gamma. \tag{6.238}$$

In the same way as the derivation of the differential Riccati equation (6.216), we can derive the algebraic Riccati equation

$$\left\langle \begin{bmatrix} \varphi \\ \psi \end{bmatrix}, A^*\Pi \begin{bmatrix} \xi \\ \eta \end{bmatrix} \right\rangle + \left\langle A^*\Pi \begin{bmatrix} \varphi \\ \psi \end{bmatrix}, \begin{bmatrix} \xi \\ \eta \end{bmatrix} \right\rangle$$

$$+\alpha \left\langle B \begin{bmatrix} \varphi \\ \psi \end{bmatrix}, \begin{bmatrix} \xi \\ \eta \end{bmatrix} \right\rangle - \int_\Gamma \frac{c^4}{\beta} \frac{\partial}{\partial \mathbf{n}} \Pi_2 \begin{bmatrix} \varphi \\ \psi \end{bmatrix} \frac{\partial}{\partial \mathbf{n}} \Pi_2 \begin{bmatrix} \xi \\ \eta \end{bmatrix} dS = 0 \qquad (6.239)$$

for all $(\varphi, \psi), (\xi, \eta) \in H_0^1(\Omega) \times L^2(\Omega)$.

We state a result about the Riccati equation from [62, 65] without proof. Consider the differential Riccati equation

$$\left\langle P' \begin{bmatrix} \varphi \\ \psi \end{bmatrix}, \begin{bmatrix} \xi \\ \eta \end{bmatrix} \right\rangle$$

$$= \left\langle \begin{bmatrix} \varphi \\ \psi \end{bmatrix}, A^*P \begin{bmatrix} \xi \\ \eta \end{bmatrix} \right\rangle + \left\langle A^*P \begin{bmatrix} \varphi \\ \psi \end{bmatrix}, \begin{bmatrix} \xi \\ \eta \end{bmatrix} \right\rangle$$

$$+\alpha \left\langle B \begin{bmatrix} \varphi \\ \psi \end{bmatrix}, \begin{bmatrix} \xi \\ \eta \end{bmatrix} \right\rangle - \int_\Gamma \frac{c^4}{\beta} \frac{\partial}{\partial \mathbf{n}} P_2 \begin{bmatrix} \varphi \\ \psi \end{bmatrix} \frac{\partial}{\partial \mathbf{n}} P_2 \begin{bmatrix} \xi \\ \eta \end{bmatrix} dS, \qquad (6.240)$$

$$\left\langle P(0) \begin{bmatrix} \varphi \\ \psi \end{bmatrix}, \begin{bmatrix} \xi \\ \eta \end{bmatrix} \right\rangle = \left\langle \begin{bmatrix} \gamma\psi \\ 0 \end{bmatrix}, \begin{bmatrix} \xi \\ \eta \end{bmatrix} \right\rangle \qquad (6.241)$$

for all $(\varphi, \psi), (\xi, \eta) \in H_0^1(\Omega) \times L^2(\Omega)$. The algebraic Riccati equation (6.239) is the steady state equation of the differential equation.

Theorem 6.16. *The differential Riccati equation (6.240)-(6.241) has a unique, non-negative, symmetric, strongly continuous solution that has the following properties:*

(1) $\left\langle P(t) \begin{bmatrix} u_0 \\ u_1 \end{bmatrix}, \begin{bmatrix} u_0 \\ u_1 \end{bmatrix} \right\rangle = \min_{\phi \in L^2(\Gamma \times (0,t))} J(\phi, u_0, u_1, 0, t)$, *where J is defined by (6.173).*

(2) If $\gamma = 0$, then the limit

$$\lim_{t \to \infty} P(t)(u_0, u_1) = \Pi(u_0, u_1) \quad \text{for all } (u_0, u_1) \in L^2(\Omega) \times H^{-1}(\Omega) \qquad (6.242)$$

exists and Π is the minimal nonnegative symmetric solution of (6.239). That is, any other solution Π_m of (6.239) satisfies

$$\left\langle \Pi \begin{bmatrix} u_0 \\ u_1 \end{bmatrix}, \begin{bmatrix} u_0 \\ u_1 \end{bmatrix} \right\rangle \le \left\langle \Pi_m \begin{bmatrix} u_0 \\ u_1 \end{bmatrix}, \begin{bmatrix} u_0 \\ u_1 \end{bmatrix} \right\rangle.$$

We define the Riccati wave energy of (6.223)-(6.225) by

$$V(u, u_t) = \left\langle \Pi \begin{bmatrix} u \\ u_t \end{bmatrix}, \begin{bmatrix} u \\ u_t \end{bmatrix} \right\rangle. \qquad (6.243)$$

In a manner analogous to Lemma 6.14, we can prove the following lemma.

Lemma 6.16. *Let Π be the nonnegative symmetric solution of (6.239). Then*

$$\frac{dV}{dt} = -\alpha \int_\Omega u^2 dV - \beta \int_\Gamma \phi^2 dS + \frac{1}{\beta} \int_\Gamma \left[c^2 \frac{\partial}{\partial \mathbf{n}} \Pi_2 \begin{bmatrix} u \\ u_t \end{bmatrix} - \beta \phi \right]^2 dS, \quad (6.244)$$

where $\Pi(u_0, u_1) = [\Pi_1(u_0, u_1), \Pi_2(u_0, u_1)]$.

The proof is left as an exercise. Integrating the identity (6.244) from 0 to t, we obtain the following lemma.

Lemma 6.17. *Let Π be the nonnegative symmetric solution of (6.239). Then*

$$J(\phi, u_0, u_1, 0, t) = \left\langle \Pi \begin{bmatrix} u_0 \\ u_1 \end{bmatrix}, \begin{bmatrix} u_0 \\ u_1 \end{bmatrix} \right\rangle - \left\langle \Pi \begin{bmatrix} u \\ u_t \end{bmatrix}, \begin{bmatrix} u \\ u_t \end{bmatrix} \right\rangle$$

$$+ \frac{1}{\beta} \int_0^t \int_\Gamma \left[c^2 \frac{\partial}{\partial \mathbf{n}} \Pi_2 \begin{bmatrix} u \\ u_t \end{bmatrix} - \beta \phi \right]^2 dS dt. \quad (6.245)$$

Theorem 6.17. *Let J be defined by (6.226). Then there exists a unique optimal control ϕ^* such that*

$$\min_{\phi \in L^2(\Gamma \times (s,T))} J(\phi; u_0, u_1) = J(\phi^*; u_0, u_1) = \left\langle \Pi \begin{bmatrix} u_0 \\ u_1 \end{bmatrix}, \begin{bmatrix} u_0 \\ u_1 \end{bmatrix} \right\rangle, \quad (6.246)$$

where Π is the nonnegative symmetric solution of (6.239) and

$$\phi^* = c^2 \frac{\partial}{\partial \mathbf{n}} \Pi_2 \begin{bmatrix} u \\ u_t \end{bmatrix} \Big|_\Gamma. \quad (6.247)$$

Proof. Since the control problem is well posed, the minimum below is finite. It follows from Theorem 6.16 that

$$\min_{\phi \in L^2(\Gamma \times (0,\infty))} J(\phi; u_0, u_1) \geq \min_{\phi \in L^2(\Gamma \times (0,\infty))} J(\phi; u_0, u_1, 0, t)$$

$$= \min_{\phi \in L^2(\Gamma \times (0,t))} J(\phi; u_0, u_1, 0, t)$$

$$= \left\langle P(t) \begin{bmatrix} u_0 \\ u_1 \end{bmatrix}, \begin{bmatrix} u_0 \\ u_1 \end{bmatrix} \right\rangle.$$

which, combined with (6.242), implies that

$$\min_{\phi \in L^2(\Gamma \times (0,\infty))} J(\phi, u_0, u_1) \geq \left\langle \Pi \begin{bmatrix} u_0 \\ u_1 \end{bmatrix}, \begin{bmatrix} u_0 \\ u_1 \end{bmatrix} \right\rangle. \quad (6.248)$$

On the other hand, by (6.245), we deduce that

$$J(\phi; u_0, u_1, 0, t) = \left\langle \Pi \begin{bmatrix} u_0 \\ u_1 \end{bmatrix}, \begin{bmatrix} u_0 \\ u_1 \end{bmatrix} \right\rangle - \left\langle \Pi \begin{bmatrix} u \\ u_t \end{bmatrix}, \begin{bmatrix} u \\ u_t \end{bmatrix} \right\rangle$$

$$+ \frac{1}{\beta} \int_\Gamma \left[c^2 \frac{\partial}{\partial \mathbf{n}} \Pi_2 \begin{bmatrix} u \\ u_t \end{bmatrix} - \beta \phi \right]^2 dS$$

$$\leq \left\langle \Pi \begin{bmatrix} u_0 \\ u_1 \end{bmatrix}, \begin{bmatrix} u_0 \\ u_1 \end{bmatrix} \right\rangle + \frac{1}{\beta} \int_\Gamma \left[c^2 \frac{\partial}{\partial \mathbf{n}} \Pi_2 \begin{bmatrix} u \\ u_t \end{bmatrix} - \beta \phi \right]^2 dS.$$

Taking $\phi = \phi^* = \frac{c^2}{\beta} \frac{\partial}{\partial \mathbf{n}} \Pi_2 \begin{bmatrix} u \\ u_t \end{bmatrix}$, we obtain

$$J(\phi^*; u_0, u_1, 0, t) \leq \left\langle \Pi \begin{bmatrix} u_0 \\ u_1 \end{bmatrix}, \begin{bmatrix} u_0 \\ u_1 \end{bmatrix} \right\rangle$$

and then

$$\min_{\phi \in L^2(\Gamma \times (0,\infty))} J(\phi, u_0, u_1) \leq J(\phi^*; u_0, u_1) \leq \left\langle \Pi \begin{bmatrix} u_0 \\ u_1 \end{bmatrix}, \begin{bmatrix} u_0 \\ u_1 \end{bmatrix} \right\rangle. \tag{6.249}$$

Putting (6.248) and (6.249) together yields (6.246). In addition, we can readily show that J is strictly convex and then the optimal control is unique.

As in the case of the reaction-convection-diffusion equations, the optimal control theory for the wave equation has been extended to the abstract dynamical control system (4.347)-(4.348). For details, we refer to [6, Part IV], [65], and [67].

Exercises 6.7

1. Verify that the operator $P(s)$ defined by (6.188) is linear.
2. Consider the optimal control problem

$$\frac{\partial^2 u}{\partial t^2} = c^2 \nabla^2 u,$$

$$u|_\Gamma = \sum_{i=1}^m b_i(\mathbf{x}) \phi_i(t),$$

$$u(\mathbf{x}, 0) = u_0(\mathbf{x}), \quad u_t(\mathbf{x}, 0) = u_1(\mathbf{x})$$

with the quadratic performance functional

$$J(\phi_1, \cdots, \phi_m) = \alpha \int_0^T \int_\Omega |u|^2 dV dt + \beta \sum_{i=1}^m \int_0^T \left(|\phi_i|^2 + |\phi_i'|^2 \right) dt + \gamma \int_\Omega |u(T)|^2 dV,$$

where $\alpha, \beta, \gamma \geq 0$ are weight constants and b_i ($i = 1, 2, \cdots, m$) are given functions. Derive the optimality system for the optimal control.

3. Prove Theorem 6.15.
4. Prove Lemma 6.16.
5. Consider the control problem

$$\frac{\partial^2 u}{\partial t^2} = \frac{\partial^2 u}{\partial x^2},$$

$$\frac{\partial u}{\partial x}(0,t) = \phi_1(t), \quad \frac{\partial u}{\partial x}(\pi,t) = \phi_2(t),$$

$$u(x,0) = u_0(x), \frac{\partial u}{\partial t}(x,0) = u_1(x)$$

with the performance functional

$$J(\phi_1, \phi_2) = \int_0^\infty \int_0^\pi u^2 dx dt + \int_0^\infty (\phi_1(t)^2 + \phi_2(t)^2) dx dt.$$

Derive the optimality system for the optimal control.

6.8 References and Notes

The material presented in this chapter is just a simplification of the material from the advanced control books [6, 24, 62, 64, 67, 69, 99] and from the references [14, 15, 16, 17, 18, 46, 47, 49, 50, 53, 66, 80]. Section 6.7 is from Example 4.3 of Chapter 3, Part 4 of [6]. The generalized Young's inequality is from [3] and Jensen's inequality from [95].

There have been numerous references on the feedback control of the wave equations. We mention some of them for further studies: Lax and Phillips [66], Morawetz [85], Strauss [100], Russell [96], Chen [14, 15, 16, 17, 18], Avalos and Lasiecka [4], Bardos, Lebeau, and Rauch [5], Carpio [11], Castro and Zuazua [12], Cox and Zuazua [22, 23], Kormornik and Zuazua [46], Lagnese [49], Lasiecka and Triggiani [52, 56, 59, 60, 61], Lions [70, 72], Liu and Williams [79, 74], Freitas and Zuazua [36], Lagnese [50], Martinez [83, 84], Lasiecka and Triggiani [51], Liu and Zuazua [80], Nakao [88, 89], Rauch, Zhang, and Zuazua [94], Wang and Chen [103], Zhang and Zuazua [108], and Zuazua [109].

References

1. Adams, R.: Sobolev Spaces. Academic Press. New York (1975)
2. Amamm, H.: Feedback stabilization of linear and semilinear parabolic systems. In: Clement, P., Invernizzi, S., Mitidieri, E. and Vrabie, I.I. (eds.) Semigroup Theory and Applications. Lecture Notes in Pure and Applied Mathematics, vol. 116, pp. 21-57. Marcel Dekker, New York (1989)
3. Arnold, V. I.: Mathematical Methods of Classical Mechanics. Springer-Verlag, New York (1989)

4. Avalos, G., Lasiecka, I.: Optimal blowup rates for the minimal energy null control of the strongly damped abstract wave equation. Ann. Sc. Norm. Super. Pisa Cl. Sci. (5) **2**, no. 3, 601-616 (2003)
5. Bardos, C., Lebeau, G., Rauch, J.: Sharp sufficient conditions for the observation, control, and stabilization of waves from the boundary. SIAM J. Control Optim. **30**, 1024-1065 (1992)
6. Bensoussan, A., Da Prato, G., Delfour, M. C., Mitter, S. K.: Representation and Control of Infinite Dimensional Systems. Birkhauser, Boston (2006)
7. Boskovic, D. M. , Krstic M. and Liu, W.: Boundary control of an unstable heat equation via measurement of domain-averaged temperature. IEEE Trans. Automatic Control **46**, 2022-2028 (2001)
8. Boskovic, D., Balogh, A., Krstic, M.: Backstepping in infinite dimension for a class of parabolic distributed parameter systems. Mathematics of Control, Signals, and Systems **16**, 44-75 (2003)
9. Burns, J. A., Rubio, D., King, B. B.: Regularity of feedback operators for boundary control of thermal processes. Proc. First International Conf. on Nonlinear Problems in Aviation and Aerospace, Daytona Beach, Florida (1996)
10. Byrnes, C. I., Priscoli, F. D., Isidori, A.: Output Regulation of Uncertain Nonlinear Systems. Birkhäuser, Boston (1997)
11. Carpio, A.: Sharp estimates of the energy decay for solutions of second order dissipative evolution equations. Potential Analysis **1**, 265-289 (1992)
12. Castro, C., Zuazua, E.: Low frequency asymptotic analysis of a string with rapidly oscillating density. SIAM J. Appl. Math. **60** 1205-1233 (2000)
13. Castro, C., Zuazua, E.: High frequency asymptotic analysis of a string with rapidly oscillating density. Eur. J. Appl. Math. **11**, 595-622 (2000)
14. Chen, G.: Energy decay estimates and exact boundary value controllability for the wave equation in a bounded domain. J. Math. Pures Appl. **58**, 249-273 (1979)
15. Chen, G.: Control and stabilization for the wave equation in a bounded domain. SIAM J. Control Optim. **17**, 66-81 (1979)
16. Chen, G.: A note on the boundary stabilization of the wave equation. SIAM J. Control Optim. **19**, 106-113 (1981)
17. Chen, G.: Control and stabilization for the wave equation in a bounded domain, part II. SIAM J. Control Optim. **19**, 114-122 (1981)
18. Chen, G.: Control and stabilization for the wave equation, part III: Domain with moving boundary. SIAM J. Control Optim. **19**, 123-138 (1981)
19. Choi Y., Chung, W. K.: PID Trajectory Tracking Control for Mechanical Systems. Springer, Berlin (2004)
20. Christofides, P.D.: Nonlinear and Robust Control of PDE Systems, Methods and Applications to Transport-Reaction Processes. Birkhäuser, Boston (2001)
21. Coppel, W. A.: Stability and Asymptotic Behavior of Differential Equations. D. C. Heath and Co., Boston (1966)
22. Cox S., Zuazua, E.: The rate at which energy decays in a damped string. Communications in Partial Differential Equations, **19**, 213-243 (1994)
23. Cox S., Zuazua, E.: The rate at which energy decays in the string damped at one end. Indiana Univ. Math. J. **44**, 545-573 (1995)
24. Curtain, R. F., Zwart, H.: An Introduction to Infinite-dimensional Linear Systems Theory. Springer-Verlag, New York (1995)
25. Datko, R., Lagnese, J., Polis, M.P.: An example on the effect of time delays in boundary feedback stabilization of wave equations. SIAM J. Control Optim. **24**, 152-156 (1986)
26. Datko, R.: Not all feedback stabilized hyperbolic systems are robust with respect to small time delays in their feedbacks. SIAM J. Control Optim. **26**, 697-713 (1988)
27. Datko, R.: The destabilizing effect of delays on certain vibrating systems. In: Advances in computing and control (Baton Rouge, LA, 1988), Lecture Notes in Control and Inform. Sci. 130, pp. 324-330. Springer, Berlin, New York (1989)
28. Datko, R., You, Y.C.: Some second-order vibrating systems cannot tolerate small time delays in their damping. J. Optim. Theory Appl. **70**, 521-537 (1991)

29. Datko, R.: Two examples of ill-posedness with respect to small time delays in stabilized elastic systems. IEEE Trans. Automat. Control **38**, 163-166 (1993)
30. Dautray, R., Lions, J.L.: Mathematical Analysis and Numerical Methods for Science and Technology, Vol.2, Functional and Variational Methods. Springer-Verlag, Berlin (1990)
31. Dautray, R., Lions, J.L.: Mathematical Analysis and Numerical Methods for Science and Technology, Vol.3, Spectral Theory and Applications. Springer-Verlag, Berlin (1990)
32. Dautray, R., Lions, J.L.: Mathematical Analysis and Numerical Methods for Science and Technology, Vol.5, Evolution Problems I. Springer-Verlag, Berlin (1992)
33. Dehman, B., Lebeau, G., Zuazua, E.: Stabilization and control for the subcritical semilinear wave equation. Annales Ecole Normale Superieure de Paris **36**, 525-551 (2003)
34. DeVito, C. L.: Functional Analysis and Linear Operator Theory. Addison-Wesley Pub. Co., Redwood City, California (1990)
35. Evans, L.S.: Partial Differential Equations. American Mathematical Society, Providence, RI (1998)
36. Freitas, P., Zuazua, E.: Stability results for the wave equation with indefinite damping. J. Diff. Equations. **132**, (1996), 338-352 (1996)
37. Fox, R. W., McDonald, A. T., Pritchard, P. J.: Introduction to Fluid Mechanics, John Wiley & Sons, Inc, Hoboken, N.J. (2004)
38. Franklin, G. F., Powell, D. J., Emami-Naeini, A.: Feedback control of dynamic systems, 4th Edition. Prentice Hall, Upper Saddle River, New Jersey (2002)
39. Gilbarg, D., Trudinger, N. S.: Elliptic Partial Differential Equations of Second Order. Springer-Verlag, Berlin (1983)
40. Grisvard, P.: Elliptic Problems in Nonsmooth Domains, Pitman, London (1985)
41. Haberman, R: Applied Partial Differential Equations with Fourier Seriers and Boundary Value Problems, 4th Edition. Prentice Hall, Upper Saddle River, New Jersey (2003)
42. Henry, D.: Geometric Theory of Semilinear Parabolic Equations, Lecture Notes in Mathematics vol. 840, Springer-Verlag, Berlin (1981)
43. Huang, J.: Nonlinear Output Regulation, Theory and Applications. Society for Industrial and Applied Mathematics, Philadelphia (2004)
44. Kato, T.: Perturbation Theory of Linear Operators. Springer-Verlag, New York (1966)
45. Khalil, H. K: Nonlinear Systems. Prentice Hall, New Jersey (2002)
46. Komornik, V., Zuazua, E.: A direct method for the boundary stabilization of the wave equation. J. Math. Pures Appl. **69**, 33-54 (1990)
47. Komornik, V.: Exact Controllability and Stabilization: The Multiplier Method. John Wiley & Sons, Masson, Paris (1994)
48. Krstic, M., Smyshlyaev, A., Boundary Control of PDEs: A Course on Backstepping Designs. SIAM, Philadelphia (2008).
49. Lagnese, J.: Decay of solutions of wave equations in a bounded region with boundary dissipation. J. Differential Equations **50**, 163-182 (1983)
50. Lagnese, J.: Control of wave process with distributed control supported on a subregion. SIAM J. Control Optim. **21**, 68-85 (1983)
51. Lasiecka, I.: Stabilization of wave and plate-like equations with nonlinear dissipation on the boundary. J. Differential Equations **79**, 340-381 (1989)
52. Lasiecka, I., Lions, J.L., Triggiani, R.: Nonhomogeneous boundary value problems for second order hyperbolic operators. J. Math. Pures Appl. **65**, 149-192 (1986)
53. Lasiecka, I., Tataru, D.: Uniform boundary stabilization of semilinear wave equations with nonlinear boundary damping. Differential & Integral Equations **6**, 507-533 (1993)
54. Lasiecka, I., Triggiani, R.: Stabilization of Neumann boundary feedback of parabolic equations: the case of trace in the feedback loop. Appl. Math. Optim. **10**, 307-350 (1983)
55. Lasiecka, I., Triggiani, R.: Stabilization and structural assignment of Dirichlet boundary feedback parabolic equations. SIAM J. Control Optim. **21**, 766-803 (1983)
56. Lasiecka, I., Triggiani, R.: Riccati equations for hyperbolic partial differential equations with $L_2(0, T; L_2(\Gamma))$ -Dirichlet boundary terms. SIAM J. Control Optim. **24**, 884-925 (1986)

57. Lasiecka, I., Triggiani, R.: The regulator problem for parabolic equations with Dirichlet boundary control. I. Riccati's feedback synthesis and regularity of optimal solution. Appl. Math. Optim. **16**, 147-168 (1987)
58. Lasiecka, I., Triggiani, R.: The regulator problem for parabolic equations with Dirichlet boundary control. II. Galerkin approximation. Appl. Math. Optim. **16**, 187-216 (1987)
59. Lasiecka, I., Triggiani, R.: Uniform exponential energy decay of wave equations in a bounded region with $L_1(0,\infty;L_2(\Gamma))$-feedback control in the Dirichlet boundary conditions. J. Differential Equations. **66**, 340-390 (1987)
60. Lasiecka, I., Triggiani, R.: Sharp regularity theory for second order hyperbolic equations of Neumann type. Atti Accad. Naz. Lincei Cl. Sci. Fis. Mat. Natur. Rend. Lincei (9) **83**, 109-113 (1989)
61. Lasiecka, I., Triggiani, R.: Uniform stabilization of the wave equation with Dirichlet-feedback control without geometrical conditions. In: Stabilization of Flexible Structures (Montpellier, 1989), Lecture Notes in Control and Information Sciences, 147, pp. 62-108. Springer-Verlag, Berlin (1990)
62. Lasiecka, I., Triggiani, R.: Differential and Algebraic Riccati Equations with Application to Boundary/Point Control problems: Continuous Theory and Approximation Theory. Lecture Notes in Control and Information Sciences, 164, springer-Verlag, Berlin (1991)
63. Lasiecka, I., Triggiani, R.: Uniform stabilization of the wave equation with Dirichlet or Neumann feedback control without geometrical conditions. Appl. Math. Optim. **25**, 189-224 (1992)
64. Lasiecka, I., Triggiani, R.: Control Theory for Partial Differential Equations, Continuous and Approximation Theories I: Abstract Parabolic Systems (Encyclopedia of Mathematics and its Applications). Cambridge University Press (2000)
65. Lasiecka, I., Triggiani, R.: Control Theory for Partial Differential Equations, Continuous and Approximation Theories II: Abstract Hyperbolic-like Systems over a Finite Time Horizon (Encyclopedia of Mathematics and its Applications). Cambridge University Press (2000)
66. Lax, P., Phillips, R. S.: Scattering theory for dissipative hyperbolic systems. J. Funct. Anal. **14**, 172-235 (1973)
67. Li, X., Yong, J.: Optimal Control Theory for Infinite Dimensional Systems, Birkhäuser, Boston (1995)
68. Li, X. J., Liu, K. S.: The effect of small time delays in the feedbacks on boundary stabilization. Science in China **36**, 1435-1443 (1993)
69. Lions, J. L.: Optimal Control of Systems Governed by Partial Differential Equations. Springer-Verlag, Berlin (1971)
70. Lions, J. L.: Contrôlabilité Exacte Perturbations et Stabilisation de Systèmes Distribués, Tome 1, Contrôlabilité Exacte. Masson, Paris Milan Barcelone Mexico (1988)
71. Lions, J. L.: Contrôlabilité Exacte Perturbations et Stabilisation de Systèmes Distribués, Tome 2, Perturbations. Masson, Paris Milan Barcelone Mexico (1988)
72. Lions, J. L.: Exact controllability, stabilization and perturbations for distributed systems. SIAM Rev. **30**, 1-68 (1988)
73. Lions, J.L., Magenes, E.: Non-homogeneous Boundary Value Problems and Applications, Vol.I. Springer-Verlag, Berlin (1972)
74. Liu, W.: Stabilization and controllability for the transmission wave equation. IEEE Transcation on Automatic Control **46**, 1900-1907 (2001)
75. Liu, W.: Boundary feedback stabilization of an unstable heat equation. SIAM J. Control Optim. **42**, 1033-1043 (2003)
76. Liu, W., Haller, G.: Strange eigenmodes and decay of variance in the mixing of diffusive tracers, Physica D **188**, 1-39 (2004)
77. Liu, W., Haller, G.: Inertial manifold and completeness of eigenmodes for unsteady magnetic dynamos. Physica D **194**, 297-319 (2004)
78. Liu, W., Krstic, M.: Boundary feedback stabilization of homogeneous equilibriums in unstable fluid mixtures. International Journal of Control **80**, 1-8 (2007)
79. Liu, W., Williams, G. H.: Exponential stability of the problem of transmission of the wave equation. Bull. Australian Math. Soc. **57**, 305-327 (1998)

80. Liu, W., Zuazua, E.: Decay rates for dissipative wave equations. Ricerche di Matematica **48**, 61-75 (1999)
81. Liu, W., Zuazua, E.: Uniform stabilization of the higher dimensional system of thermoelasticity with a nonlinear boundary feedback. Quarterly Appl. Math. **59**, 269-314 (2001)
82. Luo, Z.H., Guo, B.Z., Morgül, O.: Stability and Stabilization of Infinite Dimensional Systems with Applications. Springer, London (1999)
83. Martinez, P.: A new method to obtain decay rate estimates for dissipative systems with localized damping, Revista Matemática Complutense **12**, 251-283 (1999)
84. Martinez, P.: A new method to obtain decay rate estimates for dissipative systems. ESAIM: COCV **4**, 419-444 (1999)
85. Morawetz, C. S.: Decay for solutions of the exterior problem for the wave equation. Comm. Pure Appl. Math. **28**, 229-264 (1975)
86. Morris, K. A.: Introduction to Feedback Control. 1st edition, Academic Press (2001)
87. Muzzio, F. J., and Liu, M.: Chemical reactions in chaotic flows. Chem. Eng. J. **64** 117-127 (1996)
88. Nakao, M.: Asymptotic stability of the bounded or almost periodic solution of the wave equation with a nonlinear dissipative term. J. Math. Anal. Appl. **58**, 336-343 (1977)
89. Nakao, M.: Energy decay for the wave equation with a nonlinear weak dissipation. Differential Integral Equations, **8**, 681-688 (1995)
90. Ogata, K.: Modern Control Engineering, Fourth Edtion. Prentice Hall, Upper Saddle River, New Jersey (2002)
91. Pazy, A.: Semigroup of Linear Operators and Applications to Partial Differential Equations. Springer-Verlag, New York (1983)
92. Protter, M. H., Morrey, C. B.: A First Course in Real Analysis. Springer-Verlag, New York (1977)
93. Rauch, J., Taylor, M.: Exponential decay of solutions to hyperbolic equations in bounded domains. Indiana J. Math. **24** 79-83 (1974)
94. Rauch, J., Zhang, X., Zuazua, E.: Polynomial decay for a hyperbolic-parabolic coupled system. J. Math. Pures Appl. **84**, 407-470 (2005)
95. Rudin, W.: Functional analysis. McGraw-hill, New York (1973)
96. Russell, D. L.: Exact boundary value controllability theorems for wave and heat processes in star-complemented regions. In: Roxin, Liu, and Sternberg (eds.) Differential Games and Control Theory, pp.291-319. Marcel Dekker Inc., New York (1974)
97. Sell, G. R., You, Y.: Dynamics of Evolutionary Equations, Springer, New York (2002)
98. Smyshlyaev, A., Krstic, M.: Explicit state and output feedback boundary controllers for partial differential equations. JOURNAL OF AUTOMATIC CONTROL, UNIVERSITY OF BELGRADE **13**, 1-9 (2003)
99. Staffans, O.: Well-Posed Linear Systems. Cambridge University Press (2005)
100. Strauss, W. A.: Dispersal of waves vanishing on the boundary of an exterior domain. Comm. Pure Appl. Math., **28**, 265-278 (1975)
101. Temam, R.: Infinite Dimensonal Dynamical Systems in Mechanics and Physics, 2nd edition, Springer, New York (1997)
102. Triggiani, R.: On Nambu's boundary stabilizability problem for diffusion processes. J. Differential Equations **33**, 189-200 (1979)
103. Wang, H. K., Chen, G.: Asymptotic behavior of solutions of the one-dimensional wave equation with a nonlinear boundary stabilizer. SIAM J. Control Optim. **27**, 758-775 (1989)
104. Wonham, W. M.: Linear Multivariable Control: a Geometric Approach, Second Edition, Springer-Verlag, New York (1976)
105. Yosida, K.: Functional Analysis. Springer-Verlag, New York, 1995.
106. Zeidler, E.: Nonlinear Functional Analysis and its Applications III, Variational Methods and Optimization, Springer-Verlag, New York (1985)
107. Zhang, X., Zuazua, E.: Control, observation and polynomial decay for a coupled heat-wave system. C. R. Acad. Sci. Paris, Serie I, **336**, 823-828 (2003)
108. Zhang, X., Zuazua, E.: Polynomial decay and control of a 1-d hyperbolic-parabolic coupled system. J. Differential Equations **204**, 380-438 (2004)

109. Zuazua, E.: Uniform Stabilization of the wave equation by nonlinear boundary feedback. SIAM J. Control Optim. **28**, 466-477 (1990)
110. Zuazua, E.: Exponential decay for the semilinear wave equation with locally distributed damping. Commun. in Partial Differential Equations **15**, 205-235 (1990)
111. Zuazua, E.: Controllability and observability of partial differential equations: some results and open problems. In: Dafermos, C. M., Feireisl, E. (eds.) Handbook of Differential Equations: Evolutionary Equations, vol. 3, pp.527-621. Elsevier Science, Amsterdam, The Netherlands (2006)

Index

adjoint operator, 34
algebraic multiplicity, 54
algebraic Riccati equation, 188, 206, 284
asymptotic tracking, 112
asymptotically stable, 121, 125

backstepping, 149
Banach space, 14
boundary control, 4, 226
bounded set, 17

Cauchy problem, 44
Cauchy sequence, 14
Cauchy's integral formula, 32
Cauchy-Schwarz inequality, 10, 13
Cayley-Hamilton theorem, 57
closed operator, 26
closed set, 15
closure of a set, 16
compact set, 16
control gain, 76, 221, 226
controllability, 57
controlled output, 1, 3

decay rate, 121
derivative control, 76
detectable, 68, 140, 213
differential Riccati equation, 185, 202, 280
direct decomposition, 25
direct sum, 25
Dirichlet boundary condition, 21, 22
Dirichlet boundary control, 241
dissipative operator, 35
disturbance rejection, 112
divergence theorem, 27
dual space, 18

eigenvalue, 21

eigenvalue problem, 22, 121
energy, 215
Euclidean norm, 10
exponentially stable, 51, 121, 125

feedback control, 1, 3
feedback gain, 112
feedback gain matrix, 67
feedback matrix, 67
feedback operator, 131, 213
feedforward gain, 112
finite dimensional control systems, 1, 49
flux control, 163

Gâteaux-differential, 177
generalized Young's inequality, 261
Gronwall's inequality, 123
growth bound, 38

Hölder's inequality, 11
Hamiltonian system, 180
Hilbert space, 14
Hille-Yosida theorem, 34
Hurwitz matrix, 55

infinite dimensional system, 127
infinitesimal generator, 31
injection operator, 213
inner product, 12
inner product space, 12
integral control, 77
integral output feedback control, 87
integral state feedback control, 80
integral transform method, 149
integral-derivative (ID) feedback control, 77
integrator, 79
interior control, 4, 218, 255, 261

Jensen's inequality, 262

Kalman observability matrix, 63

Laplace operator, 4
limit and differentiation exchange theorem,
   155
limit and integration exchange theorem, 155
limit point, 15
linear operator, 17
linear quadratic optimal control problem, 173
Lumer-Phillips theorem, 35
Lyapunov design, 236
Lyapunov functional, 187

mass-spring system, 2
measured output, 1
method of energy, 221
Minkowski's inequality, 12

Neumann boundary control, 235
norm, 9
normed linear space, 10

observability inequality, 64, 252, 254
observable, 63, 252
observation system, 63
observer gain matrix, 83
open set, 15
optimal control, 173
optimal state feedback control, 184, 200, 278,
   283
optimality system, 180
output, 49
output feedback control, 84
output injection, 140
output injection matrix, 68

perturbed energy, 227
perturbed energy functional, 256
Poincaré's inequality, 28, 149, 236
point control, 135, 136
Popov-Belevitch-Hautus Test, 62, 73
proportional control, 76

proportional-integral-derivative (PID) control,
   77

reaction-convection-diffusion equations, 3
reaction-diffusion equation, 149
regulator equations, 113
regulator system, 50
resolvent set, 26
Routh-Hurwitz Criterion, 56

sectorial operator, 37
self-adjoint operator, 35
semigroup, 30
semigroup of contractions, 34
separable, 16
series uniform convergence theorem, 156
Sobolev spaces, 20
spectrum of operator, 26
stabilizable, 131, 145, 213, 219, 226, 235
stable, 51, 121, 125
stable manifold, 136
state, 1, 29, 49
state feedback, 67
state equation, 49
state observer, 83
state space, 125
strong convergence, 13
Sylvester equation, 112

term-by-term differentiation theorem, 156
term-by-term-integration theorem, 156

uniform convergence, 154
uniform convergence theorem, 154
unstable, 51, 121, 125
unstable manifold, 136

velocity feedback controller, 221

wave equation, 6
weak convergence, 91
weak solution, 44, 46
Weierstrass M-test, 157

Young's inequality, 11

Déjà parus dans la même collection

1. T. CAZENAVE, A. HARAUX : Introduction aux problèmes d'évolution semi-linéaires. 1990

2. P. JOLY : Mise en œuvre de la méthode des éléments finis. 1990

3/4. E. GODLEWSKI, P.-A. RAVIART : Hyperbolic systems of conservation laws. 1991

5/6. PH. DESTUYNDER : Modélisation mécanique des milieux continus. 1991

7. J. C. NEDELEC : Notions sur les techniques d'éléments finis. 1992

8. G. ROBIN : Algorithmique et cryptographie. 1992

9. D. LAMBERTON, B. LAPEYRE : Introduction au calcul stochastique appliqué. 1992

10. C. BERNARDI, Y. MADAY : Approximations spectrales de problèmes aux limites elliptiques. 1992

11. V. GENON-CATALOT, D. PICARD : Eléments de statistique asymptotique. 1993

12. P. DEHORNOY : Complexité et décidabilité. 1993

13. O. KAVIAN : Introduction à la théorie des points critiques. 1994

14. A. BOSSAVIT : Électromagnétisme, en vue de la modélisation. 1994

15. R. KH. ZEYTOUNIAN : Modélisation asymptotique en mécanique des fluides Newtoniens. 1994

16. D. BOUCHE, F. MOLINET : Méthodes asymptotiques en électromagnétisme. 1994

17. G. BARLES : Solutions de viscosité des équations de Hamilton-Jacobi. 1994

18. Q. S. NGUYEN : Stabilité des structures élastiques. 1995

19. F. ROBERT : Les systèmes dynamiques discrets. 1995

20. O. PAPINI, J. WOLFMANN : Algèbre discrète et codes correcteurs. 1995

21. D. COLLOMBIER : Plans d'expérience factoriels. 1996

22. G. GAGNEUX, M. MADAUNE-TORT : Analyse mathématique de modèles non linéaires de l'ingénierie pétrolière. 1996

23. M. DUFLO : Algorithmes stochastiques. 1996

24. P. DESTUYNDER, M. SALAUN : Mathematical Analysis of Thin Plate Models. 1996

25. P. ROUGEE : Mécanique des grandes transformations. 1997

26. L. HÖRMANDER : Lectures on Nonlinear Hyperbolic Differential Equations. 1997

27. J. F. BONNANS, J. C. GILBERT, C. LEMARÉCHAL, C. SAGASTIZÁBAL : Optimisation numérique. 1997

28. C. COCOZZA-THIVENT : Processus stochastiques et fiabilité des systèmes. 1997

29. B. LAPEYRE, É. PARDOUX, R. SENTIS : Méthodes de Monte-Carlo pour les équations de transport et de diffusion. 1998

30. P. SAGAUT : Introduction à la simulation des grandes échelles pour les écoulements de fluide incompressible. 1998

31. E. RIO : Théorie asymptotique des processus aléatoires faiblement dépendants. 1999

32. J. MOREAU, P.-A. DOUDIN, P. CAZES (EDS.) : L'analyse des correspondances et les techniques connexes. 1999

33. B. CHALMOND : Eléments de modélisation pour l'analyse d'images. 1999

34. J. ISTAS : Introduction aux modélisations mathématiques pour les sciences du vivant. 2000

35. P. ROBERT : Réseaux et files d'attente : méthodes probabilistes. 2000

36. A. ERN, J.-L. GUERMOND : Eléments finis : théorie, applications, mise en œuvre. 2001

37. S. SORIN : A First Course on Zero-Sum Repeated Games. 2002

38. J. F. MAURRAS : Programmation linéaire, complexité. 2002

39. B. YCART : Modèles et algorithmes Markoviens. 2002

40. B. BONNARD, M. CHYBA : Singular Trajectories and their Role in Control Theory. 2003

41. A. TSYBAKOV : Introdution à l'estimation non-paramétrique. 2003

42. J. ABDELJAOUED, H. LOMBARDI : Méthodes matricielles – Introduction à la complexité algébrique. 2004

43. U. BOSCAIN, B. PICCOLI : Optimal Syntheses for Control Systems on 2-D Manifolds. 2004

44. L. YOUNÈS : Invariance, déformations et reconnaissance de formes. 2004

45. C. BERNARDI, Y. MADAY, F. RAPETTI : Discrétisations variationnelles de problèmes aux limites elliptiques. 2004

46. J.-P. FRANÇOISE : Oscillations en biologie : Analyse qualitative et modèles. 2005

47. C. LE BRIS : Systèmes multi-échelles : Modélisation et simulation. 2005

48. A. HENROT, M. PIERRE : Variation et optimisation de formes : Une analyse géometric. 2005

49. B. BIDÉGARAY-FESQUET : Hiérarchie de modèles en optique quantique : De Maxwell-Bloch à Schrödinger non-linéaire. 2005

50. R. DÁGER, E. ZUAZUA : Wave Propagation, Observation and Control in $1 - d$ Flexible Multi-Structures. 2005

51. B. BONNARD, L. FAUBOURG, E. TRÉLAT : Mécanique céleste et contrôle des véhicules spatiaux. 2005

52. F. BOYER, P. FABRIE : Eléments d'analyse pour l'étude de quelques modèles d'écoulements de fluides visqueux incompressibles. 2005

53. E. CANCÈS, C. L. BRIS, Y. MADAY : Méthodes mathématiques en chimie quantique. Une introduction. 2006

54. J-P. DEDIEU : Points fixes, zeros et la methode de Newton. 2006

55. P. LOPEZ, A. S. NOURI : Théorie élémentaire et pratique de la commande par les régimes glissants. 2006

56. J. COUSTEIX, J. MAUSS : Analyse asympotitque et couche limite. 2006

57. J.-F. DELMAS, B. JOURDAIN : Modèles aléatoires. 2006

58. G. ALLAIRE : Conception optimale de structures. 2007

59. M. ELKADI, B. MOURRAIN : Introduction à la résolution des systèmes polynomiaux. 2007

60. N. CASPARD, B. LECLERC, B. MONJARDET : Ensembles ordonnés finis : concepts, résultats et usages. 2007

61. H. PHAM : Optimisation et contrôle stochastique appliqués à la finance. 2007

62. H. AMMARI : An Introduction to Mathematics of Emerging Biomedical Imaging. 2008

63. C. GAETAN, X. GUYON : Modélisation et statistique spatiales. 2008

64. RAKOTOSON, J.-M. : Réarrangement Relatif. 2008

65. M. CHOULLI : Une introduction aux problèmes inverses elliptiques et paraboliques. 2009

66. W. LIU : Elementary Feedback Stabilization of the Linear Reaction-Convection-Diffusion Equation and the Wave Equation. 2010